AF785334

Current State of the Field of Cell-Penetrating Peptides as an Honorific Issue for Professor Ülo Langel

Current State of the Field of Cell-Penetrating Peptides as an Honorific Issue for Professor Ülo Langel

Editors

Prisca Boisguérin
Sébastien Deshayes

MDPI

Basel • Beijing • Wuhan • Barcelona • Belgrade • Novi Sad • Cluj • Manchester

Editors

Prisca Boisguérin
PhyMedExp
Inserm U1046 - CNRS UMR
9214
University of Montpellier
Montpellier
France

Sébastien Deshayes
PhyMedExp
Inserm U1046 - CNRS UMR
9214
University of Montpellier
Montpellier
France

Editorial Office
MDPI
St. Alban-Anlage 66
4052 Basel, Switzerland

This is a reprint of articles from the Special Issue published online in the open access journal *Pharmaceutics* (ISSN 1999-4923) (available at: https://www.mdpi.com/journal/pharmaceutics/special_issues/Honorific_Issue_Professor_Ulo_Langel).

For citation purposes, cite each article independently as indicated on the article page online and as indicated below:

Lastname, A.A.; Lastname, B.B. Article Title. *Journal Name* **Year**, *Volume Number*, Page Range.

ISBN 978-3-7258-1258-5 (Hbk)
ISBN 978-3-7258-1257-8 (PDF)
doi.org/10.3390/books978-3-7258-1257-8

© 2024 by the authors. Articles in this book are Open Access and distributed under the Creative Commons Attribution (CC BY) license. The book as a whole is distributed by MDPI under the terms and conditions of the Creative Commons Attribution-NonCommercial-NoDerivs (CC BY-NC-ND) license.

Contents

pharmaceutics MDPI

Review

Cell-Penetrating Peptides and Transportan

Ülo Langel [1,2]

[1] Department of Biochemistry and Biophysics, Stockholm University, S.Arrheniusv. 16B, Room C466, SE-106 91 Stockholm, Sweden; Ulo.Langel@dbb.su; Tel.: +46-8-161-793 or +46-707-905-284; Fax: +46-8-161-371
[2] Laboratory of Molecular Biotechnology, Institute of Technology, University of Tartu, Nooruse 1, 50411 Tartu, Estonia

Abstract: In the most recent 25–30 years, multiple novel mechanisms and applications of cell-penetrating peptides (CPP) have been demonstrated, leading to novel drug delivery systems. In this review, I present a brief introduction to the CPP area with selected recent achievements. This is followed by a nostalgic journey into the research in my own laboratories, which lead to multiple CPPs, starting from transportan and paving a way to CPP-based therapeutic developments in the delivery of bio-functional materials, such as peptides, proteins, vaccines, oligonucleotides and small molecules, etc.

Keywords: transportan; cell-penetrating peptides; transfection; PepFect; NickFect

Citation: Langel, Ü. Cell-Penetrating Peptides and Transportan. *Pharmaceutics* **2021**, *13*, 987. https://doi.org/10.3390/pharmaceutics13070987

Academic Editor: Gert Fricker

Received: 28 May 2021
Accepted: 25 June 2021
Published: 29 June 2021

Publisher's Note: MDPI stays neutral with regard to jurisdictional claims in published maps and institutional affiliations.

Copyright: © 2021 by the author. Licensee MDPI, Basel, Switzerland. This article is an open access article distributed under the terms and conditions of the Creative Commons Attribution (CC BY) license (https://creativecommons.org/licenses/by/4.0/).

1. Introduction to Cell-Penetrating Peptides

The recent definition of CPPs [1,2] given below tries to summarize the diffuse diversity of a huge class of peptides with multiple bioactive properties and drug delivery abilities:

Cell-penetrating peptides (CPPs) are relatively short peptides, 4–40 aa, with the ability to gain access to the cell interior by means of different mechanisms, mainly including endocytosis, and/or with the capacity to promote the intracellular effects by these peptides themselves, or by the delivered covalently or non-covalently conjugated bioactive cargoes.

The discovery story of CPPs (also known as protein/peptide transduction domains, PTD, Trojan peptides or shuttling peptides) has been covered in detail [2], starting from the discovery of an HIV tat trans-activator protein [3,4], a membrane shuttling protein with only the portions (residues 1–72 and 37–72) necessary for cellular uptake [5]. The group of Alain Prochiantz [6] introduced the 60 aa homeodomain of Antennapedia (a *Drosophila* homeoprotein), and its short fragment, a 16 aa peptide pAntp(43–58), later named penetratin [7], which I define as a starting point for CPP research. Shortly after this, the group of Bernard Lebleu introduced the short 12 aa Tat peptide [8]. It seems that these first CPPs—penetratin and Tat peptides—are the most popular CPPs today for the trans-barrier delivery of multiple bioactive cargos. These findings introduced a breakthrough situation in cellular biochemistry back in the 1990s, breaking the traditional dogma that the cell plasma membrane was impermeable to proteins and peptides.

To date, June 2021, the website CPPsite 2.0 (http://crdd.osdd.net/raghava/cppsite/) database contains around 1700 unique, experimentally validated CPPs, together with their secondary and tertiary structures. However, in silico CPP predictions show thousands (if not millions) of such peptides awaiting confirmation and application. Most of these CPP sequences can be found in [2], and are not presented here.

Undeniably, the complexities of their mechanisms of action have rendered CPPs problematic to define, if indeed this is possible at all [9]. I have recently even suggested [2] a new way to classify CPPs, based on the multiple sides of CPP activities:

1. Protein-derived vs. designed
2. Classified by physico-chemical properties vs. classified by structural properties
3. Predicted vs. random

4. Linear vs. cyclic
5. Protein mimics vs. cargo delivery vectors
6. Nonspecific vs. targeted
7. "Direct" translocators vs. endocytosis enhancers
8. "Non-toxic" vs. antimicrobial

One can easily see from this CPP classification that the CPP subclasses often overlap, and that many CPPs can belong to several subclasses. Additionally, one can easily create one's own classification based on different CPP properties; more details are given in [2]. I hope that the work of CPP classification is still in progress today.

For the trans-barrier delivery of different cargos, diverse strategies are used in which covalent conjugation or non-covalent complex formation can be selected [10]. Many examples of CPP/cargo conjugations are available using multiple conjugation chemistries or complexation approaches [2].

CPPs have been extensively employed to transport cargo molecules in vitro and in vivo; however, the delivery uptake mechanism of the particles formed by CPPs and their cargo is poorly understood, depending on, e.g., the membrane structure, the peptide structures, the nature of the cargo, or the concentration of a particle, etc. The knowledge of these mechanisms, however, is the prerequisite for the development of drug delivery systems based on CPP technologies.

Two main types of CPP uptake mechanisms have been suggested: energy-independent ("direct penetration") and endocytotic pathways. The energy-independent uptake pathway involves CPP/membrane interactions, and can be due to, e.g., pore formation or membrane disturbance, etc. Energy-dependent pathways are usually related to endocytic mechanisms, e.g., macropinocytosis has been shown to be able to incorporate CPPs and their complexes with cargos. Usually, endocytotic uptake is initiated by the interactions of CPPs with different cell-surface receptors, e.g., anionic receptors such as neuropilin-1 and heparan sulfate proteoglycans [9]. The current understanding is that, usually, such a cellular uptake event is the consequence of the parallel action of the above pathways, depending on the conditions.

It is a paradigm in CPP research that the peptides are taken up by virtually all cells, but in vivo CPPs only target a very limited number of cells and many tissues are hardly reached at all. Today's research aims to target specifically certain cells or diseased tissues for highly efficient CPP-based targeted therapeutics. One research area fueling this research is the need for therapeutics and diagnostics in oncology [11].

All of these aspects (and more) of research of the field of cell-penetrating peptides are often studied for individual CPPs. Additionally, multiple reports are available concerning the comparison of the properties and efficacy of several CPPs in parallel, creating additional information for the development of novel drug delivery systems.

Below, I briefly summarize our work on the introduction and development of one CPP family—transportans—and their further development into a carrier of bioactive molecules as a possibility for future drug development.

2. Transportan

2.1. Discovery of Transportans, Development of PepFects and NickFects

Galparan, a chimeric peptide, was introduced in 1996 by fusing two naturally occurring aa sequences: amphipathic peptide mastoparan from wasp venom and the fragment of human neuropeptide galanin [12,13]. This fusion peptide was a logical step in our study of an exciting class of our chimeric galanin receptor ligands, developed in Prof. Tamas Bartfai's group at the beginning of 1990s [14,15]. We were studying the possible rules behind the interactions between galanin receptors and the components of these high-affinity chimeric ligands with peculiar synergistic effects, which were apparently independent of the fused constituents. The rationale behind the fusion of the membrane-active mastoparan (at C-terminus) with galanin(1–13) was the possibility to add a plasma membrane interaction to the interactions with the galanin receptors; see the sequences below.

Galanin(1–13): GWTLNSAGYLLG-P [12]
Mastoparan: INLKALAALAKKIL [12]
Galparan: GWTLNSAGYLLG-P-INLKALAALAKKIL [12]

Galanin receptors in CNS recognized galparan with high affinity, K_D = 6.4 nM; however, its bio-effects differed from those of galanin and mastoparan [16]. Galparan induced a 26-fold increase in insulin secretion from rat pancreatic islets at a distal site in the stimulus secretion coupling of the B cell [17], induced in vivo acetylcholine release when injected intracerebroventricularly into the frontal cortex of the rat by not acting at the galanin receptors or at the sites of mastoparan action [18], and modulated the activity of GTPases and Na^+,K^+-ATPase, whereas galanin does not affect these enzymes [12,19].

In the hope of understanding galparan's interactions with plasma membrane-located galanin receptors, we further modified it by the replacement of P^{13} with K^{13}. The novel peptide, called transportan, showed, to our surprise, the translocation to the cell cytosol when labelled at the K^{13} side chain with a fluorophore, which was an early experiment carried out by Margus Pooga [20]. Systematic structure-activity studies enabled the shortening of the transportan to transportan 10, which demonstrated excellent cellular translocation with decreased toxicity [21]. Hence, the discovery of transportan is a good example of serendipity in research; however, it was made possible only due to the specific exciting questions asked in the study.

Transportan, TP: GWTLNSAGYLLG-K*-INLKALAALAKKIL [20]
Transportan 10, TP10: AGYLLG-K*-INLKALAALAKKIL [21]

Later, the K^{13} side chain modification was used for the covalent modification of several cargo molecules, e.g., biotin, PNA and proteins bound to biotin, etc., see below.

Interestingly and independently, several additional chimeric peptides were reported to obtain CPP properties, e.g., MPG and Pep-1, combining the nuclear localisation signal (NLS) of SV40 T-antigen and a hydrophobic peptide with high affinity to cellular membranes [22,23], CADY [24] and Pip (PNA-internalising peptides) [25], all successful "examples for such design where cationic, hydrophobic and amphipathic sequences have been combined" [26].

The further development of transportan analogs yielded a novel CPP series, introduced by my Stockholm and Tartu groups, respectively, PepFects and NickFects, exemplified by PF6, PF14, NF51 and NF55. PepFect strategies were protected by a patent application (publication number: 20140140929, together with CePeP and General Electrics Healthcare). NickFects were protected by a patent application (PCT/EP11155275.8, WO 2012/113846 Al) and further developed by CePeP, Sweden/Estonia, PepFex, Sweden and Tartu University, leading to an additional patent application (PCT/EP2020/050524) to protect the third generation NickFects, developed by Tartu University.

In PepFect6, the side chain of Lys* was modified with branched four chloroquine analogs, hopefully enabling improved endosomal escape for PepFect/cargo. PepFect14, designed by Mattias Hällbrink, consists of the same N-terminus as PepFect6, but with an ornithine (O) containing an (educated) fantasy sequence in the C-terminus, leading also to endosomal escape similar to PF6.

PepFect 6, PF6: stearoyl-AGYLLG-K*-INLKALAALAKKIL [27]
PepFect 14, PF14: stearoyl-AGYLLG-K*-LLOOLAAAALOOLL [28]
NickFect 51, NF51: O(Nδ-stearoyl-AGYLLG)-INLKALAALAKKIL [29]
NickFect 55, NF55: O(Nδ-stearoyl-AGYLLG)-INLKALAALAKAIL [30]

While the PepFect vectors contain linear aa sequences, the NickFect peptides contain the branched structure at the side chain of ornithine, although the original sequences from galanin and mastoparan are still present. Such a branched structure in the transfection vectors yielded several CPPs with improved properties, i.e., lower toxicity and the improved tranfection of oligonucleotides, depending on different delivery systems. The mechanisms of such improvements are not clear and the current research addresses these questions intensively. One important difference between "traditional" CPPs and PepFect or NickFect vectors is that the latter form nanoparticles with oligonucleotide cargos; see below.

2.2. Antisense ON, siRNA and Plasmid Delivery by Covalent Coupling

Here, I summarize the transportan delivery of cargo (and not the delivery by ON/CPP nanoparticles); see also the earlier reviews [31,32]. The first intracellular CPP delivery of antisense ONs (ASO) was achieved in 1995 by the covalent disulfide conjugate of penetratin (then called pAntp) with phosphorothioate ONs blocking APP expression [33]. This was followed by intensive applications of CPPs in antisense ON transfection by covalent conjugation in the 1990s and, more recently, by non-covalent conjugation when the respective technologies became available; see below. Today, the transfection of "traditional" antisense ONs has been reported for multiple CPPs, e.g., penetratin, Tat, Pip, $(KFFK)_3R$, $(RXR)_4$, Pep-3, MPG, R15, TP10, PepFects, NickFects, and Chol-R_9, etc., in a covalent or non-covalent manner, in vivo and in vitro, as summarized in [2].

Our group's initial interest in vectors for the intracellular delivery of cargos was fueled by the introduction of cell-membrane-impermeable peptide nucleic acids (PNA) by a Danish group [34]. PNA oligomers with extraordinary properties, such as a high affinity to complementary DNA or RNA, and high resistance to the protease or nuclease degradation, etc., were potential intracellular regulators of DNA/RNA-initiated processes. Later, this was shown to be true after solving the main hurdles with their intracellular delivery.

After the finding of the cargo delivery properties of transportan, our group, in 1998, (patented 1997 [35], together with PerSeptive Biosystems Inc., USA, by the initiative of Michael Egholm) conjugated it (and penetratin) with a PNA oligomer via the disulfide bridge, yielding a covalent conjugate, CPP-S-S-PNA. The conjugate was designed to be reduced by intracellular glutathione, and to liberate the PNA oligomer to knock down galanin receptor 1 expression by the antisense (translational arrest) mechanism. Both constructs were successfully internalized into Bowes cells and knocked down targeted galanin receptor type 1 [36] in vitro and (intrathecally) in vivo, decreasing the galanin binding in the dorsal horn, and the inability of galanin to inhibit the C fiber stimulation-induced facilitation of the rat flexor reflex [36,37]. It seems that disulfide-based conjugates are convenient in, at least, the delivery of CPP-PNA conjugates. Several reports are available on the applications of covalent transportan-ASO and other drug conjugates.

Transportan-S-S-PNA (antisense to PTP sigma) increased the glucose-induced insulin secretion from GK rat islets, with decreased amounts of phosphatase [38]. The liposomes modified with a transportan10 analog, TH, showed enhanced cellular uptake and the delivery of paclitaxel with the inhibition of tumor cell growth in vivo [39]. The uptake into human fibroblast cytosolic compartments was seen for Tat, penetratin, R9F2 and transportan disulfide conjugates of 12mer OMe/LNA oligonucleotide conjugates targeted to the TAR RNA, in HeLa and human fibroblast cells [40]. Disulphide-conjugated penetratin, Tat, transportan, transportan-21 and transportan-22 to a 16-mer PNA, targeting the TAR region of the HIV-1 genome, showed cellular uptake and anti-HIV virucidal activity by the inhibition of HIV-1 replication in vitro [41]. Transportan10-S-S-PNA-Bpa (*p*-benzoylphenylalanine), targeting the regions of the 3′ and 5′ UTRs of ankylosis mRNA, showed intracellular crosslinking to RNA-binding proteins (RBPs) that complex with a target RNA in vivo. Several proteins were isolated and identified "in complex with or near the targeted regions of the ankylosis mRNA through UV-induced crosslinking of the annealed PNA-RNA-RBP complex" [42]. Transportan10-S-S-PNA-based antisense conjugate was used to study "the role of subtypes of the L-type voltage-gated calcium channels (LTCs), Ca(V)1.2 and Ca(V)1.3 in long-term pain sensitization in a rat model of neuropathy", showing the reverse of the neuropathy-associated mechanical hypersensitivity and the hyperexcitability, as confirmed with siRNA knock-down experiments [43]. Transportan-PNA_{TAR} internalized into the cells with the functional inhibition of HIV-1 production in chronically HIV-1-infected H9 cells [44].

An additional antisense ON application field became routinely available after R.Kole and co-workers introduced a novel antisense ON-based platform to redirect splicing by blocking aberrant splice sites in HeLa cells stably expressing luciferase containing such a site [45]. This system of splice correction by ASO is a promising therapeutic tool for a variety of diseases, e.g., Duchenne muscular dystrophy, beta-thalassemia, cystic fibrosis and certain cancers.

We applied the luciferase aberrant splice site setup in HeLa cells with known splice-correcting PNAs (PNA705), tethered to a variety of CPPs [46], Tat, penetratin and transportan, via a disulfide bridge. Transportan was the most potent vector, and it significantly restored splicing in a concentration-dependent manner, following the endocytotic cellular uptake. This suggested that CPPs can be used for the delivery of splice-correcting ASO as a potential therapeutic approach for the regulation of splicing in a variety of diseases.

Since then, the HeLa based delivery of a splice-correcting CPP-ASO assay has been in routine use in the evaluation of novel CPPs for cargo delivery and studies of internalization mechanisms, e.g., to assess different endocytic pathways and the dependence on extracellular heparan sulfates for internalization for the comparison of penetratin, Tat, transportan, TP10, MAP and pVEC [47], showing the exact endocytic internalization routes.

Disulfide-linked CPP conjugates with oligonucleotide analogues, siRNA and PNA, in the HeLa cell assay with integrated plasmid reporters showed that transportan-PNA and R6-penetratin-S-S-PNA caused the Tat-dependent trans-activation inhibition, suggesting them as potential anti-HIV agents [48].

Seven different CPPs were included in the study: transportan, R7–9, Tat, penetratin, KFF, SynB3 and NLS, conjugated by different conjugation chemistries. PNA-S-S-transportan-amide (ortho)-PNA, targeting luciferase expression correction, was the most potent conjugate, resulting in the maximum luciferase signal in the serum-free media, but it was the least sensitive to the presence of serum [49].

The development of antisense ONs (mainly PNA, LNA, mixomers, PS and PMO) in splicing redirection using CPP transfection is very active [50–52]. For example, Sarepta (Cambridge, MA, USA) works intensively with RNA-targeted therapeutic candidates for different types of RNA, using phosphorodiamidate morpholino oligomers (PMOs) as antisense ONs, attached to CPPs (PPMO) such as $(R\text{-}Ahx\text{-}R)_4$, (6-aminohexanoic acid-spaced oligo-arginine). The development of next-generation PMO-based chemistries for advanced RNA-targeted therapeutics with enhanced tissue targeting, intracellular delivery, target selectivity and drug potency is in progress [53].

PNA was designed to interact with an overlap of the NFkappaB decoy oligonucleotide consisting "of a double-stranded consensus sequence corresponding to the kappaB site localized in the IL-6 gene promoter". It was shown that the construct "blocked the effect of interleukin-1beta-induced NFkappaB activation and IL-6 gene expression" [54].

Few examples are available of applications of CPP-siRNA covalent conjugates for the knockdown of gene expression. Our group was never successful with this knock-down strategy. We often achieved the cellular internalization of the CPP-S-S-siRNA conjugates; however, the functional knockdown was never demonstrated in different cells. Interestingly, the covalent CPP (penetratin and transportan) coupling of siRNA via disulfide bond was carried out, yielding an improvement of the cellular uptake as well as the expression reduction of reporter GFP transgenes [55]. Conjugates of transportan10-S-S-siRNA showed intracellular localization and silencing by siRNA-targeted firefly luciferase GL3 in FRSK cells [56]. Transportan or transportan-r9, T9(dR), complexed with siRNA against a nucleoprotein (NP) gene segment of the influenza virus (siNP) showed the in vivo and in vitro delivery of siRNA in 293T, MDCK, RAW and A549 cells and mice. After the combined tail vein injection of siNP and T9(dR) or transportan, all of the mice "infected with PR8 influenza virus survived and showed weight recovery at 2 weeks post-infection" [57].

The improvement of intracellular plasmid delivery by CPPs has been a desirable objective for several years. Few reports are available on plasmid delivery by transportans; however, these have a low success rate, indicating the need for additional CPP vector development for efficient plasmid transfections. Transportan10 "crosslinked to a plasmid via a PNA oligomer, TP10 conjugation with polyethyleneimine (PEI), and addition of unconjugated TP10 to standard PEI transfection assay" increases the transfection efficiency several fold compared to PEI alone in Neuro-2a cells [58]. Fl-Transportan showed the maximum fluorescence among all of the tested CPPs in permebilized wheat immature embryos [59]. While Tat(2) mediated the GUS enzyme and plasmid DNA (carrying Act-1GUS) delivery to embryos, transfection success with transportan was not reported [59]. Stearoyl-TP10 was shown to form stable nanoparticles with plasmids that efficiently enter different cell-types, including primary cells, resulting in a gene expression which was almost comparable with the levels of Lipofectamine 2000 (LF2000) in vitro, and which mediates efficient gene delivery in vivo when administrated intramuscularly (i.m.) or intradermally (i.d.) in mice [60]. This work enabled us to further aim at the development of non-covalently conjugated plasmid transfection strategies.

2.3. *Delivery of Peptides and Proteins*

The conjugation of several CPPs to proteins and peptides has improved their rapid translocation into cells and through biobarriers in vivo [31,61–63]. The obvious CPP champions in protein trans-barrier delivery are Tat and penetratin; however, several reports are available on transportan conjugates with proteins and peptides, and their delivery. Besides the obvious therapeutic task with improved protein delivery, these cell-permeable proteins serve as valuable tools for the clarification of the uptake and targeting mechanisms for CPP-cargos; see below.

2.3.1. Peptides

A few examples can be found on the transportan delivery of short peptides; see below. Although efficient, the addition of transportan to the peptides yields a much longer peptide, making its production more difficult. Hence, the preferred choice for the cellular delivery of short peptides would be to use the CPP prediction-based modifications of the selected peptides.

The cellular uptake and cargo delivery kinetics were studied for penetratin, transportan, Tat and MAP, labelled with the fluorescence quencher 3-nitrotyrosine, coupled via disulfide to a pentapeptide cargo (labelled with the 2-amino benzoic acid fluorophore) [64]. The DOCK2 inhibitory peptides for protein–protein interaction conjugated to 13 different luciferin-conjugated CPPs, among them transportan, to test them as an "intracellular target for transplant rejection and inflammatory diseases" [65]. Several synthetic peptides comprised from effector caspase activational cleavage sequences fused with Tat, penetratin, transportan, and Pep1 showed the internalization and improved survival of syngeneic immortalized Schwann cells during transplantation in vivo [66]. Transportan-Aβ42 peptide showed tissue penetrating capability, and was introduced into the adult zebrafish brain [67]. Transportan (and other CPP's) conjugated with PKI and NBD peptides showed cellular uptake and its time course [68].

2.3.2. Proteins

Several initial attempts for intracellular protein delivery by CPPs have been carried out using the strong biotin–avidin interaction for the non-covalent conjugation of biotinyl-CPPs and (strept)avidin or biotin antibodies [69]. In the case of transportan, the following combinations were used: biotinyl–transportan with anti-biotin monoclonal antibodies [70]; an Nα-biotinyl-TP10 complex with fluorescently labeled streptavidin in a photo-induced endosomal escape study [71]; biotinyl–transportan, -oligoarginine and -Tat, complexed to avidin–TexasRed, showing three different populations of complexes-containing vesicles [72]; biotinyl–penetratin, Tat, transportan and pVEC complexed with avidin, showed endocytotic and clathrin-dependent and -independent mechanisms of cellular transduction [73]; colloidal gold-labeled neutravidin complexes with biotinyl-transportan and nanogold-labeled peptides showed endocytotic routes in parallel translocation [69,74,75]; biotinyl-

Tat/streptavidin conjugated to a biotinylated, pH-responsive polymer poly(propylacrylic acid) showed improved endosomal escape [74]; biotin–transportan and -TP10 showed an endocytosis-independent mechanism in colorectal cancer (CRC) cells [75]; biotinyl–penetratin, -Tat and -transportan 10 with avidin and streptavidin showed that the cellular delivery properties are dependent on the cargo used [76]; and biotinyl–transportan and -Tat complexed with avidin-β-galactosidase (ABG) showed enhanced tissue distribution in samples of freshly harvested human carcinoma or hyperplasia-containing specimens of the uterus and the cervix [77]. Colorectal cancer (CRC) cell lines HT29 and HCT116, incubated with siRNA for SASH1 mRNA in the presence of transportan and TP10, showed cellular uptake and a decreased SASH1 mRNA level [75].

A few reports with no use of biotin/avidin interaction are available on the cellular delivery of proteins by transportans. Tat, penetratin and transportan complexed with rhodamine-BSA, which showed delivery to the interior of epithelial cells, being "passive carriers that do not initiate epithelial cell-associated 'danger signals' during the process of cytoplasmic delivery of a model protein cargo" [78]. Transportan-GFP and antibodies could internalize covalently coupled molecules up to 150 kDa in Bowes cells [70]. Penetratin showed the cellular uptake of complexed BSA, while R8 and TP10 failed to deliver BSA [79]. GFP fused to penetratin, R8, Tat, transportan, Xentry and their cyclic derivatives showed cellular uptake with localization in endosomes in human cell lines HeLa, HEK, 10T1/2 and HepG2 [80].

2.4. Complexation, PepFects and NickFects

As described above, transportan and its modified versions have been demonstrated to aid efficiently the cellular internalization of a variety of covalently conjugated bioactive cargos, both in vitro and sometimes even in vivo. Often, these examples are available as comparisons with additional CPPs to transportan. However, our routine testing of multiple CPPs for their cargo delivery capacity was hindered by the need for the covalent conjugation of CPP with cargo due to the additional steps of chemistry and purification/characterization. Our attempts to attach unmodified transportan non-covalently to several cargos were mainly unsuccessful.

Hence, we further aimed to find non-covalent transportan/cargo conjugation methods which could provide strategies of the simple complexation of CPPs with different cargos yielding efficient cargo delivery strategies. The non-covalent complex formation of CPPs with ONs was an obvious choice on first sight due to the availability of cationic CPPs and anionic ONs, enabling the formation of efficient and stable aggregates or even nanoparticles, possibly yielding efficient transfections. The first CPPs for non-covalent ON complexation were stearoyl-R9, GALA, KALA, MPG, Pep-1, CADY [81–88], Chol-R9 [89], stearoyl-$(RXR)_4$ [90] and others.

Our first choice in the case of transportan was to test stearoyl–transportan [27] for ON transfection, based on the idea of Prof. Shiroh Futaki's group on stearoyl-R9 [81]. We tested multiple fatty-acid-modified transportan analogs, and now the series of PepFects (PF) and NickFects (NF) are available; see above. The PepFects and NickFects are excellent ON delivery vectors, as exemplified in multiple reports below. They form relatively stable nanocomplexes with ONs, enabling ON transfection in vitro and in vivo by the non-covalent simple formulation technology of antisense, siRNA and plasmid delivery [27–29,90–94].

Due to the formation of these nanocomplexes, it has even been questioned whether PepFects and NickFects are the "true" cell-penetrating peptides, rather than the new peptide-based detergents. My answer is a clear "yes", as they contain stearoyl-modified transportan, and the yielding CPP is a cargo delivery vector, following the rules for the CPP definition, even if they may obtain detergent-like properties. The mechanisms of such transfections are discussed below.

2.4.1. PepFects

Several reports are available on the design of the PepFect and NickFect peptides. Novel PFs and NFs were introduced using a QSAR prediction model, showing peptide-plasmid complexes and the transfection of cells with pDNA [95]. PepFect analogues with introduced His residues were introduced in order to make the peptides pH-responsive for the PepFect/SCO nanocomplexes, showing PepFect132 with high bioactivity [96]. PepFect14, double-functionalized with PEG and an MMP substrate site, complexed with pDNA, showed the efficient induction of gene expression specifically in tumors after i.v. injections [97].

Several examples of myristoyl–transportan transfection are available: NPs incorporating myristoyl–transportan and tumor-homing peptides carrying siRNA, a CpG DNA ligand of TLR9 suppressed tumor growth in several animal models of various cancers after systemic intravenous (i.v.) administration [98]; myristoyl–transportan conjugated to a transferrin receptor-targeting peptide (myr-TP-Tf) encapsulating siRNA targeted it to the brain with a functional gene silencing effect in a human glioma [99]; and a nanocomplex based on a tandem peptide of myristoyl–transportan and Lyp-1 showed the internalization of sgRNA/Cas9 ribonucleoprotein complexes and genome editing in cell lines [100].

Examples of ON cargo delivery with PFs and NFs are available. An antisense nanoprobe, ^{99m}Tc-anti-miRNA ONs/PepFect6, was used "for imaging the miRNA-21 expression in A549 lung adenocarcinoma xenografts and in vivo" [101]. The master regulator proteins in critical tumor regulation were confirmed using their lentivirus-mediated shRNA silencing with PF14 transfection, yielding sixteen master regulators which could be reproducibly silenced; of these, 94% showed reduced tumor growth/viability in vitro [102].

PF14/ASO (triplet repeat-targeting AS ON) in a muscle cell model of myotonic dystrophy yielded a dose-dependent correction of disease-typical abnormal splicing, whereas PF14 shielded the AS ON from degradation. It was shown that "intranuclear blocking-type oligonucleotide concentrations in the upper nanomolar range were required to dissolve nuclear muscleblind-like protein 1 foci" [103].

PF14/mRNA nanoparticles showed the expression of reporter protein eGFP "in two-dimensional tissue cultures and in three-dimensional cancer cell spheroids", as well as in primary ovarian cancer explants [104].

PF14/mRNA (eGFP) complexes in the glomerular endothelial cell line mGEnC, HeLa cells and SKOV-3 ovarian carcinoma cells showed uptake and protein expression with "linear correlation of dose, uptake, and expression, observed over 5 orders of magnitude in vitro and 3 orders of magnitude in vivo" [105].

The PepFects in complex with graphene oxide and plasmids, splice correction oligonucleotides and siRNA showed NPs and a >10–25 fold increase of their cell transfection [106]. Similar effects were achieved with magnetic nanoparticles [107], zeolitic imidazolate frameworks [108] and carbonized-chitosan-encapsulated hierarchical porous zeolitic imidazolate frameworks [109].

Two reports are available on the PepFect transfection of peptides and proteins. Calcium signal activity was tested following the application of a hemichannel blocking peptide, Gap19 (nine aa from connexin 43 cytoplasmic loop), complexed with PF6, showing the reduction of astrocyte response amplitudes and the proportion of ^{SE}astrocytes to the EtOH treatment in enriched astrocyte cultures [110]. Nanoparticles of PepFect14 complexed with Heat Shock Protein (HSP70), suggested first by docking [111], showed delivery into Bomirsky Hamster Melanoma cells; this protocol is shown in Falato et al., 2021 (in preparation).

2.4.2. NickFects

PepFect and NickFect supported the delivery of nanocomplexes of Fl-miRNA mimics (NF-miR-146a) into keratinocytes and dendritic cells with the downregulation of miR-146a-influenced genes by endocytosis, as well as suppressing inflammatory responses in a mouse model of irritant contact dermatitis [112].

By modifying the net charge and the helicity of the NickFects, a novel NF55 was introduced, showing in vivo DNA nanoparticle delivery with efficient gene induction in healthy mice, and showing tumor transfection in various mouse tumor models, e.g., an intracranial glioblastoma model [30,113].

The quantitative tracking of NickFect/pDNA complexes through endosomal transport was shown by the separation of endosomal vesicles by differential centrifugation and single-particle tracking using fluorescently labeled cargo and GFP expressing cells. It was shown that NF51 facilitates the rapid internalization of complexes into the cells, prolongs their stay in early endosomes and promotes the release to cytosol [30].

The cellular uptake and NickFect1- and NickFect51-mediated ON delivery were analyzed. It was shown that the "pathway for cellular uptake of peptide complexes is cargo dependent, whereas the endosomal escape efficacy depends on peptide hydrophobicity and chemical structure" [114].

NF55/pDNA nanoparticles showed promising tumor transfection in various mice tumor models, including an intracranial glioblastoma model [30,115].

3. Mechanisms

The mechanisms of the uptake of CPPs and their conjugates with bioactive cargos have not yet been clarified due to the availability of multiple diverse CPPs and examples of multiple CPP-cargos for delivery in order to introduce novel therapeutic entities for new pharmacology. This situation has caused constant interest in the characterization of CPP uptake mechanisms with all of the available strategies of chemistry and biology.

The mechanisms of the uptake of transportan and its modifications are not exceptions; one can find attempts to understand its trans-membrane delivery with multiple methods in comparison with several other CPPs [116,117]; see Table 1 and the detailed examples below.

Table 1. Summary of the mechanisms of uptake and toxicity studies of transportans and their conjugates with bioactive cargos; see the details in the text.

TP Analog Used	System	Result in Brief
Visualization		
N-ε13-biotinyl-TP/ Fl-streptavidin, ^{125}I-TP	fluorescence microscopy, gamma-counting	visualization in the fixated Bowes cells [20,36]; cellular internalization of TP is not an artefact caused by cell fixation
biotinyl-, Fl-, Abz-TP, -TP10	fluorescence microscopy, Transwell	visualization of crossing a Caco-2 human colon cancer cell layer in vitro by a transcellular pathway [118]
Fl-transportan antisense conjugates	confocal microscopy	mainly endocytic, macropinocytotic pathway for cellular uptake [46]
TP	FACS, spectrofluorimetry	protein uptake by endocytotic mechanism in HeLa cells of CPP-avidin complexes [73]
Fl-TP, -TP10	fluorescence microscopy, spectrofluorometry and FACS	visualization in different mammalian and plant cells [119]
TP	FACS	similar kinetic uptake profiles in HeLa, A549 and CHO cell lines [68]
^{111}In- or ^{68}Ga-TP	micro-PET imaging	uptake by six tumor cell lines and biodistribution in PC-3 tumor-bearing nude mice [120]
TP	fluorescence polarization, quenching and CD spectroscopy	in small phospholipid vesicles the helical penetratin and transportan lie along the vesicle surface, penetratin variants appear to penetrate deeper into the membrane [121]
AU NP-functionalized TP- and TP10-protein	TEM	cell entry mechanisms and intracellular trafficking of constructs were studied [122]
myristoylated TP-Tf	fluorescence images and functional gene silencing by siRNA	targeting of siRNA over BBB [99]

Table 1. *Cont.*

TP Analog Used	System	Result in Brief
Visualization		
TAMRA-TP10	fluorescence microscopy, cell toxicity	complexes of with cisplatin (cPt) in (HEK293, HEL299, HeLa OS143B cell lines were visualized [123]
Alexa488-TP10	flow cytometry, live-cell imaging and image analysis	showed that the glycine-phenylalanine switch was most dramatic in TP10 [79]
^{125}I-TP10 and -TP10-2	in vivo injection	BBB delivery [124]
TP10	spheroid model of the BBB, ex vivo imaging	showed increased delivery to mouse brains [125]
TP- isoniazid	Langmuir balance technique and AFM imaging	conjugates showed similar internalization rate into EBC-1 human squamous cell carcinoma in imaging of penetrated lipid layers [126]
	Structure and interactions of transportan	
TP	CD	random coil structure in water, in SDS micelles 60% induced helix [127]
TP	CD	60% alpha-helix in phospholipid vesicles [128]
TP	CD	helical structure in small phospholipid vesicles [121]
TP	NMR	in neutral bicelles alpha-helix in the C-terminus and tendency to form an alpha-helix in the N-terminus [129]
TP	CD	structure in neutral DMPC bicelles and negative DMPG-containing bicelles different from each other [130]
TP	CD	obtain amphiphilic α-helix when bound to membranes of vesicles composed of typical eukaryotic lipids [131]
TP10	solid-state ^{19}F-NMR, CD	a range of conformations in the DMPC/DMPG vesicle membrane-bound state, C-terminal α-helix is embedded in the membrane being tilted [132]
TP10 and 5 analogs	molecular dynamics simulation	forming α-helical conformation, the higher membrane disturbance yields higher cellular uptake in cells [133]
TP	CD	increased membrane affinity with DPPC and DPPC + mycolic acid mixed monolayers [126]
TP10	Gibbs energy studies	peptide-induced efflux, becoming faster with decrease of the Gibbs energies for binding and insertion [134]
TP and analogs	PepLook algorithm	peptide polymorphism showed common conformational polymorphic characteristics [135]
Cys-TP	EPR	in DMPC/cholesterol caused lipid ordering and a large increase in permeation [136]
TP10	molecular dynamics	interactions with POPC bilayer initiated α-helix with hydrophobic side facing the hydrophobic lipid core [137]
TP10	confocal microscopy	interaction of GPMVs revealed association with liquid-disordered membrane areas [138]
PF6	DLS	mean diameter of PF6/siRNA NPs was shown to be below 200 nm [27]
PF14	DLS	PF14/SCO NPs possessed a net negative charge [28]
PFs	CD, DLS, QSAR	PFs/Luc-plasmid NPs for study of structural requirements for cell penetration [95]
His-PFs, PF132	DLS, CD, calcein leakage	complexes formed were small at pH 7 and grew under acidic conditions [96]

Table 1. *Cont.*

TP Analog Used	System	Result in Brief
Visualization		
	Parallel mechanisms of endocytosis and direct translocation in cells	
biotinyl-TP	fluorescence microscopy	cellular uptake by unrelated mechanisms [139]
TP, TP10	Transwell model	translocation across a Caco-2 human colon cancer cell layer by transcellular pathway [118]
TP	CRM and AFM	entered SK-Mel-2 cells within 5 min and widespread distribution via a nonendocytic mechanism [140]
TP, TP10	confocal laser scanning microscopy, GMPV	showed the interactions with glycosaminoglycans [141]
TP, TP10	fluorescence microscopy	uptake, when conjugated to cargoes, involved both endocytosis and direct translocation [20,70]
Biotinyl-TP, -TP10/avidin	fluorescence microscopy	cellular internalization in HeLa and Bowes cellc by endocytosis with different pathways [142]
biotinyl-TP, -TP10/avidin	confocal microscopy	entered Cos-7 cells by caveolin-dependent endocytosis [72]
biotinyl-TP, -TP10/avidin	confocal microscopy	entered HeLa cells via caveolin-1-dependent pathway [122]
biotinyl-TP, -TP10/avidin	FACS analysis and spectrofluorimetry	clathrin-dependent and -independent endocytosis uptake in HeLa cells with partial depolarization [73]
TP-PNA	fluorescence microscopy	rapid cellular uptake by non-receptor-dependent endocytosis [143]
TP-PNA	splice correction assay	splice correction in HeLa/Luc cells with mainly endocytotic, particular macropinocytotic mechanism [46]
TP-, TP10-PNA	splice correction assay	showed cellular translocation by endocytosis [47]
TP/NPs, /BSA, /dextran	fluorescence microscopy	complexed with NPs showed in vitro and ex vivo cell entry via a receptor-dependent macropinocytosis process [144]
	Mechanisms of PepFects and NickFects	
PF3, PF6	interaction studies with lipid membranes	increased amphipathicity and their ability to insert into a lipid monolayer composed of zwitterionic phospholipids [145]
NFs/pDNA	membrane perturbation study	pDNA cargo inhibited membrane perturbation by NFs [114]
PF32/pDNA	Transwell model	uptake by brain endothelial cells via LRP-1 receptor-mediated endocytosis and scavenger receptors [146]
PF14/SCO	fluorescence microscopy	SR-A3 and SR-A5 recruited by PF14/SCO complexes [147]
PF14/Cy5-siRNA	FCS, FCCS	coexistence of monomers, self-aggregates of peptide/ON in complexes in solution and at the plasma membrane of live cells [148]
NF1/, NF51/pDNA	analysis of separated endosomal vesicles by differential centrifugation	NF51 facilitates rapid internalization of complexes into the cells, NF1 is less capable to induce endosomal release [149]

Table 1. *Cont.*

TP Analog Used	System	Result in Brief
Visualization		
	Delivery of the cargo	
TP-BSA	toxicity study	showed no toxicity or initiation of an immune response in epithelial cells [78]
TP6, TP7-PNA	toxicity study	significant cellular toxicity above 3−5 μM in TP-PNA conjugates [49]
St-TP10/SCO	splice-correcting assay, toxicity study	delivery of SCOs for functional splice correction with no toxicity [92]
St-TP10/plasmid	toxicity study, gene expression in vivo	entered different cells with high gene expression level with no toxicity and no nonimmunogenecity in vivo [60]
TK- and TH- camptothecin	toxicity study	TH-camptothecin showed cytotoxicity to cancer cells [150]
TP10, PFs/plasmid, /siRNA	toxicity study	peptide/plasmid and /siRNA showed no cytotoxic and immunogenic response, in vitro and in vivo [151]
TP, TP10, TP-biot1, TP-biot13, TP10-biot1	toxicity study	no significant cytotoxic effect at 0.5–5 μM [75]
T9(dR)/siRNA	cells, in vivo toxicity	showed cellular delivery of siRNA and in mice infected with PR8 influenza virus, and antiviral activity [57]
chloroquine-TP10	antimalarial activity	higher antiplasmodial activity in safe delivery of antimalarial aminoquinolines [152]
TP10-ciprofloxacin or -levofloxacin	antifungal in vitro activity	TP10-ciprofloxacin or -levofloxacin showed antifungal in vitro activity against human pathogenic yeasts [153]

3.1. Visualization

Multiple visualization and imaging methods are available to determine the distribution and translocation mechanisms of CPPs in vitro and in vivo [2]. These methods enable us to understand where and how the (drug) molecules are internalized into cells and different organs. In the case of transportan, often in comparison with other CPPs, most of these methods have been utilized; below, a brief selection is presented.

In the CPP labeling for the visualization, it should be considered that the "pharmacophoric" amino acids should not be modified due to their possible functionality, e.g., primary amino groups in the CPPs. Our initial naïve understanding of the transportan uptake mechanisms suggested the application of the side chain of Lys13 for the attachment of biotinyl and fluoresceinyl moieties, as "such modification did not compromise the transportan's internalization properties". The introduction of an extra amino acid into the CPP sequence, e.g., Cys, has been used often as a point for cargo attachment.

In the first confocal fluorescence microscopy study, we tested the uptake of N-ε13-biotinyl-transportan in live Bowes cells, then the visualization was carried out on the fixated cells after incubation with streptavidin-FITC or streptavidin-Texas red [20,36]. This method was used for a while, but it is not popular any more after a few reports indicated that, for some CPPs, the fixation could influence the cellular translocation by itself, e.g., in the case of VP22 and Tat [154–156]. Hence, the live cell visualization of CPP uptake is today usually applied, although in our hands the information obtained with the fixated cells often coincide with the later live cell studies [73,157].

For example, the ^{125}I-transportan uptake studies additionally confirmed that the cellular internalization of transportan is not an artefact caused by cell fixation [20]. The kinetics of internalization showed that "the maximal intracellular concentration is reached in about 20 min at 37 °C".

Biotinyl-, Fl-, Abz-transportan, -transportan10 and -penetratin were visualized "to cross a Caco-2 human colon cancer cell layer in vitro in Transwell model, showing that transportan peptides pass the epithelial cell layer", mainly by a transcellular pathway [118].

Fl-transportan antisense conjugates with luciferase splice-correcting ONs were shown by confocal microscopy to be mainly endocytic, in particular the macropinocytotic pathway for cellular uptake [46].

Protein uptake by penetratin, Tat, transportan, and pVEC studies by FACS and spectrofluorimetry showed the endocytotic mechanism in HeLa cells of CPP-avidin complexes [73].

We studied the internalization efficiency of Fl-CPPs, transportan, TP10, penetratin and pVEC, with fluorescence microscopy, spectrofluorometry and FACS in different mammalian and plant cells [119].

Tat, transportan and polyarginine were shown by FACS analysis to have similar kinetic uptake profiles in HeLa, A549 and CHO cell lines [68].

A ^{111}In- or ^{68}Ga-CPPs (transportan and nine additional CPPs) study by micro-PET imaging showed uptake by six tumor cell lines, biodistribution in PC-3 tumor-bearing nude mice showed accumulation in well-perfused organs, suggesting that "data reveal that CPPs do not show evidence for application in tumor targeting purposes in vivo" if they are not targeted [120].

A fluorescence polarization, quenching and CD spectroscopy study in small phospholipid vesicles showed that the helical penetratin and transportan lie along the vesicle's surface; penetratin variants appear to penetrate deeper into the membrane [121].

With TEM, the cell entry mechanisms and intracellular trafficking of gold nanoparticle-functionalized transportan- and TP10-protein constructs were studied [122].

Myristoylated transportan, modified with a transferrin receptor-targeting peptide (myr-TP-Tf), encapsulating siRNA showed the targeting of siRNA over BBB; fluorescence images indicated that the siRNA uptake in murine brain endothelioma and a human glioma cell line, and functional gene silencing in "a human glioma cell line as well as in primary murine neurons/astrocytes" [99].

Complexes of TAMRA-transportan 10 or -PTD4 with cisplatin (cPt) in HEK293, HEL299 and HeLa OS143B cell lines were visualized by fluorescence microscopy and inverted phase contrast microscopy, showing TAMRA-TP10 or TAMRA-TP10 + cPt in the interior of the HeLa cells, but not in the non-cancer cells HEK293 and HEL299. Only TP10 improved the anticancer activity of cisplatin if both compounds were used in the form of a complex [123].

Alexa488-labeled R8, penetratin and TP10, shown by flow cytometry, live-cell imaging and image analysis, demonstrated that the glycine–phenylalanine switch was most dramatic in TP10 [79].

The in vivo injection of CPPs into the right jugular externalis vein of anesthetized ICR-CD-1 mice showed the BBB delivery of ^{125}I-pVEC, -SynB3, -Tat, -transportan 10 (TP10) and -TP10-2 with a negligible-to-low brain influx by transportan analogs, while Tat, SynB3 and pVEC showed very high unidirectional influx rates; 80% of the influxed peptides reached the brain parenchyma. The CPPs (except pVEC) showed a significant efflux out of the brain [124].

The macrocyclic analogue M13 of transportan-10 showed increased delivery across a cellular spheroid model of the blood–brain barrier; the ex vivo imaging of mouse brains showed the increased penetration of the brain parenchyma following i.v. administration in mice [125].

Penetratin and transportan conjugates with isoniazid (INH, antibacterial agent against tuberculosis, Mtb) showed a similar internalization rate into EBC-1 human squamous cell carcinoma, and a markedly different subcellular localization and activity on intracellular Mtb by the Langmuir balance technique and AFM imaging of the penetrated lipid layers [124].

3.2. Structure and Interactions of Transportan and Its Modifications

Knowing the interactions of CPPs with model membranes or lipid bilayers as well as the structures helps to understand their trans-barrier translocation, as studied by different biophysical methods and model systems, e.g., large or giant phospholipid vesicles (LUVs, GUVs) or SDS micelles, etc. [2,158], which have been used to characterize even transportan and its modified versions, often in comparison.

In general, it seems that transportan shows nearly no structure in water and an induced helix in the presence of different micelles. As shown by the CD studies, transportan obtained almost a random secondary coil structure in water, turning in SDS micelles to a 60% induced helix, localized to the C-terminal part of the peptide [127]. Penetratin, pIsl and transportan showed secondary structures of up to a 60% alpha-helix in the presence of various phospholipid vesicles [128]. Transportan showed an induced helical structure in small phospholipid vesicles of varying charge densities. Penetratin interacted only with negatively charged vesicles, and the induced secondary structure depends on the membrane charge and lipid/peptide ratio [121]. The NMR solution structure and the position of the transportan in neutral bicelles showed "the alpha-helix in the C-terminus of the peptide and a weaker tendency to form an alpha-helix in the N-terminal domain, obtaining the parallel to the membrane surface structure" [129]. The structures of transportan in neutral DMPC bicelles and partly negatively charged DMPG-containing bicelles were found to be different from each other [130]. Penetratin, MSI-103, transportan, MAP, SAP, Pep-1 and AMPs were shown to obtain different structures in aqueous solutions, obtaining an amphiphilic α-helix when bound to membranes of vesicles composed of typical eukaryotic lipids [131]. Transportan 10 showed a range of conformations in the DMPC/DMPG vesicle membrane-bound state; the C-terminal α-helix is embedded in the membrane being tilted [132]. Transportan 10 and its five analogs were shown to form an α-helical conformation; the higher membrane disturbance yields a higher cellular uptake in Hela and NIH-3T3 cells [133].

A few reports are available on the characterization of the interactions of transportan with different lipid membranes. CD spectroscopy showed that the secondary structure of transportan and penetratin, isoniazid–penetratin and –transportan showed increased membrane affinity with DPPC and DPPC + mycolic acid mixed monolayers [126]. The binding of transportan 10 and its four variants to phospholipid vesicles showed the peptide-induced efflux, "becoming faster as the Gibbs energies for binding and insertion of the TP10 variants decrease" [134]. Penetratin, transportan and their variants were studied by an algorithmic method named PepLook to analyze their peptide polymorphism, showing common conformational polymorphic characteristics [135]. A cys-transportan interaction study with model DMPC membranes with moderate cholesterol concentrations by EPR showed that Cys-TP caused lipid ordering in the membranes and a large increase in the permeation of DMPC membranes. At a high cholesterol content, the effect of Cys-TP was observed, either on the membrane structure or on the membrane permeability [136]. A molecular dynamics study of the interactions between transportan 10 and a zwitterionic POPC bilayer showed the adoption of an α-helical structure on the membrane surface and binding to the membrane surface with its hydrophobic side facing the hydrophobic lipid core: "the Lys-phosphate salt bridge is a key factor in determining the orientation of the peptide in the interfacial region and the phosphate groups is also believed to be the main bottleneck for the translocation of TP10 across the membrane" [137]. The interaction of GPMVs with six Fl- CPPs (R9, Tat, penetratin, MAP, transportan and TP10) in a model system revealed that amphipathic CPPs preferentially associate with liquid-disordered membrane areas, and all of the tested CPPs accumulated into the lumen of GPMVs [138].

These studies of the structure and interactions of transportan were obtained mainly by using artificial lipid membranes, which only partly mimic the natural membranes of interest, producing the initial and necessary information for the understanding of the processes in cells and tissues.

The introduction of the PepFect (PF) and NickFect (NF) series of very efficient ON delivery vectors (see above) brought us to the need to understand their properties. To start, it was shown that PF6/siRNA [27] and PF14/SCO [28,159] formed complexes (nanoparticles) with ONs, promoting their functional cellular uptake.

The exact structures of such nanoparticles are currently being studied and, hopefully, will be characterized in order to obtain more efficient transfection vectors in the future. It is obvious that the complex nature of the components of the nanoparticles is determined by all of the possible interactions in the complex, such as electrostatic and hydrophobic interactions in the presence of a solvent (water) and its components (salts, etc.). Several examples are available concerning these studies.

The mean diameter of PF6/siRNA nanoparticles was shown to be below 200 nm [27]. DLS studies of PF14/SCO nanocomplexes as the solid formulations showed "that the particles size and particle-size distribution is highly affected by the type of excipient", and that, to our surprise, the nanocomplexes—with or without serum—possessed a net negative charge [28]. In order to improve the PepFect vectors, the peptides/Luc-plasmid complexes were studied by CD spectroscopy, DLS and a quantitative structure–activity relationship model using descriptors including hydrogen bonding, the peptide charge and the positions of nitrogen atoms. The cellular uptake data was correlated to QSAR predictions, improving the understanding of the structural requirements for cell penetration [95]. In order to make the peptides pH-responsive in an SCO assay, His-modified PepFects were introduced and characterized by DLS and CD spectroscopy. The membrane interactions in large unilammelar vesicles were studied using a calcein leakage assay. The complexes formed were small at pH 7 and grew under acidic conditions. The most promising PepFect, PepFect 132, has a significantly higher bioactivity and membrane activity [96].

3.3. Kinetics

Studies of the CPP translocation kinetics obviously add to the understanding of the internalization mechanisms. Unfortunately, only a few kinetics studies are available on CPP uptake pathways and, hence, the careful characterization of the complicated multi-step internalization is not available today. One summary of such hypothetical processes is presented in [2], in which the "free", "bound" or metabolized, etc., states of CPPs exist, likely in equilibrum. To simplify, first-order kinetics have been adopted to study CPP translocation [160]. Furthermore, the smaller cargoes seem not to influence the rate of internalization, but larger cargoes significantly slowed this rate [76,161].

Few reports are available concerning the internalization kinetics of transportan, often concerning artificial lipid membranes of different types. Hopefully, these studies will add to the general knowledge about CPP uptake mechanisms. Additionally, the difficulty in these studies is that very little is known about the kinetics of the uptake of the transportan/cargo conjugates.

3.3.1. Cells

The uptake kinetics of ^{125}I-transportan (with iodinated side-chains of Tyr by the chloramine T method) in Bowes melanoma cells "where, at 37 °C, the maximal intracellular concentration was reached in about 20 min ($t_{1/2}$ of 3–4 min)" [20].

A 2-amino benzoic acid fluorophore-modified pentapeptide conjugated by disulfide to 3-nitrotyrosine-penetratin, -transportan, -Tat and -MAP showed the cellular uptake of the cargo as an increase in fluorescence intensity when the disulfide bond of the CPP-S-S-cargo construct was reduced in the intracellular milieu [64].

Three CPPs—M918, TP10 and pVec—used a quenched fluorescence assay with the fluorophore Abz and His(DNP) as a quencher for uptake kinetics studies [162].

The uptake kinetics of PNA_{TAR}-^{125}I -transportan conjugate showed "a sigmoidal curve with a cooperativity index of 6, indicating very rapid cellular uptake" by the receptor-independent endocytotic pathway in Jurkat cells. The $[S]_{0.5}$ value and cooperativity index determined from the sigmoidal plot were 1.5 and 6 μM, respectively, with a Hill coefficient of 0.53 "suggesting that the observed cooperativity is not due to multiple conjugate binding sites on the membrane" [143].

Penetratin, Tat, transportan and polyarginine demonstrate similar kinetic uptake profiles according to FACS analysis, "being maximal at 1–3 h and independent of cell type (HeLa, A549 and CHO cell lines)". The time course of the uptake and their cellular distribution did not correlate with transferrin, a marker of clathrin-mediated endocytosis, but the peptides co-localised with a marker of the lipid raft domains, cholera toxin [68].

The uptake kinetics of ^{125}I-labeled Fl-PNA$_{Tar}$-conjugated penetratin, Tat, transportan-27, transportan-21 and transportan-22 showed a sigmoidal curve, "suggesting a cooperative interaction between the conjugate and the cellular membrane" and the possibility "that these conjugates may have more than one interaction site on the cellular membrane" [41].

3.3.2. Phospholipid Membranes

The group of Paulo Almeida has, in several reports, characterized the interactions of transportan and analogs by stopped-flow fluorescence in phospholipid membranes. These studies, although they were carried out in phospholipid vesicles and using induced dye efflux, are of high impact in the characterization of CPP interactions and translocation mechanisms in live cells. Transportan 10 induced the graded release of the contents of phospholipid vesicles, as found by an analysis of their kinetics "by directly fitting to the data the numerical solution of mathematical kinetic models". A global fit was obtained for a model in which "TP10 binds to the membrane surface and perturbs it because of the mass imbalance thus created across the bilayer". This initiates the insertion of peptide monomers "transiently into its hydrophobic core and cross the membrane, until the peptide mass imbalance is dissipated" [163]. The studies of the kinetics and thermodynamics of TP10 (and its variants) binding and induced dye efflux in phospholipid vesicles showed that "the peptide-induced efflux becomes faster as the Gibbs energies for binding and insertion of the TP10 variants decrease" [134]. Later, the kinetics of the dye efflux induced by mastoparans, masL and masX from phospholipid vesicles showed the same graded kinetic model that we previously proposed for TP10 [164]. The mechanism of TP10W in model membranes (POPC) was shown to be "determined by the thermodynamics of insertion of the peptide into the lipid bilayer from the surface-associated state" [165]. It was shown that the translocation of the TP10W is "determined by the Gibbs energy of insertion into the bilayer from the membrane interface", and that large effects on translocation are probably determined by hydrophobicity [166].

The quantitative detection of the entry of CF-TP10 into giant unilamellar vesicles (GUVs, DOPG and DOPC) containing Alexa Fluor 647 hydrazide showed first the increase of the fluorescence intensity of the GUV membrane, and at higher concentrations the leakage of AF647 was seen. It showed that "CF-TP10 can translocate across lipid bilayers without leakage of AF647, i.e., without pore formation", "but prepores formed due to thermal fluctuation of the lipid bilayers are essential" [167].

3.4. *Parallel Mechanisms of Endocytosis and Direct Translocation*

In general, CPP cellular translocation is certainly initiated by the interaction with the components of the cell surface (e.g., phospholipids or cell surface proteins such as heparan sulfate) by electrostatic/hydrophobic interactions or hydrogen bonding, often between the guanidine of arginine side chains of cationic CPPs [2]. The initial CPP interactions will probably determine the cellular uptake route of the particular CPP or CPP/cargo conjugate. The possible uptake routes are, today, mainly divided into two general types: direct translocation and endocytosis, which likely depend on the homeostasis of the cells, and occur in parallel. The direct cellular translocation pathway was suggested in the first CPP studies [20] and was later confirmed by several studies. This energy-independent route has been explained by several experimental models, e.g., the inverted micelle model and the pore-formation carpet model etc. [2]. The involvement of several endocytic pathways was shown in CPP uptake mechanisms, especially in the case of CPP/cargo conjugates, e.g., macropinocytosis, and clathrin-mediated and caveolae/lipid-raft-mediated endocytosis [168]. It is likely that these conclusions hold even for the translocation mechanisms of transportan, as summarized below, both in artificial membranes and cells.

3.4.1. Vesicles

The impact of membrane potential on the action of an AMP, lactoferricin B and TP10 in giant unilamellar vesicles (GUV) was suggested [169].

The primary amphipathicity of transportan and TP10 was shown to perturb and cause the leakage of the lipid membranes [170].

In GPMVs, the higher cholesterol content and tighter packing of the membranes reduces the accumulation of transportan, TP10 and MAP in the vesicles [171].

In model bilayers of POPC and POPG, R9, TP1, TP2 and TP3 showed that the hydrophilic and hydrophobic amino acids determine the interactions with phospholipids, and that the membrane rigidity defines the pore formation [172].

TP10 and melittin formed submicron pores in the lipid GUV membranes with the leakage of probes from the inside of the vesicles [173].

The interactions of Fl-R9, -Tat, -penetratin, -MAP, -transportan and -TP10 with giant plasma membrane vesicles (GPMVs) showed that amphipathic CPPs preferentially associate with liquid-disordered membrane areas, and that all of the tested CPPs accumulate into the lumen of GPMVs both at ambient and low temperatures, in conditions lacking endocytosis [138].

A molecular dynamics simulation study for the penetration of transportan across a DPPC bilayer yielded the free energy profile for the peptide inside the bilayer. It is energetically favorable for transportan to reside inside the bilayer because causing the higher ordering of the neighboring lipids initiates the reaching of "the other monolayer through the lysine residues". After making the connection "between the two monolayers through the peptide, the bilayer thins significantly and the formation of a pore" is likely to happen [174].

In a giant unilamellar vesicle (GUV, DOPG, DOPC) TP10 induced the leakage of fluorescent probes, inducing pore formation in lipid membranes [175].

Translocating CF-TP10 into a giant unilamellar vesicle (GUV) entered the GUV lumen before pore formation in the membrane, showing the suppression of the translocation by cholesterol, translocating "across a bilayer through transient hydrophilic prepores in the membrane" [176].

CF-TP10 translocation into the lumen of single GUVs increased the membrane potential, φ_m, up to 118 mV; it entered the GUV lumen without pore formation with an increased rate, with an increase in φ_m [177].

3.4.2. Cells

In cells, both endocytosis and direct translocation have been demonstrated for the uptake of transportan and TP10.

Biotinyl–penetratin and –transportan showed, by indirect fluorescence with fluorescein-streptavidin detection in the Bowes melanoma cell line, that penetratin and transportan enter the cells by unrelated mechanisms, and that they do not belong to the same family of translocating peptides [139]. Transportan, TP10 and a translocation study across a Caco-2 human colon cancer cell layer showed a reversible decrease of the trans-epithelial electrical resistance of the barrier model, passing the epithelial cell layer mainly by a mechanism involving a transcellular pathway; penetratin did not affect the resistance of the cell layer to the same extent [118]. Confocal Raman microscopy (CRM) and atomic force microscopy (AFM) were used to study the infiltration and physiological effects of label-free transportan in SK-Mel-2 cells, showing the rapid entry (within 5 min) and widespread distribution of the peptide throughout the cytoplasm and the nucleus after ~20 min, entering the cells via a nonendocytic mechanism with cytoskeletal changes triggered [140]. R9, Tat, transportan and TP10 showed interactions with glycosaminoglycans (GAGs) in the translocation of amphipathic CPPs [141]. The uptake of transportan and TP10 was not correlated to the presence of a receptor, energy or temperature, but when conjugated to cargoes, both by endocytosis and direct translocation [20,70], it was shown to be involved.

Several examples are available of delivery mechanism studies of cargos carried by transportan or analogs. The cellular delivery mechanisms of proteins (avidin or streptavidin) complexed with biotinylated transportan or TP10 have been studied, demonstrating cellular internalization by endocytosis. Fl-transportan and -TP10 complexed with Fl-avidin or -streptavidin-gold conjugates showed HeLa and Bowes cell transduction mostly by endocytosis with different pathways, but were also found in the perinuclear region as well as freely in the cytoplasm, suggesting endosomal escape [142]. Biotinyl–transportan, –oligoarginine and –Tat complexed to avidin–TexasRed in Cos-7 cellular uptake was induced by caveolin-dependent endocytosis [72]. Biotinyl–transportan and a biotinyl-TP10–avidin complex entered HeLa cells via the caveolin-1-dependent pathway [122]. Biotinyl–Tat, –transportan, and –pVEC complexed with avidin showed endocytotic, clathrin-dependent and independent endocytosis, a mechanism of peptide-mediated protein cellular transduction in HeLa cells with the partial depolarization of the plasma membrane [73].

Transportan-conjugated PNA oligomers also showed endocytotic cellular uptake in a few examples. A PNA_{TAR}–transportan conjugate showed rapid cellular uptake by non-receptor-dependent endocytosis [143]. Tat-, penetratin- and transportan-S-S-PNA corrected the aberrant splicing in HeLa cells stably expressing luciferase with an aberrant splice site with a mainly endocytotic—in particular macropinocytotic—mechanism [46]. PNA conjugates of M918, penetratin, Tat, transportan, TP10, MAP and pVEC showed cellular translocation by an endocytotic route, as determined by well-known endocytosis inhibitors and tracers [47].

Transportan, complexed with a variety of NPs (AgNPs, AuNPs, IONPs and QDs) and BSA or dextran, labeled by fluorescent NHS-CF555 and NHS-CF647 dyes, showed in vitro and ex vivo (live tumor slices) cell entry via a receptor-dependent macropinocytosis process; "HSPGs and scavenger receptors are likely to be the cellular receptors" for this uptake [144].

3.4.3. Mechanisms of PepFects and NickFects

PF3 and PF6 interaction studies with lipid membranes showed increased amphipathicity and the ability to insert into a lipid monolayer composed of zwitterionic phospholipids; the addition of "negatively charged phospholipids results in decreased binding and insertion of the stearylated peptides". The trifluoromethylquinoline moieties in PF6 make no significant contribution to membrane binding and insertion. Interestingly, "TP10 actively introduces pores into the bilayers of large and giant unilamellar vesicles, while PF3 and PF6 do so only at higher concentrations", suggesting their lower toxicity [145].

The membrane interactions of NF1 and NF51 with large unilamellar vesicles were studied by calcein leakage experiments showing membrane leakage by NF51, and not by NF1 and PF3.

The presence of pDNA cargo inhibited NickFect's membrane perturbation; the "peptide alone causes membrane perturbation, but the cargo complex does not" [114].

The uptake by brain endothelial cells of the PF32/pDNA nanocomplexes is initiated via LRP-1 receptor-mediated endocytosis, as well as via scavenger receptor class A and B (SR-A3, SR-A5, and SR-BI)-mediated endocytosis [146].

The SR-A3 and SR-A5 were recruited after incubation with PepFect 14/SCO complexes, initiating the translocation of SR-As to the cell surface [147].

The coexistence of distinct molecular species (monomers, self-aggregates, peptide/oligonucleotide) in complexes of Fl-PF14/cyanine5-siRNA was studied by fluorescence correlation spectroscopy (FCS) and fluorescence cross-correlation spectroscopy (FCCS) in solution and at the plasma membrane of live cells. The ratio of the complex components varied with the pH, the peptide concentration and the proximity to the plasma membrane, suggesting "that the diverse cellular uptake mechanisms, often reported for amphipathic

CPPs, might result from the synergistic effect of peptide monomers, self-aggregates and cargo complexes at the plasma membrane" [148].

The tracking of NF1 and NF51/pDNA complexes through endosomal transport was carried out by the separation of the endosomal vesicles by differential centrifugation. NF51 facilitates the rapid internalization of complexes into the cells; NF1 is less capable of inducing endosomal release, and a higher amount of complexes are routed to lysosomes for degradation [149].

3.4.4. Toxicity

The knowledge of the toxicity window for a CPP or CPP/cargo in cells or tissues is a prerequisite for the in vitro and in vivo applicability of CPPs as drug delivery vectors. Due to the (often cationic) nature of CPPs, their toxic effect is caused by the specific interaction of CPPs with the membranes of cells and organelles, or with cell ingredients. The toxicity of CPPs has been observed only in a few reports [178] and it seems that, for transportan, at least some of its bioeffects can be explained by its in vitro and in vivo side effects, and not by its drug-delivery properties.

The comparison of the degradation half-lives of DOTA-conjugated CPPs in human serum showed that different CPPs degrade in very different time ranges; for example, for MAP, TP10, NLC the $\tau_{1/2}$ was >72 h; for penetratin it was 1.2 h; and for Tat it was 8.8 h [178]. This is important to consider if the degradation products of the CPPs carry side-effect sequences, and TP10 seems to be resistant.

Transportan and analogs on their own have been tested for toxicity in few cases, sometimes in comparison with additional CPPs. It was found [76] that TP10 caused cell death in HeLa and CHO cells at a dose of 20 μM, while Tat and penetratin did not influence the cell survival at concentrations. A series of TP10 analogues showed antimicrobial activities against multidrug-resistant bacteria, showing that the toxicity could be put to service as a therapeutic benefit. TP10 killed bacteria by membrane-active and DNA-binding activities, suggesting a use as a promising antibiotic candidate [179]. Penetratin, Tat, transportan and polyarginine showed that the toxicity correlated with the mitochondrial metabolic activity and cell viability [68]. MAP and TP10 caused significant cell membrane leakage in human cancer cell lines; penetratin, Tat and pVEC showed no significant toxicity or hemolytic effect on bovine erythrocytes [180]. Molecular dynamics suggested "that higher membrane disturbance leads to higher cellular uptake of peptides" in the case of TP10 [133]. TP10 actively introduced pores into the bilayers of large and giant unilamellar vesicles, while PF3 and PF6 did so only at higher concentrations, as is consistent with the lower toxicity of PF3 and PF6 observed in previous studies [145]. The toxicity of TP10 and MAP in *Neisseria meningitides* [181] and transportan in *Mycobacterium tuberculosis* [182] has been reported.

The delivery of cargo by transportan is very efficient in many cells and also in vivo. The side effects of such delivery have been addressed in several reports. Human IFN-gamma fused with penetratin showed the decreased toxicity of IFN-gamma due to the efficient delivery [183]. Tat, penetratin and transportan, and their conjugates with BSA cargo showed no toxicity or initiation of an immune response in epithelial cells [78]. TP6, TP7 and R9 showed significant cellular toxicity above 3–5 μM in TP–PNA conjugates [49]. Stearoyl–TP10 showed the efficient delivery of a splice-correcting 2′-OMe RNA ONs for functional splice correction with no toxicity; stearoyl–R9 and –penetratin did not improve the transfection [92]. Stearoyl-TP10/plasmid nanoparticles entered different cell types with high gene expression levels with no toxicity and no nonimmunogenecity in vivo when administred intramuscularly or intradermally [60]. Analogs of TP10, TK and TH with Lys residue replaced by His showed membrane translocation with lower toxicity than TP10; the TH–camptothecin conjugate "exhibited remarkable cytotoxicity to cancer cells in a pH-dependent manner" [150]. TP10 and PepFects (PF3, PF4, and PF6) and their complexes with plasmid and siRNA showed no effect on the cytotoxic and immunogenic response, e.g., IL-1β, IL-18 and TNF-α cytokine release, cell viability, and apoptosis in vitro and in vivo [151]. TP, TP10, TP–biot1, TP–biot13 and TP10–biot1 conjugates with FITC–

streptavidin and siRNA in the HT29 and HCT116 cell lines showed no significant cytotoxic effect at 0.5–5 μM [75]. Transportan analog T9(dR) complexed with siRNA (against a nucleoprotein (NP) gene segment of the influenza virus) showed the delivery of siRNA into 293T, MDCK, RAW and A549 cells with low cellular toxicity, and in mice infected with the PR8 influenza virus when they were given a combined tail vein injection of siNP and T9(dR) or TP; all of the mice survived and showed weight recovery at 2 weeks post-infection [57]. Chloroquine–TP10 conjugates showed higher antiplasmodial activity than the parent TP10 "at the cost of an increased hemolytic activity" seeming "unsuitable for safe intracellular delivery of antimalarial aminoquinolines due to hemolysis issues" [152]. Conjugates of TP10 with ciprofloxacin or levofloxacin (fluoroquinolone antibacterial agents) showed antifungal in vitro activity against human pathogenic yeasts of the *Candida* genus, showing cytotoxicity against the HEK293, HepG2 and LLC-PK1 cells causing the intrinsic cytoplasmic membrane disruption activity [153].

3.5. Metabolomics and Transcriptomics

Genomic, transcriptomic and proteomic data is increasing in life sciences, showing growing potential for integrative proteogenomic data analyses in these areas. This analysis enables us to discover novel pathways of biochemical processes as well as the involvement of novel proteins and genes, improving our understanding of biological processes [184]. These tools are applied in a few cases in order to understand the novel CPP mechanisms, including transportan, and its modifications and analogs.

Metabolomics studies yield "the unique chemical fingerprints that specific cellular processes leave behind", identifying the small-molecule metabolite profiles [185], and could be called a functional readout of the genome, a functional genome, or a proteome [186]. We compared the alterations in the cytosolic metabolome of CHO cells caused by exposure to transportan, penetratin, Tat, R9 and MAP, showing that the "intracellular metabolome was the most affected by transportan followed by Tat and MAP"; transportan mostly affected the cellular redox potential, and depleted energy and the pools of purines and pyrimidines [187]. It remains to connect these data to the functionality of transportan in cargo delivery.

Professor Jim Eberwine's group introduced the PAIR (PNA-assisted identification of RBPs and RNA binding proteins) method for the identification and dynamics of the RNA-RBP interactions in live cells, patent number 8632972, which was further developed by the University of Pennsylvania. It applies transportan–PNA (12–18 ONs) with a photo-activatible label, targeting in live cells mRNA and, after UV light stimulation, crosslinks the PNA–Bpa–RBP complexes, which after purification can be identified by MS "in order to evaluate essential regulatory proteins that control all modes of RNA processing and regulation" [42,188–191].

Real-time transcriptome in vivo analysis (TIVA) was obtained by the application of transportan–TIVA–tag [192,193], delivered to the cytoplasm in vivo, enabling us to target and isolate cell-specific transcriptomes upon photoactivation. The method permitted us to yield the transcriptomic landscape of individual cells in mouse brains.

Following time course treatments with penetratin, PepFect14, mtCPP1 and TP10, HeLa cells were transcriptionally profiled by RNA sequencing, showing the response of "specific sets of genes related to ribosome biogenesis, microtubule dynamics and long-noncoding RNAs being differentially expressed compared to untreated controls" [194].

PF6/siRNA nanoparticles showed robust RNAi responses in various cell types with minimal associated transcriptomic or proteomic changes, and promoted strong RNAi responses in different organs following systemic delivery in mice without any associated toxicity [27]. This work is a good example where the transcriptomics analysis enables us to decide about the side effects of ON transfection.

4. Targeting

The efficient cellular uptake of CPPs in most cell lines in vitro has been demonstrated repeatedly, although some examples exist in which a certain CPP is more efficient in certain (cancer) cells, and where the PepFects vectors are able to transfect "difficult" cell lines exclusively [2,27]. In vivo, the lack of cell/organ specificity has been demonstrated, following any type of administration. In drug development, this suggests non-desired off-target side-effects and, hence, the intensive development of the targeted delivery of CPPs and their conjugated cargos is ongoing. The field is fueled by the broad interest in cancer targeting in vivo, e.g., by the application of the tumor-targeting antibodies, the targeting of specific cell surface receptors or antigens at tumor sites, the targeting of subcellular organelles with, e.g., NLS mitochondrial targeting sequences using the activatable CPP (aCPP) strategy. Below, I summarize the few reports which are available for the targeting of transportan and its modified versions using these strategies; see Figure 1.

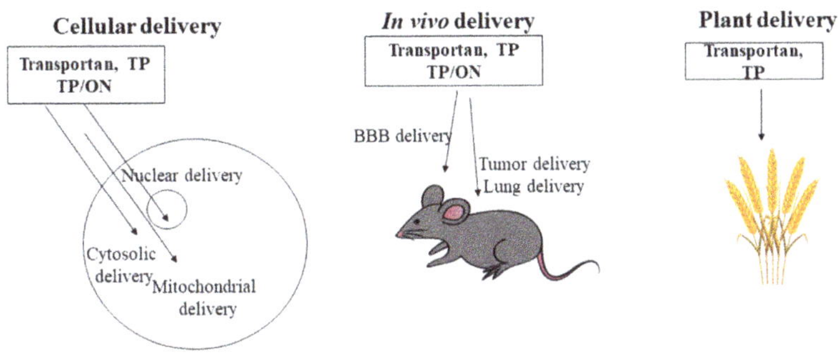

Figure 1. Schematic summary of the reported applications of transportan and its analogs with specific targeting.

Transportan, like some CPPs, showed nuclear targeting by itself [20]; see Figure 1. Later, the transportan analogues PepFects and NickFects showed powerful DNA nuclear delivery properties. The nuclear uptake of a double-stranded oligonucleotide NF-kappaB decoy ON in rat primary glial cells was facilitated by noncovalent binding to transportan 10 via a complementary PNA sequence [195,196].

Transportan showed a high translocation efficiency in plants, e.g., tobacco protoplasts [119]. The permeabilization of the cell wall increased the uptake of Fl–Tat, –Tat$_2$, -mutated–Tat, –pVEC and –transportan in immature wheat embryos [59]. Several CPPs were able to translocate into algal cells: FITC–pVEC, –Tat, –penetratin and –transportan translocated into *Chlamydomonas reinhardtii* [197], and the pVEC, Tat, penetratin and transportan translocated into algal cells [197]

Few attempts have been made to study and develop BBB-passing delivery; see Figure 1. Transportan could effectively cross the Caco-2 cell layer with a trans-cellular mechanism in a two-chambered Transwell model system [118], suggesting the possibility for a BBB-passing delivery system. The in vivo BBB transport characteristics of pVEC, SynB3, Tat, TP10 and TP10-2 showed a low brain influx for transportan, but the significant labeling of the brain parenchyma by the others, which was not correlated with their CPP properties [124]. pVEC, SynB3 and Tat 47–57 showed very high unidirectional BBB influx rates, whereas TP10 and TP10-2 showed a negligible-to-low brain influx with no significant efflux out of the brain, except for pVEC [124]. Myristoyl–transportan–Tf with encapsulated siRNA showed targeted delivery through BBB with functional gene silencing in vitro [99]. A TP10–vancomycin conjugate showed antibacterial activity and crossed the BBB in a mouse brain after i.v. administration [198]. A TP10–dopamine conjugate showed the penetration of the BBB and access to the brain tissue with anti-parkinsonian activity (higher than that of l-DOPA), "exhibiting high affinity to both dopamine D_1 and D_2 receptors" (in the case of D_1, a much higher activity than that of DA) [124].

The PepFect delivery strategy of ONs for drug delivery across the BBB has been attempted, although these studies are still at the in vitro level; see Figure 1. PF32 with the targeting ligand angiopep-2 (binding to the endothelial LDLR-related protein-1, LRP-1), showed complexed pDNA delivery in an in vitro Transwell model of the BBB through receptor-mediated endocytosis via scavenger receptors class A and B (SCARA3, SCARA5 and SR-BI) [146]. PF14 and PF28, modified covalently with BBB targeting peptides, and angiopep-2 modified by covalent conjugation and complexed with $siRNA_{Luc}$ showed specific gene-silencing efficiency in human glioblastoma cells U87 MG-luc2, as compared to non-glioma targeted cells [199].

Cancer-targeting is often the subject of CPP delivery development; see Figure 1. iRGD- and CPP (myristoyl-transportan)-grafted PEG backbone structure NPs were shown to carry siRNA into mammalian cells, silencing the RNA target in a mouse model of cancer [200]. TP10–SRC1LXXLL with the peptide from human SRC-1 nuclear receptor box 1, containing an LXXLL motif, induced the dose-dependent cell death of breast cancer cells by reducing the viability and proliferation of breast cancer MDA-MB-231 cells and dermal fibroblasts, suggesting it as a drug candidate in the treatment of cancers [201].

Due to its high efficacy, PepFect- and NickFect-assisted ON delivery has earned attention in the development of future cancer gene therapies by specific targeting. It is highly valuable that these transfection-targeting experiments are available on animal models in vivo.

Using the i.v. administration route in vivo, PEG- and MMP substrate-functionalized PF14 (aCPP approach) complexed with pDNA showed the efficient induction of gene expression specifically in tumors, avoiding normal tissues [97].

NF55/pDNA nanoparticles showed the transfection of the majority of cells and in vivo specific tumor transfection in various mouse tumor models, including an intracranial glioblastoma model [30].

MMP-2/-9 activatable PF144/pDNA nanocomplexes for anti-angiogenic gene (encoding short hairpin RNA) delivery showed the inhibition of tumor growth by silencing the vascular endothelial growth factor (VEGF) expression in orthotopic 4T1 breast tumor-bearing mice [202].

PF14 and NF55 preferentially transfect lung tissue upon their systemic administration with the complexed siRNA and pDNA encoding shRNA against cytokine TNFα in models of acute lung inflammation and asthma in mice, and showed efficient anti-inflammatory effects in both disease models [203].

PF14, covalently conjugated to mitochondrial-penetrating peptide, mtCPP1, complexed with ONs affected biological functions both in the cytoplasm and on the mitochondria, suggesting "the potential to be used as a treatment for patients with mitochondrial disorders" [204].

5. Conclusions

Transportan and its modified versions—e.g., TP10, PepFects and NickFects—are widely used as efficient delivery vectors of a wide range of cargos, such as small molecules, peptides and proteins, as well as oligonucleotides such as short ONs, siRNA, miRNA, decoy ON, plasmids and mRNA. These various examples of applications have been used in studies of CPP mechanisms as well as for the development of therapies and the diagnosis of diseases.

Remarkably, PepFects and NickFects have demonstrated the ability to form stable nanoparticles with the very efficient transfection of ONs in vitro and in vivo, paving a way to future gene therapy. The addition of these CPPs to the available nanoparticle platforms may, in the future, contribute to novel, improved drug delivery systems.

Transportan and its versions have been modified in order to achieve the controlled targeted delivery of bioactive cargos, especially for future cancer gene therapy. For that, the detailed knowledge of CPP mechanisms, toxicity, immunogenicity, efficiency and kinetics should be achieved, and this work is ongoing.

Funding: The writing of this review was supported by the Swedish Research Council (VR-NT, 2017-03691), by the Estonian Ministry of Education and Research (project IUT20-26), and by the EU (project 2014–2020.4.01.15-0013).

Data Availability Statement: The study did not report any original data.

Acknowledgments: I acknowledge Matjaž Zorko, Kaido Kurrikoff and Piret Arukuusk for reading and commenting on this manuscript.

Conflicts of Interest: The authors declare no conflict of interest.

Abbreviations

Abbreviations

aa	amino acid(s)
Abz	aminobenzoic acid
AFM	atomic force microscopy
AMP	antimicrobial peptide
AS	antisense
ASO	antisense oligonucleotide
BBB	blood-brain-barrier
BSA	bovine serum albumin
CD	circular dichroism
CF-	carboxyfluoresceinyl
Chol	cholesteryl
CPP	Cell-penetrating peptides, e.g., pVEC, SynB3, CADY etc.
CRM	confocal Raman microscopy
DLS	dynamic light scattering
DMPC	1,2-dimyristoyl-sn-glycero-3-phosphocholine
DMPG	1,2-dimyristoyl-sn-glycero-3-phosphoglycerol
DPPC	dipalmitoylphosphatidylcholine
FACS	Fluorescence-activated cell sorting
Fl-	fluorescently labelled
GFP	green fluorescent protein
GPMV	giant plasma membrane vesicles
GUV	giant unilamellar phospholipid vesicles
LNA	locked nucleic acid
Luc	luciferase
LUVs	large unilamellar phospholipid vesicles
MAP	membrane active peptide, CPP
miRNA	microRNA
NF	NickFect, e.g., NF51 and NF55
NLS	nuclear localisation signal
NMR	nuclear magnetic resonance
NP	nanoparticle
ON	oligonucleotide
O	ornithine
pDNA	plasmid
PEI	polyethyleneimine
PET	Positron emission tomography
PF	PepFects, e.g., PF6, PF14
PMO	phosphorodiamidate morpholino ON
PNA	peptide nucleic acids
POPC	1-palmitoyl-2-oleoyl-sn-glycero-3-phosphocholine
PS	phosphothioate ON
QSAR	quantitative structure-activity relationhip
R9	Arg_9

SCO	splice correctin ON
shRNA	short hairpin RNA
siRNA	short interfering RNA
TAMRA	5-(and-6)-carboxytetramethylrhodamine
TEM	transmission electron microscopy
TP	transportan
TP10	transportan 10
Tf	transferrin

References

1. Langel, Ü. Cell Penetrating Peptides. In *Methods and Protocols*, 2nd ed.; Humana Press: New York, NY, USA; Heidelberg, Germany; Dordrecht, The Netherland; London, UK, 2015; Volume 1324.
2. Langel, Ü. *CPP, Cell-Penetrating Peptides*; Springer Nature Singapore Ltd.: Singapore, 2019; ISBN 9789811387470.
3. Frankel, A.D.; Pabo, C.O. Cellular uptake of the tat protein from human immunodeficiency virus. *Cell* **1988**, *55*, 1189–1193. [CrossRef]
4. Green, M.; Loewenstein, P.M. Autonomous functional domains of chemically synthesized human immunodeficiency virus tat trans-activator protein. *Cell* **1988**, *55*, 1179–1188. [CrossRef]
5. Fawell, S.; Seery, J.; Daikh, Y.; Moore, C.; Chen, L.L.; Pepinsky, B.; Barsoum, J. Tat-mediated delivery of heterologous proteins into cells. *Proc. Natl. Acad. Sci. USA* **1994**, *91*, 664–668. [CrossRef]
6. Joliot, A.; Pernelle, C.; Deagostini-Bazin, H.; Prochiantz, A. Antennapedia homeobox peptide regulates neural morphogenesis. *Proc. Natl. Acad. Sci. USA* **1991**, *88*, 1864–1868. [CrossRef] [PubMed]
7. Derossi, D.; Joliot, A.H.; Chassaing, G.; Prochiantz, A. The third helix of the Antennapedia homeodomain translocates through biological membranes. *J. Biol. Chem.* **1994**, *269*, 10444–10450. [CrossRef]
8. Vives, E.; Brodin, P.; Lebleu, B. A truncated HIV-1 Tat protein basic domain rapidly translocates through the plasma membrane and accumulates in the cell nucleus. *J. Biol. Chem.* **1997**, *272*, 16010–16017. [CrossRef]
9. Gestin, M.; Dowaidar, M.; Langel, Ü. Uptake Mechanism of Cell-Penetrating Peptides. *Adv. Exp. Med. Biol.* **2017**, *1030*, 255–264.
10. Sánchez-Navarro, M. Advances in peptide-mediated cytosolic delivery of proteins. *Adv. Drug Deliv. Rev.* **2021**, *171*, 187–198. [CrossRef] [PubMed]
11. Timur, S.S.; Gürsoy, R.N. The role of peptide-based therapeutics in oncotherapy. *J. Drug Target.* **2021**. [CrossRef]
12. Langel, Ü.; Pooga, M.; Kairane, C.; Zilmer, M.; Bartfai, T. A galanin-mastoparan chimeric peptide activates the Na+,K(+)-ATPase and reverses its inhibition by ouabain. *Regul. Pept.* **1996**, *62*, 47–52. [CrossRef]
13. Soomets, U.; Hällbrink, M.; Zorko, M.; Langel, Ü. From galanin and mastoparan to galparan and transportan. *Curr. Top. Pept. Protein Res.* **1997**, *2*, 83–113.
14. Langel, Ü.; Land, T.; Bartfai, T. Design of chimeric peptide ligands to galanin receptors and substance P receptors. *Int. J. Pept Protein Res.* **1992**, *39*, 516–522. [CrossRef]
15. Bartfai, T.; Bedecs, K.; Land, T.; Langel, Ü.; Bertorelli, R.; Girotti, P.; Consolo, S.; Xu, X.J.; Wiesenfeld-Hallin, Z.; Nilsson, S.; et al. M-15: High-affinity chimeric peptide that blocks the neuronal actions of galanin in the hippocampus, locus coeruleus, and spinal cord. *Proc. Natl. Acad. Sci. USA* **1991**, *88*, 10961–10965. [CrossRef]
16. Pooga, M.; Lindgren, M.; Hällbrink, M.; Bråkenhielm, E.; Langel, Ü. Galanin-based peptides, galparan and transportan, with receptor-dependent and independent activities. *Ann. N. Y. Acad. Sci.* **1998**, *863*, 450–453. [CrossRef]
17. Östenson, C.-G.; Zaitsev, S.; Berggren, P.O.; Efendic, S.; Langel, Ü.; Bartfai, T. Galparan: A powerful insulin-releasing chimeric peptide acting at a novel site. *Endocrinology* **1997**, *138*, 3308–3313. [CrossRef]
18. Consolo, S.; Baldi, G.; Nannini, L.; Uboldi, M.C.; Pooga, M.; Langel, Ü.; Bartfai, T. Galparan induces in vivo acetylcholine release in the frontal cortex. *Brain Res.* **1997**, *756*, 174–178. [CrossRef]
19. Zorko, M.; Pooga, M.; Saar, K.; Rezaei, K.; Langel, Ü. Differential regulation of GTPase activity by mastoparan and galparan. *Arch. Biochem. Biophys.* **1998**, *349*, 321–328. [CrossRef] [PubMed]
20. Pooga, M.; Hällbrink, M.; Zorko, M.; Langel, Ü. Cell penetration by transportan. *FASEB J.* **1998**, *12*, 67–77. [CrossRef] [PubMed]
21. Soomets, U.; Lindgren, M.; Gallet, X.; Hällbrink, M.; Elmquist, A.; Balaspiri, L.; Zorko, M.; Pooga, M.; Brasseur, R.; Langel, Ü. Deletion analogues of transportan. *Biochim. Biophys. Acta* **2000**, *1467*, 165–176. [CrossRef]
22. Morris, M.C.; Vidal, P.; Chaloin, L.; Heitz, F.; Divita, G. A new peptide vector for efficient delivery of oligonucleotides into mammalian cells. *Nucleic Acids Res.* **1997**, *25*, 2730–2736. [CrossRef] [PubMed]
23. Morris, M.C.; Depollier, J.; Mery, J.; Heitz, F.; Divita, G. A peptide carrier for the delivery of biologically active proteins into mammalian cells. *Nat. Biotechnol.* **2001**, *19*, 1173–1176. [CrossRef] [PubMed]
24. Crombez, L.; Aldrian-Herrada, G.; Konate, K.; Nguyen, Q.N.; McMaster, G.K.; Brasseur, R.; Heitz, F.; Divita, G. A new potent secondary amphipathic cell-penetrating peptide for siRNA delivery into mammalian cells. *Mol. Ther.* **2009**, *17*, 95–103. [CrossRef]
25. Abes, R.; Arzumanov, A.; Moulton, H.; Abes, S.; Ivanova, G.; Gait, M.J.; Iversen, P.; Lebleu, B. Arginine-rich cell penetrating peptides: Design, structure-activity, and applications to alter pre-mRNA splicing by steric-block oligonucleotides. *J. Pept. Sci.* **2008**, *14*, 455–460. [CrossRef] [PubMed]

26. Talaei, B.; Fathi Vavsari, V.; Balalaie, S.; Arabanian, A.; Bijanzadeh, H.R. Synthesis of Novel Peptides Using Unusual Amino Acids. *Iran. J. Pharm. Res.* **2020**, *19*, 370–382. [PubMed]
27. El-Andaloussi, S.; Lehto, T.; Mäger, I.; Rosenthal-Aizman, K.; Oprea, I.I.; Simonson, O.E.; Sork, H.; Ezzat, K.; Copolovici, D.M.; Kurrikoff, K.; et al. Design of a peptide-based vector, PepFect6, for efficient delivery of siRNA in cell culture and systemically in vivo. *Nucleic Acids Res.* **2011**, *39*, 3972–3987. [CrossRef] [PubMed]
28. Ezzat, K.; Andaloussi, S.E.; Zaghloul, E.M.; Lehto, T.; Lindberg, S.; Moreno, P.M.; Viola, J.R.; Magdy, T.; Abdo, R.; Guterstam, P.; et al. PepFect 14, a novel cell-penetrating peptide for oligonucleotide delivery in solution and as solid formulation. *Nucleic Acids Res.* **2011**, *39*, 5284–5298. [CrossRef]
29. Arukuusk, P.; Pärnaste, L.; Oskolkov, N.; Copolovici, D.M.; Margus, H.; Padari, K.; Moll, K.; Maslovskaja, J.; Tegova, R.; Kivi, G.; et al. New generation of efficient peptide-based vectors, NickFects, for the delivery of nucleic acids. *Biochim. Biophys. Acta* **2013**, *1828*, 1365–1373. [CrossRef]
30. Freimann, K.; Arukuusk, P.; Kurrikoff, K.; Vasconcelos, L.D.F.; Veiman, K.L.; Uusna, J.; Margus, H.; Garcia-Sosa, A.T.; Pooga, M.; Langel, U. Optimization of in vivo DNA delivery with NickFect peptide vectors. *J. Control. Release* **2016**, *241*, 135–143. [CrossRef]
31. Lindsay, M.A. Peptide-mediated cell delivery: Application in protein target validation. *Curr. Opin. Pharmacol.* **2002**, *2*, 587–594. [CrossRef]
32. El-Andaloussi, S.; Holm, T.; Langel, Ü. Cell-penetrating peptides: Mechanisms and applications. *Curr. Pharm. Des.* **2005**, *11*, 3597–3611. [CrossRef]
33. Allinquant, B.; Hantraye, P.; Mailleux, P.; Moya, K.; Bouillot, C.; Prochiantz, A. Downregulation of amyloid precursor protein inhibits neurite outgrowth in vitro. *J. Cell Biol.* **1995**, *128*, 919–927. [CrossRef]
34. Nielsen, P.E.; Egholm, M.; Berg, R.H.; Buchardt, O. Sequence-selective recognition of DNA by strand displacement with a thymine-substituted polyamide. *Science* **1991**, *254*, 1497–1500. [CrossRef]
35. Langel, Ü.; Bartfai, T.; Pooga, M.; Valkna, A.; Saar, K.; Hällbrink, M. Membrane-Permeable Constructs for Transport across a Lipid Membrane. U.S. Patent No. US6025140, 24 July 1997.
36. Pooga, M.; Soomets, U.; Hällbrink, M.; Valkna, A.; Saar, K.; Rezaei, K.; Kahl, U.; Hao, J.X.; Xu, X.J.; Wiesenfeld-Hallin, Z.; et al. Cell penetrating PNA constructs regulate galanin receptor levels and modify pain transmission in vivo. *Nat. Biotechnol.* **1998**, *16*, 857–861. [CrossRef]
37. Kilk, K.; Elmquist, A.; Saar, K.; Pooga, M.; Land, T.; Bartfai, T.; Soomets, U.; Langel, Ü. Targeting of antisense PNA oligomers to human galanin receptor type 1 mRNA. *Neuropeptides* **2004**, *38*, 316–324. [CrossRef]
38. Ostenson, C.G.; Sandberg-Nordqvist, A.C.; Chen, J.; Hällbrink, M.; Rotin, D.; Langel, U.; Efendic, S. Overexpression of protein-tyrosine phosphatase PTP sigma is linked to impaired glucose-induced insulin secretion in hereditary diabetic Goto-Kakizaki rats. *Biochem. Biophys. Res. Commun.* **2002**, *291*, 945–950. [CrossRef]
39. Zhang, Q.; Tang, J.; Fu, L.; Ran, R.; Liu, Y.; Yuan, M.; He, Q. A pH-responsive alpha-helical cell penetrating peptide-mediated liposomal delivery system. *Biomaterials* **2013**, *34*, 7980–7993. [CrossRef] [PubMed]
40. Turner, J.J.; Arzumanov, A.A.; Gait, M.J. Synthesis, cellular uptake and HIV-1 Tat-dependent trans-activation inhibition activity of oligonucleotide analogues disulphide-conjugated to cell-penetrating peptides. *Nucleic Acids Res.* **2005**, *33*, 27–42. [CrossRef] [PubMed]
41. Tripathi, S.; Chaubey, B.; Ganguly, S.; Harris, D.; Casale, R.A.; Pandey, V.N. Anti-HIV-1 activity of anti-TAR polyamide nucleic acid conjugated with various membrane transducing peptides. *Nucleic Acids Res.* **2005**, *33*, 4345–4356. [CrossRef] [PubMed]
42. Zielinski, J.; Kilk, K.; Peritz, T.; Kannanayakal, T.; Miyashiro, K.Y.; Eiriksdottir, E.; Jochems, J.; Langel, Ü.; Eberwine, J. In vivo identification of ribonucleoprotein-RNA interactions. *Proc. Natl. Acad. Sci. USA* **2006**, *103*, 1557–1562. [CrossRef]
43. Fossat, P.; Dobremez, E.; Bouali-Benazzouz, R.; Favereaux, A.; Bertrand, S.S.; Kilk, K.; Leger, C.; Cazalets, J.R.; Langel, Ü.; Landry, M.; et al. Knockdown of L calcium channel subtypes: Differential effects in neuropathic pain. *J. Neurosci.* **2010**, *30*, 1073–1085. [CrossRef]
44. Kaushik, N.; Basu, A.; Palumbo, P.; Myers, R.L.; Pandey, V.N. Anti-TAR polyamide nucleotide analog conjugated with a membrane-permeating peptide inhibits human immunodeficiency virus type 1 production. *J. Virol.* **2002**, *76*, 3881–3891. [CrossRef]
45. Dominski, Z.; Kole, R. Restoration of correct splicing in thalassemic pre-mRNA by antisense oligonucleotides. *Proc. Natl. Acad. Sci. USA* **1993**, *90*, 8673–8677. [CrossRef]
46. El-Andaloussi, S.; Johansson, H.J.; Lundberg, P.; Langel, Ü. Induction of splice correction by cell-penetrating peptide nucleic acids. *J. Gene Med.* **2006**, *8*, 1262–1273. [CrossRef] [PubMed]
47. Lundin, P.; Johansson, H.; Guterstam, P.; Holm, T.; Hansen, M.; Langel, Ü. Distinct uptake routes of cell-penetrating peptide conjugates. *Bioconjug. Chem.* **2008**, *19*, 2535–2542. [CrossRef]
48. Turner, J.J.; Jones, S.; Fabani, M.M.; Ivanova, G.; Arzumanov, A.A.; Gait, M.J. RNA targeting with peptide conjugates of oligonucleotides, siRNA and PNA. *Blood Cells Mol. Dis.* **2007**, *38*, 1–7. [CrossRef] [PubMed]
49. Bendifallah, N.; Rasmussen, F.W.; Zachar, V.; Ebbesen, P.; Nielsen, P.E.; Koppelhus, U. Evaluation of cell-penetrating peptides (CPPs) as vehicles for intracellular delivery of antisense peptide nucleic acid (PNA). *Bioconjug. Chem.* **2006**, *17*, 750–758. [CrossRef] [PubMed]
50. Boisguérin, P.; Deshayes, S.; Gait, M.J.; O'Donovan, L.; Godfrey, C.; Betts, C.A.; Wood, M.J.A.; Lebleu, B. Delivery of therapeutic oligonucleotides with cell penetrating peptides. *Adv. Drug Deliv. Rev.* **2015**, *87*, 52–67. [CrossRef]

51. Boisguerin, P.; O'Donovan, L.; Gait, M.J.; Lebleu, B. In Vitro Assays to Assess Exon Skipping in Duchenne Muscular Dystrophy. *Methods Mol. Biol.* **2015**, *1324*, 317–329.
52. Pooga, M.; Land, T.; Bartfai, T.; Langel, Ü. PNA oligomers as tools for specific modulation of gene expression. *Biomol. Eng.* **2001**, *17*, 183–192. [CrossRef]
53. Sheikh, O.; Yokota, T. Developing DMD therapeutics: A review of the effectiveness of small molecules, stop-codon readthrough, dystrophin gene replacement, and exon-skipping therapies. *Expert Opin. Investig. Drugs* **2021**, *30*, 1–10. [CrossRef]
54. Fisher, L.; Soomets, U.; Cortes Toro, V.; Chilton, L.; Jiang, Y.; Langel, Ü.; Iverfeldt, K. Cellular delivery of a double-stranded oligonucleotide NFkappaB decoy by hybridization to complementary PNA linked to a cell-penetrating peptide. *Gene Ther.* **2004**, *11*, 1264–1272. [CrossRef] [PubMed]
55. Muratovska, A.; Eccles, M.R. Conjugate for efficient delivery of short interfering RNA (siRNA) into mammalian cells. *FEBS Lett.* **2004**, *558*, 63–68. [CrossRef]
56. Ishihara, T.; Goto, M.; Kodera, K.; Kanazawa, H.; Murakami, Y.; Mizushima, Y.; Higaki, M. Intracellular delivery of siRNA by cell-penetrating peptides modified with cationic oligopeptides. *Drug Deliv.* **2009**, *16*, 153–159. [CrossRef]
57. Zhang, C.; Ren, W.; Liu, Q.; Tan, Z.; Li, J.; Tong, C. Transportan-derived cell-penetrating peptide delivers siRNA to inhibit replication of influenza virus in vivo. *Drug Des. Devel. Ther.* **2019**, *13*, 1059–1068. [CrossRef] [PubMed]
58. Kilk, K.; El Andaloussi, S.; Järver, P.; Meikas, A.; Valkna, A.; Bartfai, T.; Kogerman, P.; Metsis, M.; Langel, Ü. Evaluation of transportan 10 in PEI mediated plasmid delivery assay. *J. Control. Release* **2005**, *103*, 511–523. [CrossRef] [PubMed]
59. Chugh, A.; Eudes, F. Study of uptake of cell penetrating peptides and their cargoes in permeabilized wheat immature embryos. *FEBS J.* **2008**, *275*, 2403–2414. [CrossRef]
60. Lehto, T.; Simonson, O.E.; Mager, I.; Ezzat, K.; Sork, H.; Copolovici, D.M.; Viola, J.R.; Zaghloul, E.M.; Lundin, P.; Moreno, P.M.; et al. A peptide-based vector for efficient gene transfer in vitro and in vivo. *Mol. Ther.* **2011**, *19*, 1457–1467. [CrossRef]
61. Van den Berg, A.; Dowdy, S.F. Protein transduction domain delivery of therapeutic macromolecules. *Curr. Opin. Biotechnol.* **2011**, *22*, 888–893. [CrossRef]
62. Lönn, P.; Dowdy, S.F. Cationic PTD/CPP-mediated macromolecular delivery: Charging into the cell. *Expert Opin. Drug Deliv.* **2015**, *12*, 1627–1636. [CrossRef] [PubMed]
63. Lukanowska, M.; Howl, J.; Jones, S. Bioportides: Bioactive cell-penetrating peptides that modulate cellular dynamics. *Biotechnol. J.* **2013**, *8*, 918–930. [CrossRef]
64. Hällbrink, M.; Floren, A.; Elmquist, A.; Pooga, M.; Bartfai, T.; Langel, Ü. Cargo delivery kinetics of cell-penetrating peptides. *Biochim. Biophys. Acta* **2001**, *1515*, 101–109. [CrossRef]
65. Adachi, Y.; Sakamoto, K.; Umemoto, T.; Fukuda, Y.; Tani, A.; Asami, T. Investigation on cellular uptake and pharmacodynamics of DOCK2-inhibitory peptides conjugated with cell-penetrating peptides. *Bioorg. Med. Chem.* **2017**, *25*, 2148–2155. [CrossRef] [PubMed]
66. Mikhailov, A.; Sankai, Y. Cell penetrating caspase substrates promote survival of the transplanted cells. *BMC Res. Notes* **2019**, *12*, 440. [CrossRef] [PubMed]
67. Bhattarai, P.; Thomas, A.K.; Cosacak, M.I.; Papadimitriou, C.; Mashkaryan, V.; Zhang, Y.; Kizil, C. Modeling Amyloid-beta42 Toxicity and Neurodegeneration in Adult Zebrafish Brain. *J. Vis. Exp.* **2017**, *128*, 56014.
68. Jones, S.W.; Christison, R.; Bundell, K.; Voyce, C.J.; Brockbank, S.M.; Newham, P.; Lindsay, M.A. Characterisation of cell-penetrating peptide-mediated peptide delivery. *Br. J. Pharmacol.* **2005**, *145*, 1093–1102. [CrossRef] [PubMed]
69. Margus, H.; Padari, K.; Pooga, M. Insights into cell entry and intracellular trafficking of peptide and protein drugs provided by electron microscopy. *Adv. Drug Deliv. Rev.* **2013**, *65*, 1031–1038. [CrossRef]
70. Pooga, M.; Kut, C.; Kihlmark, M.; Hällbrink, M.; Fernaeus, S.; Raid, R.; Land, T.; Hallberg, E.; Bartfai, T.; Langel, Ü. Cellular translocation of proteins by transportan. *FASEB J.* **2001**, *15*, 1451–1453. [CrossRef]
71. Räägel, H.; Hein, M.; Kriiska, A.; Säälik, P.; Floren, A.; Langel, Ü.; Pooga, M. Cell-penetrating peptide secures an efficient endosomal escape of an intact cargo upon a brief photo-induction. *Cell Mol. Life Sci.* **2013**, *70*, 4825–4839. [CrossRef]
72. Räägel, H.; Säälik, P.; Hansen, M.; Langel, Ü.; Pooga, M. CPP-protein constructs induce a population of non-acidic vesicles during trafficking through endo-lysosomal pathway. *J. Control. Release* **2009**, *139*, 108–117. [CrossRef]
73. Säälik, P.; Elmquist, A.; Hansen, M.; Padari, K.; Saar, K.; Viht, K.; Langel, Ü.; Pooga, M. Protein cargo delivery properties of cell-penetrating peptides. A comparative study. *Bioconjug. Chem.* **2004**, *15*, 1246–1253. [CrossRef]
74. Rinne, J.; Albarran, B.; Jylhava, J.; Ihalainen, T.O.; Kankaanpaa, P.; Hytonen, V.P.; Stayton, P.S.; Kulomaa, M.S.; Vihinen-Ranta, M. Internalization of novel non-viral vector TAT-streptavidin into human cells. *BMC Biotechnol.* **2007**, *7*, 1. [CrossRef]
75. Wierzbicki, P.M.; Kogut-Wierzbicka, M.; Ruczynski, J.; Siedlecka-Kroplewska, K.; Kaszubowska, L.; Rybarczyk, A.; Alenowicz, M.; Rekowski, P.; Kmiec, Z. Protein and siRNA delivery by transportan and transportan 10 into colorectal cancer cell lines. *Folia Histochem. Cytobiol.* **2014**, *52*, 270–280. [CrossRef]
76. El-Andaloussi, S.; Järver, P.; Johansson, H.J.; Langel, Ü. Cargo-dependent cytotoxicity and delivery efficacy of cell-penetrating peptides: A comparative study. *Biochem. J.* **2007**, *407*, 285–292. [CrossRef] [PubMed]
77. Saar, K.; Saar, H.; Hansen, M.; Langel, Ü.; Pooga, M. Distribution of CPP-Protein Complexes in Freshly Resected Human Tissue Material. *Pharmaceuticals* **2010**, *3*, 621–635. [CrossRef] [PubMed]
78. Carter, E.; Lau, C.Y.; Tosh, D.; Ward, S.G.; Mrsny, R.J. Cell penetrating peptides fail to induce an innate immune response in epithelial cells in vitro: Implications for continued therapeutic use. *Eur. J. Pharm. Biopharm.* **2013**, *85*, 12–19. [CrossRef]

79. Sayers, E.J.; Cleal, K.; Eissa, N.G.; Watson, P.; Jones, A.T. Distal phenylalanine modification for enhancing cellular delivery of fluorophores, proteins and quantum dots by cell penetrating peptides. *J. Control. Release* **2014**, *195*, 55–62. [CrossRef]
80. Patel, S.G.; Sayers, E.J.; He, L.; Narayan, R.; Williams, T.L.; Mills, E.M.; Allemann, R.K.; Luk, L.Y.P.; Jones, A.T.; Tsai, Y.H. Cell-penetrating peptide sequence and modification dependent uptake and subcellular distribution of green florescent protein in different cell lines. *Sci. Rep.* **2019**, *9*, 6298. [CrossRef] [PubMed]
81. Futaki, S.; Ohashi, W.; Suzuki, T.; Niwa, M.; Tanaka, S.; Ueda, K.; Harashima, H.; Sugiura, Y. Stearylated Arginine-Rich Peptides: A New Class of Transfection Systems. *Bioconjug. Chem.* **2001**, *12*, 1005–1011. [CrossRef]
82. Crombez, L.; Morris, M.C.; Deshayes, S.; Heitz, F.; Divita, G. Peptide-based nanoparticle for ex vivo and in vivo drug delivery. *Curr. Pharm. Des.* **2008**, *14*, 3656–3665. [CrossRef]
83. Deshayes, S.; Konate, K.; Aldrian, G.; Crombez, L.; Heitz, F.; Divita, G. Structural polymorphism of non-covalent peptide-based delivery systems: Highway to cellular uptake. *Biochim. Biophys. Acta* **2010**, *1798*, 2304–2314. [CrossRef]
84. Deshayes, S.; Morris, M.C.; Divita, G.; Heitz, F. Cell-penetrating peptides: Tools for intracellular delivery of therapeutics. *Cell Mol. Life Sci.* **2005**, *62*, 1839–1849. [CrossRef] [PubMed]
85. Gottschalk, S.; Sparrow, J.T.; Hauer, J.; Mims, M.P.; Leland, F.E.; Woo, S.L.; Smith, L.C. A novel DNA-peptide complex for efficient gene transfer and expression in mammalian cells. *Gene Ther.* **1996**, *3*, 448–457. [PubMed]
86. Wyman, T.B.; Nicol, F.; Zelphati, O.; Scaria, P.V.; Plank, C.; Szoka, F.C., Jr. Design, synthesis, and characterization of a cationic peptide that binds to nucleic acids and permeabilizes bilayers. *Biochemistry* **1997**, *36*, 3008–3017. [CrossRef]
87. Konate, K.; Crombez, L.; Deshayes, S.; Decaffmeyer, M.; Thomas, A.; Brasseur, R.; Aldrian, G.; Heitz, F.; Divita, G. Insight into the cellular uptake mechanism of a secondary amphipathic cell-penetrating peptide for siRNA delivery. *Biochemistry* **2010**, *49*, 3393–3402. [CrossRef]
88. Konate, K.; Rydstrom, A.; Divita, G.; Deshayes, S. Everything you always wanted to know about CADY-mediated siRNA delivery* (* but afraid to ask). *Curr. Pharm. Des.* **2013**, *19*, 2869–2877. [CrossRef]
89. Kim, W.J.; Christensen, L.V.; Jo, S.; Yockman, J.W.; Jeong, J.H.; Kim, Y.H.; Kim, S.W. Cholesteryl oligoarginine delivering vascular endothelial growth factor siRNA effectively inhibits tumor growth in colon adenocarcinoma. *Mol. Ther.* **2006**, *14*, 343–350. [CrossRef]
90. Lehto, T.; Abes, R.; Oskolkov, N.; Suhorutsenko, J.; Copolovici, D.M.; Mäger, I.; Viola, J.R.; Simonson, O.E.; Ezzat, K.; Guterstam, P.; et al. Delivery of nucleic acids with a stearylated (RxR)4 peptide using a non-covalent co-incubation strategy. *J. Control. Release* **2010**, *141*, 42–51. [CrossRef] [PubMed]
91. Kurrikoff, K.; Gestin, M.; Langel, Ü. Recent in vivo advances in cell-penetrating peptide-assisted drug delivery. *Expert Opin. Drug Deliv.* **2016**, *13*, 373–387. [CrossRef] [PubMed]
92. Mäe, M.; El Andaloussi, S.; Lundin, P.; Oskolkov, N.; Johansson, H.J.; Guterstam, P.; Langel, Ü. A stearylated CPP for delivery of splice correcting oligonucleotides using a non-covalent co-incubation strategy. *J. Control. Release* **2009**, *134*, 221–227. [CrossRef] [PubMed]
93. Oskolkov, N.; Arukuusk, P.; Copolovici, D.-M.; Lindberg, S.; Margus, H.; Padari, K.; Pooga, M.; Langel, Ü. NickFects, phosphorylated derivatives of transportan 10 for cellular delivery of oligonucleotides. *Int. J. Pep. Res.Therap.* **2011**, *17*, 147–157. [CrossRef]
94. Arukuusk, P.; Pärnaste, L.; Hällbrink, M.; Langel, Ü. PepFects and NickFects for the Intracellular Delivery of Nucleic Acids. *Methods Mol. Biol.* **2015**, *1324*, 303–315. [PubMed]
95. Regberg, J.; Srimanee, A.; Erlandsson, M.; Sillard, R.; Dobchev, D.A.; Karelson, M.; Langel, Ü. Rational design of a series of novel amphipathic cell-penetrating peptides. *Int. J. Pharm.* **2014**, *464*, 111–116. [CrossRef]
96. Regberg, J.; Vasconcelos, L.; Madani, F.; Langel, Ü.; Hällbrink, M. pH-responsive PepFect cell-penetrating peptides. *Int. J. Pharm.* **2016**, *501*, 32–38. [CrossRef]
97. Veiman, K.L.; Kunnapuu, K.; Lehto, T.; Kiisholts, K.; Pärn, K.; Langel, Ü.; Kurrikoff, K. PEG shielded MMP sensitive CPPs for efficient and tumor specific gene delivery in vivo. *J. Control. Release* **2015**, *209*, 238–247. [CrossRef] [PubMed]
98. Buss, C.G.; Bhatia, S.N. Nanoparticle delivery of immunostimulatory oligonucleotides enhances response to checkpoint inhibitor therapeutics. *Proc. Natl. Acad. Sci. USA* **2020**, *117*, 13428–13436. [CrossRef]
99. Youn, P.; Chen, Y.; Furgeson, D.Y. A myristoylated cell-penetrating peptide bearing a transferrin receptor-targeting sequence for neuro-targeted siRNA delivery. *Mol. Pharm.* **2014**, *11*, 486–495. [CrossRef]
100. Jain, P.K.; Lo, J.H.; Rananaware, S.; Downing, M.; Panda, A.; Tai, M.; Raghavan, S.; Fleming, H.E.; Bhatia, S.N. Non-viral delivery of CRISPR/Cas9 complex using CRISPR-GPS nanocomplexes. *Nanoscale* **2019**, *11*, 21317–21323. [CrossRef]
101. Yang, G.; Zhao, Y.; Gong, A.; Miao, W.; Yan, L.; Nie, P.; Wang, Z. Improved Cellular Delivery of Antisense Oligonucleotide for miRNA-21 Imaging In Vivo Using Cell-Penetrating Peptide-Based Nanoprobes. *Mol. Pharm.* **2021**, *18*, 787–795. [CrossRef]
102. Alvarez, M.J.; Subramaniam, P.S.; Tang, L.H.; Grunn, A.; Aburi, M.; Rieckhof, G.; Komissarova, E.V.; Hagan, E.A.; Bodei, L.; Clemons, P.A.; et al. A precision oncology approach to the pharmacological targeting of mechanistic dependencies in neuroendocrine tumors. *Nat. Genet.* **2018**, *50*, 979–989. [CrossRef]
103. Van der Bent, M.L.; Paulino da Silva Filho, O.; Willemse, M.; Hallbrink, M.; Wansink, D.G.; Brock, R. The nuclear concentration required for antisense oligonucleotide activity in myotonic dystrophy cells. *FASEB J.* **2019**, *33*, 11314–11325. [CrossRef] [PubMed]

104. Van den Brand, D.; Gorris, M.A.J.; van Asbeck, A.H.; Palmen, E.; Ebisch, I.; Dolstra, H.; Hallbrink, M.; Massuger, L.; Brock, R. Peptide-mediated delivery of therapeutic mRNA in ovarian cancer. *Eur. J. Pharm. Biopharm.* **2019**, *141*, 180–190. [CrossRef] [PubMed]
105. Van Asbeck, A.H.; Dieker, J.; Oude Egberink, R.; van den Berg, L.; van der Vlag, J.; Brock, R. Protein Expression Correlates Linearly with mRNA Dose over Up to Five Orders of Magnitude In Vitro and In Vivo. *Biomedicines* **2021**, *9*, 511. [CrossRef]
106. Dowaidar, M.; Abdelhamid, H.; Hällbrink, M.; Zou, X.; Langel, Ü. Graphene oxide Mediated Cell Penetrating Peptides for Oligonucleotides Delivery. *Biochim. Biophys. Acta Gen. Subj.* **2017**, *1861*, 2334–2341. [CrossRef] [PubMed]
107. Dowaidar, M.; Abdelhamid, H.; Hällbrink, M.; Zou, X.; Langel, Ü. Magnetic nanoparticles Assist Self-assembly of Cell Penetrating Peptides for Oligonucleotides Delivery. *Sci. Rep.* **2017**, *7*, 9159. [CrossRef]
108. Abdelhamid, H.N.; Dowaidar, M.; Hallbrink, M.; Langel, U. Gene delivery using cell penetrating peptides-zeolitic imidazolate frameworks. *Microporous Mesoporous Mat.* **2020**, *300*, 10. [CrossRef]
109. Abdelhamid, H.N.; Dowaidar, M.; Langel, U. Carbonized chitosan encapsulated hierarchical porous zeolitic imidazolate frameworks nanoparticles for gene delivery. *Microporous Mesoporous Mat.* **2020**, *302*, 8. [CrossRef]
110. Kim, H.B.; Morris, J.; Miyashiro, K.; Lehto, T.; Langel, Ü.; Eberwine, J.; Sul, J.Y. Astrocytes promote ethanol-induced enhancement of intracellular Ca(2+) signals through intercellular communication with neurons. *Science* **2021**, *24*, 102436.
111. Dowaidar, M.; Gestin, M.; Cerrato, C.P.; Jafferali, M.H.; Margus, H.; Kivistik, P.A.; Ezzat, K.; Hallberg, E.; Pooga, M.; Hällbrink, M.; et al. Role of autophagy in cell-penetrating peptide transfection model. *Sci. Rep.* **2017**, *7*, 12635. [CrossRef] [PubMed]
112. Carreras-Badosa, G.; Maslovskaja, J.; Periyasamy, K.; Urgard, E.; Padari, K.; Vaher, H.; Tserel, L.; Gestin, M.; Kisand, K.; Arukuusk, P.; et al. NickFect type of cell-penetrating peptides present enhanced efficiency for microRNA-146a delivery into dendritic cells and during skin inflammation. *Biomaterials* **2020**, *262*, 120316. [CrossRef]
113. Park, K. In vivo DNA delivery with NickFect peptide vectors. *J. Control. Release Soc.* **2016**, *241*, 242. [CrossRef]
114. Vasconcelos, L.; Madani, F.; Arukuusk, P.; Pärnaste, L.; Gräslund, A.; Langel, Ü. Effects of cargo molecules on membrane perturbation caused by transportan10 based cell-penetrating peptides. *Biochim. Biophys. Acta* **2014**, *1838*, 3118–3129. [CrossRef] [PubMed]
115. Freimann, K.; Arukuusk, P.; Kurrikoff, K.; Parnaste, L.; Raid, R.; Piirsoo, A.; Pooga, M.; Langel, U. Formulation of Stable and Homogeneous Cell-Penetrating Peptide NF55 Nanoparticles for Efficient Gene Delivery In Vivo. *Mol. Ther. Nucleic Acids* **2018**, *10*, 28–35. [CrossRef] [PubMed]
116. Kurrikoff, K.; Vunk, B.; Langel, Ü. Status update in the use of cell-penetrating peptides for the delivery of macromolecular therapeutics. *Expert Opin. Biol. Ther.* **2021**, *21*, 361–370. [CrossRef] [PubMed]
117. Kurrikoff, K.; Langel, U. Recent CPP-based applications in medicine. *Expert Opin. Drug Deliv.* **2019**, *16*, 1183–1191. [CrossRef] [PubMed]
118. Lindgren, M.E.; Hällbrink, M.M.; Elmquist, A.M.; Langel, Ü. Passage of cell-penetrating peptides across a human epithelial cell layer in vitro. *Biochem. J.* **2004**, *377*, 69–76. [CrossRef]
119. Mäe, M.; Myrberg, H.; Jiang, Y.; Paves, H.; Valkna, A.; Langel, Ü. Internalisation of cell-penetrating peptides into tobacco protoplasts. *Biochim. Biophys. Acta* **2005**, *1669*, 101–107. [CrossRef]
120. Sarko, D.; Beijer, B.; Garcia Boy, R.; Nothelfer, E.M.; Leotta, K.; Eisenhut, M.; Altmann, A.; Haberkorn, U.; Mier, W. The pharmacokinetics of cell-penetrating peptides. *Mol. Pharm.* **2010**, *7*, 2224–2231. [CrossRef] [PubMed]
121. Magzoub, M.; Eriksson, L.E.; Graslund, A. Comparison of the interaction, positioning, structure induction and membrane perturbation of cell-penetrating peptides and non-translocating variants with phospholipid vesicles. *Biophys. Chem.* **2003**, *103*, 271–288. [CrossRef]
122. Säälik, P.; Padari, K.; Niinep, A.; Lorents, A.; Hansen, M.; Jokitalo, E.; Langel, Ü.; Pooga, M. Protein delivery with transportans is mediated by caveolae rather than flotillin-dependent pathways. *Bioconjug. Chem.* **2009**, *20*, 877–887.
123. Izabela, R.; Jarosław, R.; Magdalena, A.; Piotr, R.; Ivan, K. Transportan 10 improves the anticancer activity of cisplatin. *Naunyn-Schmiedeberg's Arch. Pharmacol.* **2016**, *389*, 485–497. [CrossRef] [PubMed]
124. Stalmans, S.; Bracke, N.; Wynendaele, E.; Gevaert, B.; Peremans, K.; Burvenich, C.; Polis, I.; De Spiegeleer, B. Cell-Penetrating Peptides Selectively Cross the Blood-Brain Barrier In Vivo. *PLoS ONE* **2015**, *10*, e0139652. [CrossRef]
125. Fadzen, C.M.; Wolfe, J.M.; Cho, C.F.; Chiocca, E.A.; Lawler, S.E.; Pentelute, B.L. Perfluoroarene-Based Peptide Macrocycles to Enhance Penetration Across the Blood-Brain Barrier. *J. Am. Chem. Soc.* **2017**, *139*, 15628–15631. [CrossRef] [PubMed]
126. Pári, E.; Horváti, K.; Bősze, S.; Biri-Kovács, B.; Szeder, B.; Zsila, F.; Kiss, É. Drug Conjugation Induced Modulation of Structural and Membrane Interaction Features of Cationic Cell-Permeable Peptides. *Int. J. Mol. Sci.* **2020**, *21*, 2197. [CrossRef] [PubMed]
127. Lindberg, M.; Jarvet, J.; Langel, Ü.; Gräslund, A. Secondary structure and position of the cell-penetrating peptide transportan in SDS micelles as determined by NMR. *Biochemistry* **2001**, *40*, 3141–3149. [CrossRef]
128. Magzoub, M.; Kilk, K.; Eriksson, L.E.; Langel, Ü.; Gräslund, A. Interaction and structure induction of cell-penetrating peptides in the presence of phospholipid vesicles. *Biochim. Biophys. Acta* **2001**, *1512*, 77–89. [CrossRef]
129. Barany-Wallje, E.; Andersson, A.; Gräslund, A.; Mäler, L. NMR solution structure and position of transportan in neutral phospholipid bicelles. *FEBS Lett.* **2004**, *567*, 265–269. [CrossRef] [PubMed]
130. Barany-Wallje, E.; Andersson, A.; Gräslund, A.; Mäler, L. Dynamics of transportan in bicelles is surface charge dependent. *J. Biomol. NMR* **2006**, *35*, 137–147. [CrossRef] [PubMed]

131. Wadhwani, P.; Reichert, J.; Burck, J.; Ulrich, A.S. Antimicrobial and cell-penetrating peptides induce lipid vesicle fusion by folding and aggregation. *Eur. Biophys. J.* **2012**, *41*, 177–187. [CrossRef]
132. Fanghanel, S.; Wadhwani, P.; Strandberg, E.; Verdurmen, W.P.; Burck, J.; Ehni, S.; Mykhailiuk, P.K.; Afonin, S.; Gerthsen, D.; Komarov, I.V.; et al. Structure analysis and conformational transitions of the cell penetrating peptide transportan 10 in the membrane-bound state. *PLoS ONE* **2014**, *9*, e99653. [CrossRef]
133. Song, J.; Kai, M.; Zhang, W.; Zhang, J.; Liu, L.; Zhang, B.; Liu, X.; Wang, R. Cellular uptake of transportan 10 and its analogs in live cells: Selectivity and structure-activity relationship studies. *Peptides* **2011**, *32*, 1934–1941. [CrossRef]
134. Yandek, L.E.; Pokorny, A.; Almeida, P.F. Small changes in the primary structure of transportan 10 alter the thermodynamics and kinetics of its interaction with phospholipid vesicles. *Biochemistry* **2008**, *47*, 3051–3060. [CrossRef] [PubMed]
135. Deshayes, S.; Decaffmeyer, M.; Brasseur, R.; Thomas, A. Structural polymorphism of two CPP: An important parameter of activity. *Biochim. Biophys. Acta* **2008**, *1778*, 1197–1205. [CrossRef]
136. Arsov, Z.; Nemec, M.; Schara, M.; Johansson, H.; Langel, Ü.; Zorko, M. Cholesterol prevents interaction of the cell-penetrating peptide transportan with model lipid membranes. *J. Pept. Sci.* **2008**, *14*, 1303–1308. [CrossRef] [PubMed]
137. Dunkin, C.M.; Pokorny, A.; Almeida, P.F.; Lee, H.S. Molecular dynamics studies of transportan 10 (tp10) interacting with a POPC lipid bilayer. *J. Phys. Chem. B* **2011**, *115*, 1188–1198. [CrossRef] [PubMed]
138. Säälik, P.; Niinep, A.; Pae, J.; Hansen, M.; Lubenets, D.; Langel, Ü.; Pooga, M. Penetration without cells: Membrane translocation of cell-penetrating peptides in the model giant plasma membrane vesicles. *J. Control. Release* **2011**, *153*, 117–125. [CrossRef]
139. Lindgren, M.; Gallet, X.; Soomets, U.; Hällbrink, M.; Brakenhielm, E.; Pooga, M.; Brasseur, R.; Langel, Ü. Translocation properties of novel cell penetrating transportan and penetratin analogues. *Bioconjug. Chem.* **2000**, *11*, 619–626. [CrossRef] [PubMed]
140. Cosme, P.J.; Ye, J.; Sears, S.; Wojcikiewicz, E.P.; Terentis, A.C. Label-Free Confocal Raman Mapping of Transportan in Melanoma Cells. *Mol. Pharm.* **2018**, *15*, 851–860. [CrossRef]
141. Pae, J.; Liivamagi, L.; Lubenets, D.; Arukuusk, P.; Langel, U.; Pooga, M. Glycosaminoglycans are required for translocation of amphipathic cell-penetrating peptides across membranes. *Biochim. Biophys. Acta* **2016**, *23*, 30137. [CrossRef]
142. Padari, K.; Säälik, P.; Hansen, M.; Koppel, K.; Raid, R.; Langel, Ü.; Pooga, M. Cell transduction pathways of transportans. *Bioconjug. Chem.* **2005**, *16*, 1399–1410. [CrossRef]
143. Chaubey, B.; Tripathi, S.; Ganguly, S.; Harris, D.; Casale, R.A.; Pandey, V.N. A PNA-transportan conjugate targeted to the TAR region of the HIV-1 genome exhibits both antiviral and virucidal properties. *Virology* **2005**, *331*, 418–428. [CrossRef]
144. Li, Y.X.; Wei, Y.; Zhong, R.; Li, L.; Pang, H.B. Transportan Peptide Stimulates the Nanomaterial Internalization into Mammalian Cells in the Bystander Manner through Macropinocytosis. *Pharmaceutics* **2021**, *13*, 552. [CrossRef] [PubMed]
145. Anko, M.; Majhenc, J.; Kogej, K.; Sillard, R.; Langel, Ü.; Anderluh, G.; Zorko, M. Influence of stearyl and trifluoromethylquinoline modifications of the cell penetrating peptide TP10 on its interaction with a lipid membrane. *Biochim. Biophys. Acta* **2012**, *1818*, 915–924. [CrossRef] [PubMed]
146. Srimanee, A.; Regberg, J.; Hallbrink, M.; Vajragupta, O.; Langel, U. Role of scavenger receptors in peptide-based delivery of plasmid DNA across a blood-brain barrier model. *Int. J. Pharm.* **2016**, *500*, 128–135. [CrossRef] [PubMed]
147. Juks, C.; Lorents, A.; Arukuusk, P.; Langel, U.; Pooga, M. Cell-penetrating peptides recruit type A scavenger receptors to the plasma membrane for cellular delivery of nucleic acids. *FASEB J.* **2017**, *31*, 975–988. [CrossRef]
148. Vasconcelos, L.; Lehto, T.; Madani, F.; Radoi, V.; Hällbrink, M.; Vukojević, V.; Langel, Ü. Simultaneous membrane interaction of amphipathic peptide monomers, self-aggregates and cargo complexes detected by fluorescence correlation spectroscopy. *Biochim. Biophys. Acta Biomembr.* **2018**, *1860*, 491–504. [CrossRef]
149. Pärnaste, L.; Arukuusk, P.; Zagato, E.; Braeckmans, K.; Langel, Ü. Methods to follow intracellular trafficking of cell-penetrating peptides. *J. Drug Target.* **2016**, *24*, 508–519. [CrossRef]
150. Zhang, W.; Song, J.; Zhang, B.; Liu, L.; Wang, K.; Wang, R. Design of acid-activated cell penetrating peptide for delivery of active molecules into cancer cells. *Bioconjug. Chem.* **2011**, *22*, 1410–1415. [CrossRef]
151. Suhorutsenko, J.; Oskolkov, N.; Arukuusk, P.; Kurrikoff, K.; Eriste, E.; Copolovici, D.M.; Langel, Ü. Cell-penetrating peptides, PepFects, show no evidence of toxicity and immunogenicity in vitro and in vivo. *Bioconjug. Chem.* **2011**, *22*, 2255–2262. [CrossRef]
152. Aguiar, L.; Machado, M.; Sanches-Vaz, M.; Prudencio, M.; Vale, N.; Gomes, P. Coupling the cell-penetrating peptides transportan and transportan 10 to primaquine enhances its activity against liver-stage malaria parasites. *MedChemComm* **2019**, *10*, 221–226. [CrossRef]
153. Ptaszyńska, N.; Gucwa, K.; Olkiewicz, K.; Heldt, M.; Serocki, M.; Stupak, A.; Martynow, D.; Dębowski, D.; Gitlin-Domagalska, A.; Lica, J.; et al. Conjugates of Ciprofloxacin and Levofloxacin with Cell-Penetrating Peptide Exhibit Antifungal Activity and Mammalian Cytotoxicity. *Int. J. Mol. Sci.* **2020**, *21*, 4696. [CrossRef]
154. Lundberg, M.; Johansson, M. Is VP22 nuclear homing an artifact? *Nat. Biotechnol.* **2001**, *19*, 713–714. [CrossRef]
155. Lundberg, M.; Johansson, M. Positively Charged DNA-Binding Proteins Cause Apparent Cell Membrane Translocation. *Biochem. Biophys. Res. Commun.* **2002**, *291*, 367–371. [CrossRef]
156. Richard, J.P.; Melikov, K.; Vives, E.; Ramos, C.; Verbeure, B.; Gait, M.J.; Chernomordik, L.V.; Lebleu, B. Cell-penetrating peptides. A reevaluation of the mechanism of cellular uptake. *J. Biol. Chem.* **2003**, *278*, 585–590. [CrossRef]
157. Console, S.; Marty, C.; Garcia-Echeverria, C.; Schwendener, R.; Ballmer-Hofer, K. Antennapedia and HIV transactivator of transcription (TAT) "protein transduction domains" promote endocytosis of high molecular weight cargo upon binding to cell surface glycosaminoglycans. *J. Biol. Chem.* **2003**, *278*, 35109–35114. [CrossRef]

158. Madani, F.; Lindberg, S.; Langel, Ü.; Futaki, S.; Gräslund, A. Mechanisms of cellular uptake of cell-penetrating peptides. *J. Biophys.* **2011**, *2011*, 414729. [CrossRef] [PubMed]
159. El-Andaloussi, S.; Lehto, T.; Lundin, P.; Langel, Ü. Application of PepFect peptides for the delivery of splice-correcting oligonucleotides. *Methods Mol. Biol.* **2011**, *683*, 361–373.
160. Zorko, M.; Langel, Ü. Cell-penetrating peptides: Mechanism and kinetics of cargo delivery. *Adv. Drug Deliv Rev.* **2005**, *57*, 529–545. [CrossRef] [PubMed]
161. Ma, D.X.; Shi, N.Q.; Qi, X.R. Distinct transduction modes of arginine-rich cell-penetrating peptides for cargo delivery into tumor cells. *Int. J. Pharm.* **2011**, *419*, 200–208. [CrossRef] [PubMed]
162. Mäger, I.; Eiriksdottir, E.; Langel, K.; El Andaloussi, S.; Langel, Ü. Assessing the uptake kinetics and internalization mechanisms of cell-penetrating peptides using a quenched fluorescence assay. *Biochim. Biophys. Acta* **2010**, *1798*, 338–343. [CrossRef] [PubMed]
163. Yandek, L.E.; Pokorny, A.; Floren, A.; Knoelke, K.; Langel, Ü.; Almeida, P.F. Mechanism of the cell-penetrating peptide transportan 10 permeation of lipid bilayers. *Biophys. J.* **2007**, *92*, 2434–2444. [CrossRef]
164. Yandek, L.E.; Pokorny, A.; Almeida, P.F. Wasp mastoparans follow the same mechanism as the cell-penetrating peptide transportan 10. *Biochemistry* **2009**, *48*, 7342–7351. [CrossRef] [PubMed]
165. McKeown, A.N.; Naro, J.L.; Huskins, L.J.; Almeida, P.F. A thermodynamic approach to the mechanism of cell-penetrating peptides in model membranes. *Biochemistry* **2011**, *50*, 654–662. [CrossRef]
166. Ablan, F.D.O.; Spaller, B.L.; Abdo, K.I.; Almeida, P.F. Charge Distribution Fine-Tunes the Translocation of α-Helical Amphipathic Peptides across Membranes. *Biophys. J.* **2016**, *111*, 1738–1749. [CrossRef]
167. Moghal, M.M.R.; Islam, M.Z.; Sharmin, S.; Levadnyy, V.; Moniruzzaman, M.; Yamazaki, M. Continuous detection of entry of cell-penetrating peptide transportan 10 into single vesicles. *Chem. Phys. Lipids* **2018**, *212*, 120–129. [CrossRef]
168. Pae, J.; Pooga, M. Peptide-mediated delivery: An overview of pathways for efficient internalization. *Ther. Deliv.* **2014**, *5*, 1203–1222. [CrossRef]
169. Moghal, M.M.R.; Hossain, F.; Yamazaki, M. Action of antimicrobial peptides and cell-penetrating peptides on membrane potential revealed by the single GUV method. *Biophys. Rev.* **2020**, *12*, 339–348. [CrossRef]
170. Guterstam, P.; Madani, F.; Hirose, H.; Takeuchi, T.; Futaki, S.; El Andaloussi, S.; Gräslund, A.; Langel, Ü. Elucidating cell-penetrating peptide mechanisms of action for membrane interaction, cellular uptake, and translocation utilizing the hydrophobic counter-anion pyrenebutyrate. *Biochim. Biophys. Acta* **2009**, *1788*, 2509–2517. [CrossRef] [PubMed]
171. Pae, J.; Saalik, P.; Liivamagi, L.; Lubenets, D.; Arukuusk, P.; Langel, Ü.; Pooga, M. Translocation of cell-penetrating peptides across the plasma membrane is controlled by cholesterol and microenvironment created by membranous proteins. *J. Control. Release* **2014**, *192*, 103–113. [CrossRef]
172. Hu, Y.; Patel, S. Structural and Thermodynamic Insight into Spontaneous Membrane-Translocating Peptides Across Model PC/PG Lipid Bilayers. *J. Membr. Biol.* **2015**, *248*, 505–515. [CrossRef] [PubMed]
173. Karal, M.A.S.; Ahamed, M.K.; Ahmed, M.; Ahamed, S.; Mahbub, Z.B. Location of Peptide-Induced Submicron Discontinuities in the Membranes of Vesicles Using Image. *J. Fluoresc.* **2020**, *30*, 735–740. [CrossRef]
174. Pourmousa, M.; Wong-ekkabut, J.; Patra, M.; Karttunen, M. Molecular dynamic studies of transportan interacting with a DPPC lipid bilayer. *J. Phys. Chem. B* **2013**, *117*, 230–241. [CrossRef]
175. Islam, M.Z.; Ariyama, H.; Alam, J.M.; Yamazaki, M. Entry of cell-penetrating peptide transportan 10 into a single vesicle by translocating across lipid membrane and its induced pores. *Biochemistry* **2014**, *53*, 386–396. [CrossRef]
176. Islam, M.Z.; Sharmin, S.; Levadnyy, V.; Alam Shibly, S.U.; Yamazaki, M. Effects of Mechanical Properties of Lipid Bilayers on the Entry of Cell-Penetrating Peptides into Single Vesicles. *Langmuir* **2017**, *33*, 2433–2443. [CrossRef] [PubMed]
177. Moghal, M.M.R.; Islam, M.Z.; Hossain, F.; Saha, S.K.; Yamazaki, M. Role of Membrane Potential on Entry of Cell-Penetrating Peptide Transportan 10 into Single Vesicles. *Biophys. J.* **2020**, *118*, 57–69. [CrossRef] [PubMed]
178. Shi, N.Q.; Qi, X.R.; Xiang, B.; Zhang, Y. A survey on "Trojan Horse" peptides: Opportunities, issues and controlled entry to "Troy". *J. Control. Release* **2014**, *194*, 53–70. [CrossRef]
179. Xie, J.; Gou, Y.; Zhao, Q.; Li, S.; Zhang, W.; Song, J.; Mou, L.; Li, J.; Wang, K.; Zhang, B.; et al. Antimicrobial activities and action mechanism studies of transportan 10 and its analogues against multidrug-resistant bacteria. *J. Pept. Sci.* **2015**, *21*, 599–607. [CrossRef] [PubMed]
180. Saar, K.; Lindgren, M.; Hansen, M.; Eiriksdottir, E.; Jiang, Y.; Rosenthal-Aizman, K.; Sassian, M.; Langel, Ü. Cell-penetrating peptides: A comparative membrane toxicity study. *Anal. Biochem.* **2005**, *345*, 55–65. [CrossRef]
181. Eriksson, O.S.; Georg, M.; Sjolinder, H.; Sillard, R.; Lindberg, S.; Langel, Ü.; Jonsson, A.B. Identification of cell-penetrating peptides that are bactericidal to Neisseria meningitidis and prevent inflammatory responses upon infection. *Antimicrob. Agents Chemother.* **2013**, *57*, 3704–3712. [CrossRef] [PubMed]
182. Horvati, K.; Bacsa, B.; Mlinko, T.; Szabo, N.; Hudecz, F.; Zsila, F.; Bosze, S. Comparative analysis of internalisation, haemolytic, cytotoxic and antibacterial effect of membrane-active cationic peptides: Aspects of experimental setup. *Amino Acids* **2017**, *49*, 1053–1067. [CrossRef]
183. Lee, J.; Jung, E.; Park, J.; Park, D. Transdermal delivery of interferon-gamma (IFN-gamma) mediated by penetratin, a cell-permeable peptide. *Biotechnol. Appl. Biochem.* **2005**, *42*, 169–173.
184. Tsang, O.; Wong, J.W.H. Proteogenomic interrogation of cancer cell lines: An overview of the field. *Expert Rev. Proteom.* **2021**, *18*, 221–232. [CrossRef] [PubMed]

185. Daviss, B. Growing pains for metabolomics. *Scientist* **2005**, *19*, 25–28.
186. Fan, T.W.M.; Lorkiewicz, P.K.; Sellers, K.; Moseley, H.N.B.; Higashi, R.M.; Lane, A.N. Stable isotope-resolved metabolomics and applications for drug development. *Pharmacol. Ther.* **2012**, *133*, 366–391. [CrossRef]
187. Kilk, K.; Mahlapuu, R.; Soomets, U.; Langel, Ü. Analysis of in vitro toxicity of five cell-penetrating peptides by metabolic profiling. *Toxicology* **2009**, *265*, 87–95. [CrossRef] [PubMed]
188. Bell, T.J.; Eberwine, J. Live Cell Genomics: RNA Exon-Specific RNA-Binding Protein Isolation. *Methods Mol. Biol.* **2015**, *1324*, 457–468.
189. Bell, T.J.; Eiriksdottir, E.; Langel, Ü.; Eberwine, J. PAIR technology: Exon-specific RNA-binding protein isolation in live cells. *Methods Mol. Biol.* **2011**, *683*, 473–486.
190. Peritz, T.; Zeng, F.; Kannanayakal, T.J.; Kilk, K.; Eiriksdottir, E.; Langel, Ü.; Eberwine, J. Immunoprecipitation of mRNA-protein complexes. *Nat. Protoc.* **2006**, *1*, 577–580. [CrossRef] [PubMed]
191. Zeng, F.; Peritz, T.; Kannanayakal, T.J.; Kilk, K.; Eiriksdottir, E.; Langel, Ü.; Eberwine, J. A protocol for PAIR: PNA-assisted identification of RNA binding proteins in living cells. *Nat. Protoc.* **2006**, *1*, 920–927. [CrossRef]
192. Bell, T.J.; Eberwine, J. Live Cell Genomics: Cell-Specific Transcriptome Capture in Live Tissues and Cells. *Methods Mol. Biol.* **2015**, *1324*, 447–456. [PubMed]
193. Lovatt, D.; Ruble, B.K.; Lee, J.; Dueck, H.; Kim, T.K.; Fisher, S.; Francis, C.; Spaethling, J.M.; Wolf, J.A.; Grady, M.S.; et al. Transcriptome in vivo analysis (TIVA) of spatially defined single cells in live tissue. *Nat. Methods* **2014**, *11*, 190–196. [CrossRef]
194. Venit, T.; Dowaidar, M.; Gestin, M.; Mahmood, S.R.; Langel, Ü.; Percipalle, P. Transcriptional Profiling Reveals Ribosome Biogenesis, Microtubule Dynamics and Expression of Specific lncRNAs to be Part of a Common Response to Cell-Penetrating Peptides. *Biomolecules* **2020**, *10*, 1567. [CrossRef] [PubMed]
195. Fisher, L.; Samuelsson, M.; Jiang, Y.; Ramberg, V.; Figueroa, R.; Hallberg, E.; Langel, Ü.; Iverfeldt, K. Targeting cytokine expression in glial cells by cellular delivery of an NF-kappaB decoy. *J. Mol. Neurosci.* **2007**, *31*, 209–219.
196. El-Andaloussi, S.; Johansson, H.; Magnusdottir, A.; Järver, P.; Lundberg, P.; Langel, Ü. TP10, a delivery vector for decoy oligonucleotides targeting the Myc protein. *J. Control. Release* **2005**, *110*, 189–201. [CrossRef]
197. Suresh, A.; Kim, Y.C. Translocation of cell penetrating peptides on Chlamydomonas reinhardtii. *Biotechnol. Bioeng.* **2013**, *110*, 2795–2801. [CrossRef]
198. Ruczynski, J.; Rusiecka, I.; Turecka, K.; Kozlowska, A.; Alenowicz, M.; Gagalo, I.; Kawiak, A.; Rekowski, P.; Waleron, K.; Kocic, I. Transportan 10 improves the pharmacokinetics and pharmacodynamics of vancomycin. *Sci. Rep.* **2019**, *9*, 3247. [CrossRef]
199. Srimanee, A.; Arvanitidou, M.; Kim, K.; Hallbrink, M.; Langel, U. Cell-penetrating peptides for siRNA delivery to glioblastomas. *Peptides* **2018**, *104*, 62–69. [CrossRef]
200. Rosenke, K.; Leventhal, S.; Moulton, H.M.; Hatlevig, S.; Hawman, D.; Feldmann, H.; Stein, D.A. Inhibition of SARS-CoV-2 in Vero cell cultures by peptide-conjugated morpholino-oligomers. *bioRxiv* **2020**, *76*, 413–417. [CrossRef] [PubMed]
201. Tints, K.; Prink, M.; Neuman, T.; Palm, K. LXXLL peptide converts transportan 10 to a potent inducer of apoptosis in breast cancer cells. *Int. J. Mol. Sci.* **2014**, *15*, 5680–5698. [CrossRef]
202. Künnapuu, K.; Veiman, K.L.; Porosk, L.; Rammul, E.; Kiisholts, K.; Langel, Ü.; Kurrikoff, K. Tumor gene therapy by systemic delivery of plasmid DNA with cell-penetrating peptides. *FASEB Bioadv.* **2019**, *1*, 105–114. [CrossRef] [PubMed]
203. Kurrikoff, K.; Freimann, K.; Veiman, K.L.; Peets, E.M.; Piirsoo, A.; Langel, Ü. Effective lung-targeted RNAi in mice with peptide-based delivery of nucleic acid. *Sci. Rep.* **2019**, *9*, 19926. [CrossRef]
204. Cerrato, C.P.; Kivijärvi, T.; Tozzi, R.; Lehto, T.; Gestin, M.; Langel, Ü. Intracellular delivery of therapeutic antisense oligonucleotides targeting mRNA coding mitochondrial proteins by cell-penetrating peptides. *J. Mater. Chem. B* **2020**, *8*, 10825–10836. [CrossRef] [PubMed]

pharmaceutics

MDPI

Review

Cell-Penetrating Peptides with Unexpected Anti-Amyloid Properties

Nicklas Österlund [1], Sebastian K. T. S. Wärmländer [2,3] and Astrid Gräslund [1,3,*]

[1] Department of Biochemistry and Biophysics, Arrhenius Laboratories, Stockholm University, 10691 Stockholm, Sweden; nicklas.osterlund@dbb.su.se
[2] Department of Archaeology and Classical Studies, Stockholm University, 10691 Stockholm, Sweden; seb@student.su.se
[3] CellPept Sweden AB, Kvarngatan 10B, 11847 Stockholm, Sweden
* Correspondence: astrid@dbb.su.se

Abstract: Cell-penetrating peptides (CPPs) with sequences derived originally from a prion protein (PrP) have been shown to exhibit both anti-prion and anti-amyloid properties particularly against prion proteins and the amyloid-β (Aβ) peptide active in Alzheimer's disease. These disease-modifying properties are so far observed in cell cultures and in vitro. The CPP sequences are composed of a hydrophobic signal sequence followed by a highly positively charged hexapeptide segment. The original signal sequence of the prion protein can be changed to the signal sequence of the NCAM1 protein without losing the anti-prion activity. Although the detailed molecular mechanisms of these CPP peptides are not fully understood, they do form amyloid aggregates by themselves, and molecular interactions between the CPPs and PrP/Aβ can be observed in vitro using various spectroscopic techniques. These initial intermolecular interactions appear to re-direct the aggregation pathways for prion/amyloid formation to less cell-toxic molecular structures (i.e., co-aggregates), which likely is why the disease-inducing PrP/Aβ aggregation is counteracted in vivo.

Keywords: protein aggregation; secretion signal peptide; peptide engineering; drug design

Citation: Österlund, N.; Wärmländer, S.K.T.S.; Gräslund, A. Cell-Penetrating Peptides with Unexpected Anti-Amyloid Properties. *Pharmaceutics* **2022**, *14*, 823. https://doi.org/10.3390/pharmaceutics14040823

Academic Editors: Prisca Boisguérin and Sébastien Deshayes

Received: 21 February 2022
Accepted: 31 March 2022
Published: 9 April 2022

Publisher's Note: MDPI stays neutral with regard to jurisdictional claims in published maps and institutional affiliations.

Copyright: © 2022 by the authors. Licensee MDPI, Basel, Switzerland. This article is an open access article distributed under the terms and conditions of the Creative Commons Attribution (CC BY) license (https://creativecommons.org/licenses/by/4.0/).

1. Introduction

We here describe designed cell-penetrating peptides (CPPs) which have been shown to display anti-prion and anti-amyloid properties in cell cultures [1,2]. In general, CPPs are relatively short peptides which can transport a molecular cargo across a cellular membrane and into the cell. Such peptides typically display high transport efficiency and low cell toxicity [3]. CPPs can generally be classified by their specific physico-chemical properties such as hydrophobicity and charge. They may also cross the blood–brain barrier, which makes them interesting as possible drug candidates [4].

The CPPs discussed in this review belong to a group of primary amphipathic CPPs, with a hydrophobic N-terminus and a hydrophilic positively charged C-terminus. The peptides were derived starting from the peptide sequences of the N-terminal segments of the mouse prion protein ($mPrP_{1-28}$). This segment consists of a hydrophobic signal peptide ($mPrP_{1-22}$) and a highly cationic hexapeptide sequence ($mPrP_{23-28}$, KKRPKP). The mPrP signal peptide segment was later exchanged into the slightly less hydrophobic signal peptide for a type-I membrane protein found in neuronal cells, i.e., the mouse neural cell adhesion molecule 1 (NCAM1) protein ($mNCAM1_{1-19}$). This yielded the chimeric CPP construct $mNCAM1_{1-19}$-$mPrP_{23-28}$, which we refer to as NCAM-PrP. The peptides' variants and their properties are summarized in Figure 1A. Both CPP constructs are predicted to be cleaved by signal peptidase I in vivo, resulting in release of the signal peptide and the $mPrP_{23-28}$ peptide.

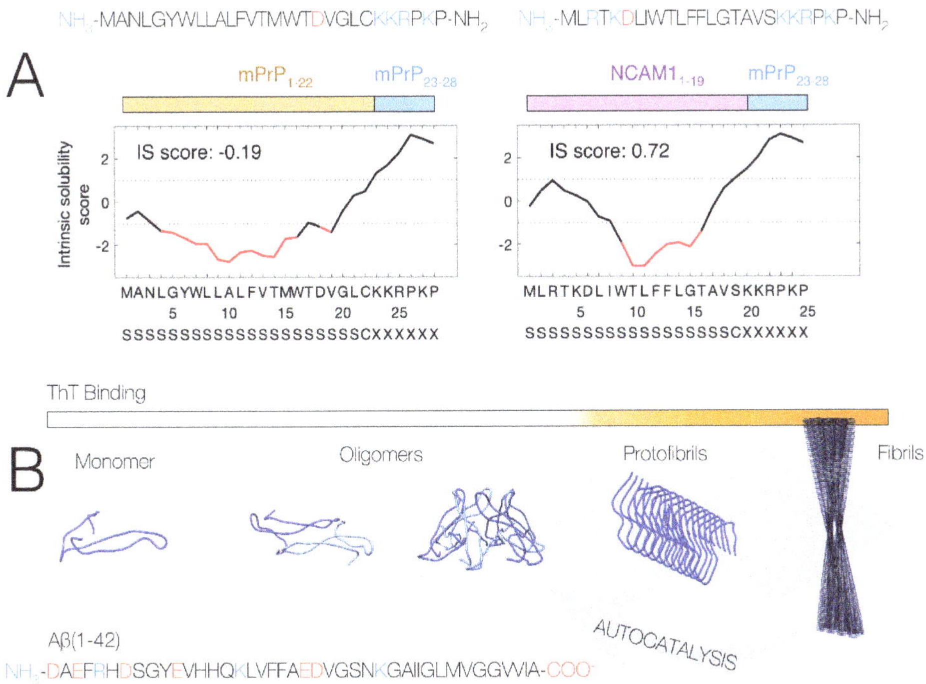

Figure 1. (**A**) Overview of the sequences for mPrP$_{1-28}$ and the chimeric NCAM-PrP construct. Positive charges are colored blue, and negative charges are colored red. The intrinsic solubility score according to the CamSol method [5] over the peptide sequences are shown. Poorly soluble regions are shown in red, indicating the higher solubility of NCAM-PrP compared to the original mPrP$_{1-28}$ peptide. Eukaryotic signal peptide predictions by the SignalP 5.0 method [6,7] are shown at the bottom, with predicted signal peptide segments marked as "S" and predicted cleavage sites by signal peptidase I marked as "C". The figure is reprinted/adapted from Król, S. et al. The amyloid-inhibiting NCAM-PrP peptide targets Aβ peptide aggregation in membrane-mimetic environments. Science 2021, 24, 102852. https://doi.org/10.1016/j.isci.2021.102852. Ref. [8] under the terms of the Creative Commons CC-BY license. (**B**) Overview of the Aβ aggregation process where soluble monomeric peptides self-assemble into mature amyloid fibrils via intermediate states often termed "oligomers" and "protofibrils". Oligomers are believed to primarily form in an autocatalytic cycle involving the fibril surface. The fluorescent dye Thioflavin T (ThT) is commonly used to report on the formation of amyloid structures in time-dependent fluorescence spectroscopy experiments.

2. Amyloid Processes in Alzheimer's Disease Involving the Amyloid-β Peptide

Aggregation of the Amyloid-β (Aβ) peptide into amyloid material is associated with neurodegeneration in Alzheimer's disease (AD). This so-called amyloid cascade hypothesis has been extensively reviewed elsewhere [9]. The Aβ peptide is an amphiphilic peptide with an overall negatively charged N-terminal segment, followed by a hydrophobic segment that originates from a membrane-embedded portion of the amyloid-β precursor protein. Self-interaction between the C-terminal part of Aβ and its central KLVFF motif is believed to be important in the self-assembly process [10]. Figure 1B shows a schematic view of the amyloid aggregation pathway of the Aβ peptide, and the proposed intermediate stages on the way from soluble monomeric peptide into insoluble fibrillar aggregates. Intermediate oligomeric aggregates have been shown to be particularly neurotoxic [11,12]. One suggested mechanism is that such oligomers are capable of disturbing the integrity of cell membranes, for example by formation of transient pore structures [13,14]. Formation of oligomers is greatly enhanced by the presence of fibrillar aggregates, in an autocatalytic cycle [15] (Figure 1B).

Aβ aggregation has been the target for development of therapeutic drugs for AD. Most such efforts have failed, and there are today no generally approved therapies for AD which target Aβ peptides. There has however been some recent success with drug candidates involving the antibodies BAN2401 (Bioarctic AB) and aducanumab (Biogen Inc.), as well as with the small molecule ALZ-801 (Alzheon Inc.). They are all in ongoing or finished clinical trials [16]. All these drug candidates target, at least in part, smaller Aβ aggregates, which further highlights the suspected importance of such intermediate structures in AD.

In vitro studies of Aβ are generally performed on purified recombinant or synthetic Aβ peptides, either the full lengths variants $A\beta_{1-40}$ or $A\beta_{1-42}$ or smaller fragments. The time-dependent aggregation from monomeric peptides into mature amyloid fibrils is often followed using fluorescence spectroscopy with the amyloid-specific dye Thioflavin T (ThT), which significantly increases its intrinsic fluorescence when bound to amyloid fibrils [17]. Other techniques can similarly be used to follow the aggregation process over time, for example by monitoring changes in the Aβ secondary structure with circular dichroism (CD) or infrared (IR) spectroscopy, or by monitoring the size distribution of the Aβ peptide ensemble with dynamic light scattering (DLS) or fluorescence correlation spectroscopy (FCS). Figure 2A,B show examples of kinetic data of $A\beta_{1-42}$, measured using ThT fluorescence and CD spectroscopy in vitro. Such time-dependent studies are commonly used to obtain information about the underlying kinetics of the aggregation processes [18,19]. Kinetic parameters can be supplemented with molecular structures of the peptides and their aggregates. Solution state NMR can reveal the structures of monomeric peptides, while atomic force microscopy (AFM) and electron microscopy (EM) images can reveal the structures of larger aggregates, which might be either fibrillar or non-fibrillar (see examples in Figure 2D).

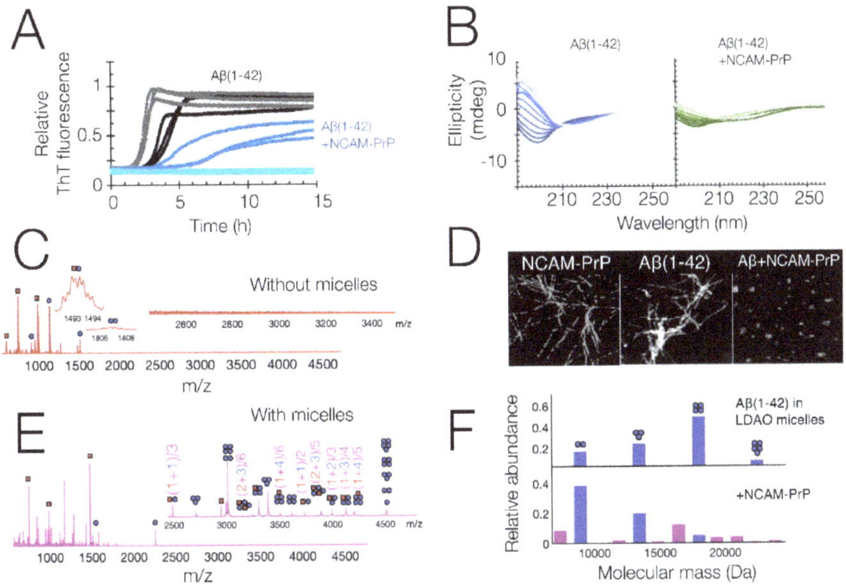

Figure 2. Overview of experimental in vitro results on the NCAM-PrP peptide and its interactions with $A\beta_{1-42}$. (**A**) Time-dependent fluorescence spectroscopy using the amyloid specific ThT dye. Grey/Black: $A\beta_{1-42}$ (with/without 150 mM NaF salt) Light blue/Dark blue: $A\beta_{1-42}$ + NCAM-PrP at 1:1 molar ratio (with/without 150 mM NaF salt). 5 µM peptides in 20 mM NaP buffer pH 8. (**B**) CD spectra recorded at multiple time-points (lightest trace = 0 h, darkest trace = 4 h), of $A\beta_{1-42}$ alone (blue) and $A\beta_{1-42}$ + NCAM-PrP at 1:1 molar ratio (green); (**C**) electrospray ionization mass spectrum of $A\beta_{1-42}$ + NCAM-PrP at 1:1 molar ratio, at native conditions. $A\beta_{1-42}$ species are shown as blue circles, NCAM-PrP species are shown as red squares; (**D**) Left: 1 × 1 µm AFM image of 10 µM NCAM-PrP peptide incubated for 26 h in PBS buffer, pH 7.4 at 42 °C [20]. Reprinted (adapted) with permission from Pansieri, J. et al. Pro-Inflammatory S100A9 Protein Aggregation Promoted by NCAM1 Peptide Constructs. ACS Chem. Biol. 2019, 14, 1410–1417; Copyright 2019 American Chemical Society. Middle, and Right: 2 × 2 µm AFM images of 5 µM $A\beta_{42}$ peptide, incubated without (middle) or together with 5 µM NCAM-PrP peptide (right), for 15 hrs in 20 mM NaP buffer, pH 8 at 37 °C [8]. (**E**) electrospray mass ionization mass spectrum of: $A\beta_{1-42}$ + NCAM-PrP at 1:1 molar ratio, at native conditions, similar conditions as in C, but with 4 mM LDAO micelles to mimic a membrane environment. The peptide species were removed from the micelles using collision induced dissociation; (**F**) overview of how the distribution of $A\beta_{1-42}$ oligomers in the LDAO micelles shifts upon co-incubation with NCAM-PrP. All results are adapted and reprinted from reference [8], under the terms of the Creative Commons CC-BY license, unless otherwise is stated.

Detailed studies of the small oligomeric Aβ structures have proven to be more difficult, due to their low abundance and transient nature under normal in vitro conditions. Studies have shown that the maximum amount of $A\beta_{1-42}$ oligomers is reached in the middle of the transition phase during amyloid formation, and that their relative abundance never reaches more than approximately 1.5% of the total peptide molar concentration [15]. Such small populations are usually not distinguishable by methods that detect the average of the entire peptide ensemble. One method that can detect oligomers at low concentrations is mass spectrometry (MS), as in a mass spectrum every molecular species gives rise to a separate signal. The use of soft electrospray ionization and gentle instrument conditions to study protein samples in aqueous solution is usually called native MS, analogous to native gel electrophoresis. Native MS has successfully been used to study Aβ oligomers, their structure, their ligand binding, and their stability upon co-incubation with other molecules [21–23].

3. Anti-Prion Properties of Signal Peptide Derived CPPs

The discovery of the anti-amyloid CPPs discussed here was partly by serendipity. It had been observed that the $mPrP_{1-28}$ peptide displayed striking sequence similarities with CPPs. In a collaboration with Ülo Langel and his group, we showed that prion peptides with such unprocessed signal sequences are indeed CPPs which can carry large cargoes into live cells [24]. The $mPrP_{1-28}$ peptide could, for example, carry the 67 kDa hydrophilic protein avidin into cells. This was, however, associated with large precipitation as the $mPrP_{1-28}$ peptide was prone to aggregation, especially in the presence of negatively charged lipids. The internalized peptide–complex also seemed to stress the cell, as cell morphology was altered.

It was later observed that the corresponding bovine prion peptide, $bPrP_{1-30}$, was internalized into live mammalian cells by macropinocytosis, initiated by peptide binding to cell surface proteoglycans [25]. The peptide was found to localize to the cytoplasm and not to the nucleus, which could otherwise be suggested by the similarity of the KKRPKP hexapeptide to a nuclear localization signal (NLS). $bPrP_{1-30}$ was found to alter the studied cells just like $mPrP_{1-28}$, indicating some toxicity.

The PrP-derived CPPs were then found to be efficient anti-prion peptides [1,26]. Both $mPrP_{1-28}$ and $bPrP_{1-30}$ could inhibit the conversion between the cellular prion protein PrP^{c} and the misfolded disease (scrapie) form PrP^{Sc}, and efficiently counteracted prion propagation in cell cultures. The levels of PrP^{c} in healthy cells were unaffected, indicating a specific effect on the PrP^{Sc} form. The effect was retained, and even enhanced, upon exchanging the PrP signal peptide into the NCAM1 signal peptide [26]. A study based on AFM imaging showed that co-aggregation between our constructs and the prion protein into non-fibrillar/non-amyloid aggregates appears to be an important molecular interaction underlying these results [27]. The NCAM1 signal peptide is shorter and less hydrophobic than its PrP counterpart, which leads to higher solubility and a lower aggregation propensity. This is likely beneficial since aggregation of the PrP-derived signal peptides had been shown to stress the cells.

The anti-prion effect was lost when five or more N-terminal residues were removed from the peptide [26]. There was also no effect for the $mPrP_{23-28}$ segment alone, without any of the signal peptides linked. All this points towards the importance of protein translocation for the CPP constructs to have their anti-prion function. The anti-prion effect was not retained upon exchanging the signal peptide into various other common CPPs, such as the chimeric penetratin-$mPrP_{23-28}$, TAT-$mPrP_{23-28}$, and Transportan-10-$mPrP_{23-28}$ constructs, indicating some specificity for the signal sequence.

The synthetic CPP peptides counteracted prion infection equally well in chiral D-form as in L-form, suggesting that no chiral recognition such as by a receptor is involved in the observed cellular uptake or anti-amyloid function [26]. The D-form of the peptide should promote a longer biological lifetime, which is beneficial for therapeutic purposes. So far, all experiments with our CPP constructs have been carried out in vitro or in cell cultures. Finding out how the constructs perform in vivo, i.e., in animal models or in human patients, is an important task for future research.

No exact molecular mechanism for the anti-prion effect of the CPP constructs has been pinpointed, although co-aggregation between the constructs and the amyloidogenic proteins and peptides appears to be one important aspect [8,20,27]. We hypothesize based on the above-mentioned experimental findings that the signal peptide sequence causes translocation of the peptides into cells where it co-localizes with and binds specifically to PrP^{Sc}, through the KKRPKP recognition sequence. The CPP-PrP^{Sc} complex is then unable to bind additional PrP^{c} units, which hinders the conversion from PrP^{c} to PrP^{Sc}.

4. Anti-Amyloid Properties of Signal Peptide Derived CPPs

Our studies continued from prion infection to more general amyloid effects on living cells. We found that the NCAM-PrP construct could efficiently counteract the cellular toxicity caused by the Aβ peptide, when both peptides were added extracellularly to cell cultures [2]. We also showed that the cationic PrP segment could be exchanged with another cationic hexapeptide (KKLVFF), derived from the $A\beta_{16\text{–}20}$ sequence, with similar anti-amyloid effects [2]. The idea behind this new peptide variant, which we labelled NCAM-Aβ, was that the recognition sequence $A\beta_{16\text{–}20}$ (KLVFF) is previously known to bind full-length Aβ peptides and affect their amyloid formation in vitro as well as in vivo [28,29]. The KKLVFF sequence also has a strong cationic net charge, similar to the prion-derived KKRPKP sequence previously examined.

The NCAM-PrP and NCAM-Aβ constructs display similar properties in terms of translocation into cells, decreasing Aβ toxicity in cell studies, and decreasing the concentration of Aβ oligomers in vitro as detected by ELISA and dot–blot assays [2]. NCAM-Aβ, however, displayed slightly worse solubility and stability, probably due to the smaller positive net charge of the construct. The NCAM-PrP peptide was also able to suppress Aβ fibrillation at lower CPP/Aβ ratios compared to NCAM-Aβ, further indicating an importance of the positive net charge for the anti-amyloid effect.

In vitro studies of $A\beta_{1\text{–}40}$ and $A\beta_{1\text{–}42}$ peptide amyloid formation showed that the amyloid process was affected by the presence of the CPPs, as seen by time-dependent ThT fluorescence spectroscopy [8] (Figure 2A). AFM imaging of the aggregation end-point state indicated that the Aβ aggregation process was seemingly redirected to non-fibrillar states by the presence of the CPP (Figure 2D). No clear interactions with the monomeric Aβ peptide were observed in ^{15}N HSQC NMR, which instead showed a general drop in signal intensity. A loss in spectroscopic signal upon co-incubation of Aβ and NCAM-PrP was also observed in CD spectroscopy (Figure 2B). No abundant heterooligomer formation was observed using native MS, with only a weak heterodimer being observed (Figure 2C). All this seems to point towards a rapid co-aggregation process where large co-aggregates are formed which precipitate out of solution. Similar non-fibrillar co-aggregates were also observed for the PrP protein together with the NCAM-Aβ construct [27].

A different behavior was observed upon introduction of a membrane-mimicking environment, where loss of NMR, CD, and MS signals was not observed when micelles were included in the samples [8]. Native MS showed that Aβ and NCAM-PrP formed an abundance of heterooligomers in this membrane-mimicking environment, indicating that the two peptides can also interact in a membrane (Figure 2E). Aggregates formed in the membrane are on a size-scale compatible with detection by native MS. This is in contrast to the situation in a pure aqueous solution, where no hetero-oligomerization is observed (Figure 2C). Co-aggregates formed in pure aqueous solution (Figure 2D, right) instead seem to be larger and therefore not detectable by native MS. They also seem to be less soluble, and perhaps also less specific, as seen by the large general losses in signal intensity observed in CD spectroscopy, NMR spectroscopy, and native MS. The co-oligomerization shifted the Aβ oligomer population towards lower oligomerization states (Figure 2F), indicating that the binding of NCAM-PrP interferes with the oligomerization process.

5. Discussion

We have here described how CPPs derived from the PrP and NCAM1 signal peptide sequences, together with added cationic hexamer sequences, can decrease the toxicity of PrP and Aβ in cell cultures, as well as modulate their aggregation in vitro. The CPP peptides are made up of a hydrophobic signal peptide part that likely is responsible for translocation, and a cationic short sequence that likely is most responsible for the interaction with the toxic peptides. The exact molecular mechanisms are not known, for example, if the interaction in vivo takes place between Aβ and the intact CPP construct, or if the CPP construct is cleaved upon entry to the cell, thereby releasing the free cationic peptide for further interactions with Aβ and PrP. It is also possible that the interaction could take place in the membrane itself, as both Aβ and the signal peptides are membrane-interacting molecules. Our in vitro results demonstrate direct interactions between small Aβ oligomers and NCAM-PrP in membrane-mimicking micelles. Such interactions could inhibit the formation of e.g., pore-complexes, which have been suggested as particularly toxic states of Aβ aggregation [13,30,31].

The nature of the interaction between Aβ and the CPPs is of interest, as both the PrP and Aβ derived cationic sequences decrease Aβ toxicity. It was even observed that the PrP-sequences were more efficient at counteracting Aβ aggregation compared to the Aβ-sequence, which perhaps indicates that the net positive charge is more important than the specific sequence. One would otherwise expect that the $A\beta_{16-21}$-recognition sequence would interact more specifically with Aβ peptide compared to the mPrP sequence. These speculations on the importance of the positive charge are in agreement with a previous study showing that modifications to the KLVFF segment by adding additional positive K residues increased the segment's ability to inhibit Aβ aggregation [32]. Another study has similarly shown how increasing the positive charge of hydrophilic proteins lead to enhancement of their ability to inhibit Aβ aggregation [33]. In addition, cationic amphiphiles have been shown to interact strongly with Aβ and inhibit amyloid aggregation [34].

There are recent reports about a relatively short (12 amino acid residues) positively charged peptide, named RD2, which targets Aβ oligomers and shows similar effects as the CPP constructs described here [35,36]. The RD2 peptide has a C-terminal 6xR repeat similar to the cationic hexapeptides presented above, and the D-form of the RD2 peptide showed a beneficial anti-AD effect in animal studies [37]. The suggested model of action is that the RD2 peptide may interact with unstructured Aβ monomers, thereby stabilizing the monomeric state of Aβ [38].

Currently, there is a growing interest in finding sequences of relatively short peptides that may be active against amyloid formation [39,40], where amyloid structures generally are defined as having characteristic cross-β-sheet molecular structures [41]. The Aβ peptide is a typical example of an amyloid-forming pathological peptide. It has a hydrophilic and charged N-terminal segment, followed by a more hydrophobic C-terminal section, which is the basis for formation of the β hairpin structure in the fibrils [10]. The described anti-amyloid peptides have few common features in terms of sequences and potential structures, but certain features seem to be related to the anti-amyloid activity, such as one region having mainly hydrophilic and charged residues and another region mainly having hydrophobic residues [40]. This pattern is in fact similar to generally amyloid-forming sequences. The designed anti-amyloid peptides described here seem to follow this pattern, with the N-terminal segment being mainly hydrophobic and the C-terminal segment being mainly hydrophilic and positively charged. Other short peptides have been found to display anti-cancer activities, with sequences derived from e.g., the azurin protein [42,43]. Here, the ability to enter cancer cells seems to be important for its activity. In general, short peptides, possibly with CPP activities, may be very interesting for covalent or non-covalent addition in the strategic development of therapeutic molecules such as antibodies [44].

6. Conclusions

The studies here reviewed, about anti-prion and anti-amyloid effects of certain CPP constructs, suggest that there may be a general principle for how to stop or at least counteract cellular amyloid formation (involving both cell location and cell chemistry). The signal sequence in these constructs may be important for localization of the peptide construct, and the cationic hexapeptide segment may be most important for the anti-amyloid effects on the amyloid and prion aggregation processes.

Author Contributions: All authors contributed to writing the text and making the figures. All authors have read and agreed to the published version of the manuscript.

Funding: This work was supported by grants from the Swedish Brain Foundation, the Swedish Research Council, and Region Stockholm to AG.

Institutional Review Board Statement: Not applicable.

Informed Consent Statement: Not applicable.

Data Availability Statement: Not applicable.

Acknowledgments: We wish to thank Ulo Langel for many interesting discussions about the properties of Cell-Penetrating peptides.

Conflicts of Interest: The authors declare no conflict of interest.

References

1. Löfgren, K.; Wahlström, A.; Lundberg, P.; Langel, U.; Gräslund, A.; Bedecs, K. Antiprion properties of prion protein-derived cell-penetrating peptides. *FASEB J.* **2008**, *22*, 2177–2184. [CrossRef] [PubMed]
2. Henning-Knechtel, A.; Kumar, S.; Wallin, C.; Król, S.; Wärmländer, S.K.T.S.; Jarvet, J.; Esposito, G.; Kirmizialtin, S.; Gräslund, A.; Hamilton, A.D.; et al. Designed Cell-Penetrating Peptide Inhibitors of Amyloid-beta Aggregation and Cytotoxicity. *Cell Rep. Phys. Sci.* **2020**, *1*, 100014. [CrossRef]
3. Gräslund, A.; Eriksson, L.E.G. Biophysical Studies of Cell-Penetrating Peptides. In *Cell-Penetrating Peptides*; CRC Press: Boca Raton, FL, USA, 2002; pp. 223–244. [CrossRef]
4. Xie, J.; Shen, Z.; Anraku, Y.; Kataoka, K.; Chen, X. Nanomaterial-based blood-brain-barrier (BBB) crossing strategies. *Biomaterials* **2019**, *224*, 119491. [CrossRef]
5. Sormanni, P.; Vendruscolo, M. Protein Solubility Predictions Using the CamSol Method in the Study of Protein Homeostasis. *Cold Spring Harb. Perspect. Biol.* **2019**, *11*, a033845. [CrossRef]
6. Nielsen, H.; Engelbrecht, J.; Brunak, S.; Von Heijne, G. Identification of prokaryotic and eukaryotic signal peptides and prediction of their cleavage sites. *Protein Eng.* **1997**, *10*, 1–6. [CrossRef]
7. Almagro Armenteros, J.J.; Tsirigos, K.D.; Sønderby, C.K.; Petersen, T.N.; Winther, O.; Brunak, S.; Von Heijne, G.; Nielsen, H. SignalP 5.0 improves signal peptide predictions using deep neural networks. *Nat. Biotechnol.* **2019**, *37*, 420–423. [CrossRef]
8. Król, S.; Österlund, N.; Vosough, F.; Jarvet, J.; Wärmländer, S.; Barth, A.; Ilag, L.L.; Magzoub, M.; Gräslund, A.; Mörman, C. The amyloid-inhibiting NCAM-PrP peptide targets Aβ peptide aggregation in membrane-mimetic environments. *Science* **2021**, *24*, 102852. [CrossRef]
9. Selkoe, D.J.; Hardy, J. The amyloid hypothesis of Alzheimer's disease at 25 years. *EMBO Mol. Med.* **2016**, *8*, 595–608. [CrossRef]
10. Abelein, A.; Abrahams, J.P.; Danielsson, J.; Gräslund, A.; Jarvet, J.; Luo, J.; Tiiman, A.; Wärmländer, S.K.T.S. The hairpin conformation of the amyloid β peptide is an important structural motif along the aggregation pathway. *J. Biol. Inorg. Chem.* **2014**, *19*, 623–634. [CrossRef]
11. Mroczko, B.; Groblewska, M.; Litman-Zawadzka, A.; Kornhuber, J.; Lewczuk, P. Amyloid β oligomers (AβOs) in Alzheimer's disease. *J. Neural Transm.* **2018**, *125*, 177–191. [CrossRef]
12. Sandberg, A.; Luheshi, L.M.; Sollvander, S.; de Barros, T.P.; Macao, B.; Knowles, T.P.J.; Biverstal, H.; Lendel, C.; Ekholm-Petterson, F.; Dubnovitsky, A.; et al. Stabilization of neurotoxic Alzheimer amyloid-oligomers by protein engineering. *Proc. Natl. Acad. Sci. USA* **2010**, *107*, 15595–15600. [CrossRef] [PubMed]
13. Sciacca, M.F.M.; Kotler, S.A.; Brender, J.R.; Chen, J.; Lee, D.K.; Ramamoorthy, A. Two-step mechanism of membrane disruption by Aβ through membrane fragmentation and pore formation. *Biophys. J.* **2012**, *103*, 702–710. [CrossRef] [PubMed]
14. Österlund, N.; Moons, R.; Ilag, L.L.; Sobott, F.; Gräslund, A. Native Ion Mobility-Mass Spectrometry Reveals the Formation of β-Barrel Shaped Amyloid-β Hexamers in a Membrane-Mimicking Environment. *J. Am. Chem. Soc.* **2019**, *141*, 10440–10450. [CrossRef] [PubMed]
15. Michaels, T.C.T.; Šarić, A.; Curk, S.; Bernfur, K.; Arosio, P.; Meisl, G.; Dear, A.J.; Cohen, S.I.A.; Dobson, C.M.; Vendruscolo, M.; et al. Dynamics of oligomer populations formed during the aggregation of Alzheimer's Aβ42 peptide. *Nat. Chem.* **2020**, *12*, 445–451. [CrossRef]
16. Tolar, M.; Abushakra, S.; Hey, J.A.; Porsteinsson, A.; Sabbagh, M. Aducanumab, gantenerumab, BAN2401, and ALZ-801—The first wave of amyloid-targeting drugs for Alzheimer's disease with potential for near term approval. *Alzheimer's Res. Ther.* **2020**, *12*, 95. [CrossRef]
17. Biancalana, M.; Koide, S. Molecular mechanism of Thioflavin-T binding to amyloid fibrils. *Biochim. Biophys. Acta Proteins Proteom.* **2010**, *1804*, 1405–1412. [CrossRef]
18. Linse, S.; Thalberg, K.; Knowles, T.P.J. The unhappy chaperone. *QRB Discov.* **2021**, *2*, e7. [CrossRef]
19. Meisl, G.; Kirkegaard, J.B.; Arosio, P.; Michaels, T.C.T.; Vendruscolo, M.; Dobson, C.M.; Linse, S.; Knowles, T.P. Molecular mechanisms of protein aggregation from global fitting of kinetic models. *Nat. Protoc.* **2016**, *11*, 252–272. [CrossRef]
20. Pansieri, J.; Ostojić, L.; Iashchishyn, I.A.; Magzoub, M.; Wallin, C.; Wärmländer, S.K.T.S.; Gräslund, A.; Ngoc, M.N.; Smirnovas, V.; Svedruzic, Z.; et al. Pro-Inflammatory S100A9 Protein Aggregation Promoted by NCAM1 Peptide Constructs. *ACS Chem. Biol.* **2019**, *14*, 1410–1417. [CrossRef]
21. Lieblein, T.; Zangl, R.; Martin, J.; Hoffmann, J.; Hutchison, M.J.; Stark, T.; Stirnal, E.; Schrader, T.; Schwalbe, H.; Morgner, N. Structural rearrangement of amyloid-β upon inhibitor binding suppresses formation of Alzheimer's disease related oligomers. *eLife* **2020**, *9*, e59306. [CrossRef]

22. Österlund, N.; Lundqvist, M.; Ilag, L.L.; Gräslund, A.; Emanuelsson, C. Amyloid-β oligomers are captured by the DNAJB6 chaperone: Direct detection of interactions that can prevent primary nucleation. *J. Biol. Chem.* **2020**, *295*, 8135–8144. [CrossRef] [PubMed]
23. Puig, E.; Tolchard, J.; Riera, A.; Carulla, N. Somatostatin, an In Vivo Binder to Aβ Oligomers, Binds to βPFOAβ(1–42) Tetramers. *ACS Chem. Neurosci.* **2020**, *11*, 3358–3365. [CrossRef] [PubMed]
24. Lundberg, P.; Magzoub, M.; Lindberg, M.; Hällbrink, M.; Jarvet, J.; Eriksson, L.E.G.; Langel, U.; Gräslund, A. Cell membrane translocation of the N-terminal (1–28) part of the prion protein. *Biochem. Biophys. Res. Commun.* **2002**, *299*, 85–90. [CrossRef]
25. Magzoub, M.; Sandgren, S.; Lundberg, P.; Oglęcka, K.; Lilja, J.; Wittrup, A.; Eriksson, L.E.G.; Langel, U.; Belting, M.; Gräslund, A. N-terminal peptides from unprocessed prion proteins enter cells by macropinocytosis. *Biochem. Biophys. Res. Commun.* **2006**, *348*, 379–385. [CrossRef]
26. Söderberg, K.L.; Guterstam, P.; Langel, U.; Gräslund, A. Targeting prion propagation using peptide constructs with signal sequence motifs. *Arch. Biochem. Biophys.* **2014**, *564*, 254–261. [CrossRef]
27. Gielnik, M.; Zhukova, L.; Zhukov, I.; Gräslund, A.; Kozak, M.; Wärmländer, S.K.T.S. The engineered peptide construct NCAM1-Aβ inhibits aggregation of the human prion protein (PrP). *Acta Biochim. Pol.* **2022**, *69*, 257–261.
28. Tjernberg, L.O.; Näslund, J.; Lindqvist, F.; Johansson, J.; Karlström, A.R.; Thyberg, J.; Terenius, L.; Nordstedt, C. Arrest of -Amyloid Fibril Formation by a Pentapeptide Ligand. *J. Biol. Chem.* **1996**, *271*, 8545–8548. [CrossRef]
29. Lowe, T.L.; Strzelec, A.; Kiessling, A.L.L.; Murphy, R.M. Structure-Function Relationships for Inhibitors of β-Amyloid Toxicity Containing the Recognition Sequence KLVFF. *Biochemistry* **2001**, *40*, 7882–7889. [CrossRef]
30. Ciudad, S.; Puig, E.; Botzanowski, T.; Meigooni, M.; Arango, A.S.; Do, J.; Mayzel, M.; Bayoumi, M.; Chaignepain, S.; Maglia, G.; et al. Aβ(1–42) tetramer and octamer structures reveal edge conductivity pores as a mechanism for membrane damage. *Nat. Commun.* **2020**, *11*, 3014. [CrossRef]
31. Österlund, N.; Luo, J.; Wärmländer, S.K.T.S.; Gräslund, A. Membrane-mimetic systems for biophysical studies of the amyloid-β peptide. *Biochim. Biophys. Acta Proteins Proteom.* **2018**, *1867*, 492–501. [CrossRef]
32. Bett, C.K.; Serem, W.K.; Fontenot, K.R.; Hammer, R.P.; Garno, J.C. Effects of Peptides Derived from Terminal Modifications of the Aβ Central Hydrophobic Core on Aβ Fibrillization. *ACS Chem. Neurosci.* **2010**, *1*, 661–678. [CrossRef] [PubMed]
33. Assarsson, A.; Hellstrand, E.; Cabaleiro-Lago, C.; Linse, S. Charge dependent retardation of amyloid β aggregation by hydrophilic proteins. *ACS Chem. Neurosci.* **2014**, *5*, 266–274. [CrossRef] [PubMed]
34. Österlund, N.; Kulkarni, Y.S.; Misiaszek, A.D.; Wallin, C.; Krüger, D.M.; Liao, Q.; Rad, F.M.; Jarvet, J.; Strodel, B.; Wärmländer, S.K.T.S.; et al. Amyloid-β Peptide Interactions with Amphiphilic Surfactants: Electrostatic and Hydrophobic Effects. *ACS Chem. Neurosci.* **2018**, *9*, 1680–1692. [CrossRef]
35. Van Groen, T.; Schemmert, S.; Brener, O.; Gremer, L.; Ziehm, T.; Tusche, M.; Nagel-Steger, L.; Kadish, I.; Schartmann, E.; Elfgen, A.; et al. The Aβ oligomer eliminating D-enantiomeric peptide RD2 improves cognition without changing plaque pathology. *Sci. Rep.* **2017**, *7*, 16275. [CrossRef]
36. Olubiyi, O.O.; Frenzel, D.; Bartnik, D.; Gluck, J.M.; Brener, O.; Nagel-Steger, L.; Funke, S.A.; Willbold, D.; Strodel, B. Amyloid aggregation inhibitory mechanism of arginine-rich D-peptides. *Curr. Med. Chem.* **2014**, *21*, 1448–1457. [CrossRef] [PubMed]
37. Willbold, D.; Kutzsche, J.; Willuweit, A.; Windisch, M.; Jürgens, D. Clinical phase I data of the first orally available anti-aβ-prionic drug PRI-002 that reverses behavioral and cognitive deficits, and decelerates neurodegeneration in AD animal models. *Alzheimer's Dement.* **2020**, *16*, e12001. [CrossRef]
38. Zhang, T.; Gering, I.; Kutzsche, J.; Nagel-Steger, L.; Willbold, D. Toward the Mode of Action of the Clinical Stage All-d-Enantiomeric Peptide RD2 on Aβ42 Aggregation. *ACS Chem. Neurosci.* **2019**, *10*, 4800–4809. [CrossRef]
39. Neddenriep, B.; Calciano, A.; Conti, D.; Sauve, E.; Paterson, M.; Bruno, E.; Moffet, D.A. Short Peptides as Inhibitors of Amyloid Aggregation. *Open Biotechnol. J.* **2011**, *5*, 39–46. [CrossRef]
40. Roterman, I.; Banach, M.; Konieczny, L. Towards the design of anti-amyloid short peptide helices. *Bioinformation* **2018**, *14*, 1–7. [CrossRef]
41. Riek, R.; Eisenberg, D.S. The activities of amyloids from a structural perspective. *Nature* **2016**, *539*, 227–235. [CrossRef]
42. Taylor, B.N.; Mehta, R.R.; Yamada, T.; Lekmine, F.; Christov, K.; Chakrabarty, A.M.; Green, A.; Bratescu, L.; Shilkaitis, A.; Beattie, C.W.; et al. Noncationic Peptides Obtained from Azurin Preferentially Enter Cancer Cells. *Cancer Res.* **2009**, *69*, 537–546. [CrossRef] [PubMed]
43. Yaghoubi, A.; Khazaei, M.; Avan, A.; Hasanian, S.M.; Cho, W.C.; Soleimanpour, S. p28 Bacterial Peptide, as an Anticancer Agent. *Front. Oncol.* **2020**, *10*, 1303. [CrossRef] [PubMed]
44. Horsley, J.R.; Jovcevski, B.; Wegener, K.L.; Yu, J.; Pukala, T.L.; Abell, A.D. Rationally designed peptide-based inhibitor of Aβ42 fibril formation and toxicity: A potential therapeutic strategy for Alzheimer's disease. *Biochem. J.* **2019**, *477*, 2039–2054. [CrossRef] [PubMed]

pharmaceutics

Article

Stem Cell Bioengineering with Bioportides: Inhibition of Planarian Head Regeneration with Peptide Mimetics of Eyes Absent Proteins

Sarah Jones [1,*], Bárbara Matos [2], Sarah Dennison [3], Margarida Fardilha [2] and John Howl [1]

1 Research Institute in Healthcare Science, Faculty of Science & Engineering, University of Wolverhampton, Wulfruna Street, Wolverhampton WV1 1LY, UK; john_871@hotmail.co.uk
2 Laboratory of Signal Transduction, Department of Medical Sciences, Institute of Biomedicine—iBiMED, University of Aveiro, 3810-193 Aveiro, Portugal; barbaracostamatos@ua.pt (B.M.); mfardilha@ua.pt (M.F.)
3 School of Pharmacy and Biomedical Sciences, University of Central Lancashire, Preston PR1 2HE, UK; srdennison1@uclan.ac.uk
* Correspondence: s.jones4@wlv.ac.uk; Tel.: +44-1902-322748

Abstract: Djeya1 (RKLAFRYRRIKELYNSYR) is a very effective cell penetrating peptide (CPP) that mimics the α5 helix of the highly conserved Eya domain (ED) of eyes absent (Eya) proteins. The objective of this study was to bioengineer analogues of Djeya1 that, following effective translocation into planarian tissues, would reduce the ability of neoblasts (totipotent stem cells) and their progeny to regenerate the anterior pole in decapitated *S. mediterranea*. As a strategy to increase the propensity for helix formation, molecular bioengineering of Djeya1 was achieved by the mono-substitution of the helicogenic aminoisobutyric acid (Aib) at three species-variable sites: 10, 13, and 16. CD analyses indicated that Djeya1 is highly helical, and that Aib-substitution had subtle influences upon the secondary structures of bioengineered analogues. Aib-substituted Djeya1 analogues are highly efficient CPPs, devoid of influence upon cell viability or proliferation. All three peptides increase the migration of PC-3 cells, a prostate cancer line that expresses high concentrations of Eya. Two peptides, [Aib^{13}]Djeya1 and [Aib^{16}]Djeya1, are bioportides which delay planarian head regeneration. As neoblasts are the only cell population capable of division in planaria, these data indicate that bioportide technologies could be utilised to directly manipulate other stem cells in situ, thus negating any requirement for genetic manipulation.

Keywords: cell penetrating peptide; bioportide; eyes absent protein; neoblast; planarian; confocal microscopy; stem cell

Citation: Jones, S.; Matos, B.; Dennison, S.; Fardilha, M.; Howl, J. Stem Cell Bioengineering with Bioportides: Inhibition of Planarian Head Regeneration with Peptide Mimetics of Eyes Absent Proteins. *Pharmaceutics* **2023**, *15*, 2018. https://doi.org/10.3390/pharmaceutics15082018

Academic Editors: Prisca Boisguérin and Sébastien Deshayes

Received: 15 June 2023
Revised: 17 July 2023
Accepted: 24 July 2023
Published: 26 July 2023

Copyright: © 2023 by the authors. Licensee MDPI, Basel, Switzerland. This article is an open access article distributed under the terms and conditions of the Creative Commons Attribution (CC BY) license (https://creativecommons.org/licenses/by/4.0/).

1. Introduction

Cellular and tissue permeability barriers are significant caveats for bioengineering strategies to develop bioactive agents and exploit novel intracellular drug modalities. As recently reviewed [1–3], cell penetrating peptides (CPPs), often polycationic linear sequences of 12–24 amino acids, are a versatile technology that can overcome the common biophysical constraint of ineffective intracellular access. Bioportides, CPPs with intrinsic bioactivities and so partially distinct from conventional inert vectors, accrete within eukaryotic cells to influence protein function and impact cell biology [4]. The molecular organisation of bioportides commonly includes mimetic sequences derived from functional protein domains to serve as selective modulators of intracellular protein-protein interactions (PPIs; [4–6]). Hence, bioportide technologies, presumably acting by a dominant-negative mechanism, enable both the understanding and discrete manipulation of intracellular signalling pathways that regulate cellular biology [5,6].

The triploblastic bilateral planarian *Schmidtea mediterranea* is a very common model organism employed to address fundamental cellular processes that include tissue regeneration and the differentiation of neoblasts, relatively small pluripotent stem cells of mesenchymal origin [7,8]. Indeed, a total of 20–30% of *S. mediterranea* cells are neoblasts, the only planarian cell type capable of mitotic division [7,8]. A comprehensive *S. mediterranea* genomic database (SmedGD 2.0) confirmed that planaria are genetically more like vertebrates than both *Drosophila melanogaster* and *Caenorhabditis elegans* [9,10]. *S. mediterranea* is also a rigorous three-dimensional model to analyse the import of CPPs and bioportides into complex tissues presenting both physical and metabolic barriers [11].

In common with studies of arginine-rich peptides derived from both RNA- and DNA-binding proteins [12], planarian proteins that collectively control head remodelling and eye regeneration following decapitation are a viable source of cationic CPP vectors [11]. Djeya1 (RKLAFRYRRIKELYNSYR), an octadecapeptide sequence mimicking a highly conserved domain of eyes absent (Eya) proteins, is a particularly efficient and seemingly inert example of such a CPP vector sequence. Three days after head amputation, fluorescent Djeya1 effectively enters the unpigmented *S. mediterranea* blastema, a transient and heterogenous cell mass responsible for head morphogenesis, to penetrate deeper along the dorsal ventral axis [11]. Thus, CPPs such as Djeya1 provide the means to target bioactive agents to differentiate post-mitotic neoblast progeny [13] in addition to epithelial precursor cells or neoblast-derived mesenchymal cells.

Djeya1 mimics part of the $\alpha5$ helix within the evolutionary conserved C-terminal Eya Domain (ED) of Eya proteins [14,15]. The ED domain acts as a transcriptional regulator known to bind proteins such as Dachshund and Sine Oculus [14,15]. Hence, the major objective of this study was to determine whether the bioengineering of Djeya1 analogues could provide rhegnylogically organised bioportides, CPPs in which the pharmacophores that enable cellular penetration and those essential for bioactivity are discontinuously organised. This is in contrast to a sychnologic organisation in which the pharmacophores for penetration and bioactivity are distinct. Moreover, given that Arg is the quantitatively dominant amino acid at PPI interfaces, there will clearly be some pharmacophores which are both cell penetrant and bioactive within a rhegnylogic organisation [4,5]. We hypothesized that analogues of Djeya1 with enhanced helicity would more effectively mimic ED to modulate PPIs and inhibit anterior pole and eye regeneration in *S. mediterranea* [11,14–16]. To promote helicity, we introduced the helicogenic amino acid α-aminoisobutyric acid (Aib) at three sites within Djeya1 (I^{10}, L^{13}, S^{16}) where there is species-specific heterogeneity of the protein sequence in Eya proteins. All three bioengineered Djeya1 analogues are highly efficient CPPs. Two of these, [Aib^{13}]Djeya1 and [Aib^{16}]Djeya1, inhibited planarian head regeneration whilst [Aib^{10}]Djeya1 was inert. As neoblast populations are the only cell type capable of division in planarians [7,8,13,16–18], our data indicate that both [Aib^{13}]Djeya1 and [Aib^{16}]Djeya1 directly influence the biology of stem cells. Furthermore, cellular assays confirmed that this action was unlikely the consequence of a detrimental influence of bioportides upon cellular viability, proliferation, or motility. Hence, the same bioportides might modulate the morphogenesis of mammalian stem cells regulated by eyes absent proteins [19].

2. Materials and Methods

2.1. Peptide Synthesis and Secondary Structure Analysis

2.1.1. Microwave Enhanced Solid Phase Peptide Synthesis

Aib-substituted Djeya1 analogues (Figure 1) were synthesized using a CEM Liberty Blue Synthesizer, equipped with a UV analyser (CEM Microwave Technology Ltd., Buckingham, UK) in dimethylformamide (DMF; Cambridge Reagents, Barton upon Humber, UK) on Rink-amide 4-methylbenzhydrylamine (MBHA) resin (Novabiochem, Beeston, UK) to generate peptide amides [11,20]. N,N′-diisopropylcarbodiimide (DIC; Sigma-Aldrich, Gillingham, UK) was used as a condensation reagent with the additive ethyl 2-cyano-2-(hydroxyimino)acetate (Oxyma; [21]; Novabiochem, Beeston, UK) plus a trace concentration of N,N-diisopropylyethylamine (DIPEA; Sigma-Aldrich, Gillingham, UK). Syntheses on a 0.1 mmole scale were performed using a 5-fold molar excess of AA/DIC/Oxyma with

0.1 molar equivalent of DIPEA in a final volume of 4 mL. Single coupling of most Fmoc-AAs (Novabiochem, Beeston, UK) was achieved at 90 °C/120 s. To overcome any steric hindrance, both the dialkylated Aib and following AA were double coupled using an identical protocol. Arg was routinely double coupled at 75 °C/300 s. A standard deprotection cycle with 20% (*v*/*v*) Piperidine (Sigma-Aldrich, Gillingham, UK) in DMF was 90 °C/60 s routinely monitored by determining the UV absorption at 301 nm of dibenzofulvene-piperidine adducts [20]. Amino-terminal acylation of Aib-substituted Djeya1 analogues with 6-carboxytetramethylrhodamine (Novabiochem, Beeston, UK) yielded fluorescent analogues with excellent photo-stability and pH-insensitive fluorescence. All peptides described herein were purified to apparent homogeneity by reverse phase HPLC [4,6,22].

2.1.2. Circular Dichroism

Secondary structure analyses of Djeya1 peptides were performed on a Jasco J-815 CD spectrometer (Jasco Ltd., Heckmondwike, UK). Peptides (0.01 mg mL^{-1}) were dissolved in either 1 × phosphate buffered saline (PBS) at pH 7.4 or in a 2,2,2-tetrafluoroethanol (TFE) and 1 × PBS mixture (50.0% *v*/*v*). All buffers were prepared using ultra-pure water (resistivity 18 MΩ cm), PBS and TFE (Scientific Laboratory Supplies Ltd., Nottingham, UK). CD experiments were also performed at peptide:lipid ratios of 1:100, whilst 1,2-dimyristoyl-sn-glycero-3-phosphatidylcholine (DMPC) and 1,2-dimyristoyl-sn-glycero-3-phosphatidylserine (DMPS) (5 mg mL^{-1}) (Avanti Polar Lipids, Alabaster, AL, USA) were dissolved separately in chloroform (HPLC grade, VWR International Ltd Lutterworth, UK) and dried under N_2 gas. The lipid film was rehydrated using 1 × PBS, pH 7.5 for 1 h or until the solution was no longer turbid. The solution then underwent 5 cycles of freeze thaw before the lipid/peptide samples were prepared by the addition of peptide stock solution (final concentration 0.01 mg mL^{-1}). Far UV CD spectra (180 nm to 260 nm) were recorded at 20 °C using 0.5 nm intervals, a bandwidth of 1 nm, a scan speed of 50 nm min^{-1} and a 10 mm path-length cell. Ten accumulations were selected to average a single spectrum. Samples without peptide were prepared for background spectral subtraction. Data, baseline-subtracted and averaged CD, were analyzed using the Dichroweb server CDSSTR and SELCON3 reference dataset 3 [23–25]. To generate percentage structure averages, these analyses were repeated four times.

2.2. *Schmidtea mediterranea* Culture and Head Morphogenesis

2.2.1. Planarian Maintenance

S. mediterranea, hermaphroditic sexual strain, were a kind gift from Kerstin Bartscherer (Max Planck Institute for Molecular Medicine, Münster, Germany). As previously described [11], animals were maintained at 20 °C in planarian artificial medium (PAM, [26]) containing NaCl (1.6 mM), $MgSO_4$ (1 mM), $MgCl_2$ (0.1 mM), KCl (0.1 mM), $NaHCO_3$ (1.2 mM), and $CaCl_2$ (1 mM) in ultrapure water with gentamycin (3 μg mL^{-1}). For maintenance, animals were routinely fed finely minced calf's liver at two-weekly intervals.

2.2.2. Anterior Pole Regeneration and Eye Development Assay

Adult planaria, 4–6 mm in length, and selected for experimentation were starved 5 days prior to amputation. Transverse amputation of planarian heads, post-auricle and pre-pharynx, induced blastema formation in the trunk section, leading to eye regeneration and head remodeling [11].

Post amputation, planaria were immediately treated with exogenously added Djeya1 analogues in 35 mm diameter × 10 mm depth culture dishes (VWR International Ltd., Lutterworth, UK) to a final volume of 4 mL PAM and maintained at 20 °C. PAM, containing Djeya1 analogues, was replenished 4 days post amputation. Animals were allowed to regenerate for 7–8 days post amputation, whilst observations were performed daily to assess any morphological variations. To document the influence of bioportides upon head morphogenesis, living specimens were observed with a Swift SM series stereo microscope equipped with Moticam BTU assembly for image acquisition (Swift Microscope World, Carlsbad, CA, USA).

2.3. Biological Characterization of Aib-Substituted Djeya1 Analogues

In the absence of a transformed neoblast cell line or a readily isolated primary culture, U373MG astrocytic tumour cells were employed as a robust model to determine CPP internalization and discount any unwanted cytotoxic effects [20,27]. Transformed cell lines known to overexpress Eya proteins were used to quantify the impact of Djeya1 analogues on cell migration (PC-3 prostate cancer cells [28]) and proliferation (U251 glioblastoma cells [29,30]).

2.3.1. Cell Culture Maintenance

U373MG and U251 cells were both maintained in Dulbecco's Modified Eagle's Medium (DMEM) containing L-glutamine (0.1 mg mL^{-1}) (Sigma-Aldrich, Gillingham, UK) and PC-3 in Roswell Park Memorial Institute media-1640 (RPMI-1640) with L-glutamine (Gibco, Life Technologies, Carlsbad, CA, USA). Both media were supplemented with foetal bovine serum (FBS) 10% (v/v), penicillin (100 U mL^{-1}), and streptomycin (100 μg mL^{-1}), and all cell lines were maintained in a humidified atmosphere of 5% CO_2 at 37 °C.

2.3.2. Qualitative Peptide Uptake Analyses

Live confocal cell imaging analyses were employed to avoid fixation artefacts and establish the intracellular distributions of tetramethylrhodamine (TAMRA)-conjugated Djeya1 analogues [22]. U373MG cells were maintained as above, transferred to 35 mm sterile glass base dishes (NuncTM, Fisher Scientific, Loughborough, UK), and grown to ~75% confluence. Cells were washed in phenol red-free DMEM prior to treatment with (TAMRA)-conjugated Djeya1 analogues diluted to a final concentration of 5 μM in phenol red free media. Treated cells were maintained at 37 °C in a humidified atmosphere of 5% CO_2 for the designated time periods. For the 1 h incubations, cells were also treated with CellMaskTM (5 μg mL^{-1}; Molecular Probes, Thermo Fisher Scientific, Waltham, MA, USA) for an additional 5 min prior to observation using a Zeiss LSM 880 microscope equipped with live cell imaging chamber (Zeiss, Cambridge, UK). A total of 5 h incubations were also observed using a photomultiplier for transmitted light (T-PMT).

2.3.3. Quantitative Peptide Uptake Analyses

U373MG and U251 cells were maintained as above and then transferred to 6-well plates and grown to 80% confluence. Cells were washed and maintained in phenol red-free DMEM and subsequently treated with TAMRA-conjugated Djeya1 analogues at final concentrations of 1 μM, 2.5 μM, and 5 μM for 1 h in culture conditions as above. Cells were washed four times, detached with 300 μL of 1% (w/v) trypsin (without phenol red) at 37 °C, collected by centrifugation, and lysed in 300 μL 0.1 M NaOH for 2 h on ice. A total of 250 μL of each sample cell lysate was transferred to a black 96-well plate and analysed using a Thermo Fischer Scientific (Loughborough, UK) Fluoroskan Ascent FL fluorescence spectrophotometer (λAbs 544 nm/λEm 590 nm).

2.3.4. Cytotoxicity Assays

When employing U373MG cells, cytotoxicity was quantitatively assessed using the 3-(4,5-dimethylthazol-2-yl)-2,5-diphenyl tetrazolium bromide (MTT; Sigma-Aldrich, Gillingham, UK) conversion assay [31,32]. U373MG cells were cultured as above in 96-well plates and treated with peptides (0.1–30 μM) in DMEM without FBS. At 4 h post treatment, cells were incubated for a further 3 h with MTT (0.5 mg mL^{-1}). The insoluble formazan product was solubilised with DMSO (Sigma-Aldrich, Gillingham, UK) and MTT conversion determined by colorimetric analysis at 540 nm (Labsystems Multiskan Ascent 354 Microplate Reader, Thermo Fischer Scientific, Loughborough, UK). Cellular viability was expressed as a percentage of those cells treated with vehicle (medium) alone.

PC-3 prostate cancer cells were seeded in supplemented RPMI-1640 (as described above) into 96-well plates (1.0×10^4 cells/well) and maintained in a humidified atmosphere of 5% CO_2 at 37 °C for 24 h. Thereafter, cells were treated with 3 µM, 10 µM, and 25 µM of Djeyal analogues in RPMI-1640 medium without FBS for 24 h, whilst maintaining culture conditions. Untreated (medium alone) cells were used as a control representing 100% viability. For the final hour of incubation, 10 µL of PrestoBlue™ Cell Viability reagent, resazurin-based reagent (Thermo Fisher Scientific, Waltham, MA, USA) was added to each well. A total of 100 µL of culture medium from each well was transferred to a black bottomed 96-well plate and the fluorescence at λ_{abs} = 560 nm and λ_{em} = 590 nm was measured using a microplate reader (Tecan Infinite® 200 PROseries, Mannedorf, Switzerland). Cellular viability was expressed as a percentage of those cells treated with vehicle (medium) alone. Three independent experiments with five replicates for each condition were performed.

2.3.5. Cellular Proliferation

Cellular proliferation was determined using measurements of cell viability and employed the MTT conversion assay as described above. U251 cells were cultured as previously described, grown to anastomose in 96-well plates, and treated from 4–72 h with Djeyal analogues (3 µM, 10 µM, 25 µM) in DMEM supplemented with FBS 10% (*v*/*v*) and maintained in a humidified atmosphere of 5% CO_2 at 37 °C. Cell viability was expressed as absorbance minus background at 540 nm. Cells treated with medium alone were also included at each designated time point and acted as a comparator when constructing growth curves.

2.3.6. Cell Migration Assays

PC-3 cells (1.00×10^5) were seeded in supplemented RPMI-1640 (as described above) in 24-well plates and incubated for 24 h at 37 °C in a humidified atmosphere of 5% CO_2. Confluent cells monolayers were wounded by scratching lines with a 200 µL pipette tip. Cells were washed in phosphate-buffered saline (PBS) and incubated with fresh medium without FBS containing 3 µM or 25 µM of Djeyal analogues for 48 h. Untreated cells were also included in the assay. Photographs were taken under ×40 magnification for EVOS™ M5000 imaging (Thermo Fisher Scientific, Waltham, MA, USA), immediately after wound incision and after 48 h. Results were expressed as a percentage of wound closure, relative to the time at 0 h. Three independent replicates of each condition were performed.

2.4. Graphical Representations and Statistical Analyses

These were performed using GraphPad Prism 9 software. For cytotoxicity and cellular proliferation assays, statistical analyses of changes in cellular viability employed an unpaired, 2-tailed, non-parametric Mann–Whitney test. Statistical analyses of changes in wound closure used a paired, 2-tailed, non-parametric Wilcoxon matched pairs signed rank test.

3. Results

3.1. Site-Directed Bioengineering of the Djeya1 CPP

The entire ED of Eya proteins is exceptionally well conserved to provide a unique tyrosine phosphatase activity within a multi-domain protein classified as a haloacid dehalogenase [13–15,33]. The transcriptional activity of Eya proteins, dependent upon PPIs notably with Dachshund and SIX family members, appears to be regulated by the dephosphorylation activity of the ED [14,15,33,34]. The CPP Djeya1 (Table 1; [11]), identical in planarians *Dugesia juponica* and *S. mediterranea*, mimics helix 5 of the ED conserved within plants, fungi, and animals, a region that could participate in PPIs required for trancriptional activity [33,34]. Within a consensus sequence of this helical domain, three variable positions (10,13,16) were selected for Aib-substitution as a bioengineering strategy to promote helicity. Very conservative changes are indicated at positions 10 (I/V) and

13 (L/I) with more chemical variability at site 16 (S/T/A/Q). We did not introduce Aib at position 18 (R/K) as the loss of cationic charge could impact on the uptake efficacy of the modified octadecapeptide. The sequences of these peptides and others used in this study are presented in Table 2.

Table 1. Sequence conservation within the α5 helix of ED.

Species	Sequence
Dugesia japonica (Djeya 1)	RKLAFRYRRIKELYNSYR
Lingula unguis	RKLAFRYRRIKEIYNSYR
Euprymna scolopes	RKLAFRYRRIKEIYNSYR
Euperipatoides kanangrensis	RKLAFRYRRIKEIYNSYR
Dryobates pubescens	RKLAFRYRRVKELYNTYR
Helobdella robusta	RKLAFRYRRIKELYSAYR
Branchiostoma floridae	RKLAFRYRRIKEIYNSYK
Branchiostoma lanceolatum	RKLAFRYRRIKEIYNSYK
Capitella teleta	RKLAFRYRRIKEIYSSYR
Melanaphis sacchari	RKLAFRYRRVKELYNQYR
Schizaphis graminum	RKLAFRYRRVKELYNQYR
Felis catus	RKLAFRYRRVKELYNTYK
Canus familiarus	RKLAFRYRRVKELYNTYK
Bos Taurus	RKLAFRYRRVKELYNTYK
Homo sapiens	RKLAFRYRRVKELYNTYK
Ornithodoros moubata	RKLAFRYRRIKEIYNQYR

Consensus—RKLAFRYRR(I/V)KE(L/I)YN(S/T/A/Q)Y(R/K).

Table 2. Primary sequences of CPPs and bioportides.

Peptide	Sequence
Djeya	*H*-RKLAFRYRRIKELYNSYR-*NH*$_2$
[Aib^{10}]Djeya1	*H*-RKLAFRYRR(Aib)KELYNSYR-*NH*$_2$
[Aib^{13}]Djeya1	*H*-RKLAFRYRRIKE(Aib)YNSYR-*NH*$_2$
[Aib^{16}]Djeya1	*H*-RKLAFRYRRIKELYN(Aib)YR-*NH*$_2$
Tat	*H*-GRKKRRQRRRPPQ-*NH*$_2$
C105Y	*H*-CSIPPEVKFNKPFVYLI-*NH*$_2$
Mitoparan	*H*-INLKKLAKL(Aib)KKIL-*NH*$_2$

The synthesis and purification of Djeya1, Tat (denotes the Tat peptide, Tat $^{48-60}$), C105Y, and mitoparan have been described elsewhere [11]. All peptides were synthesized as C-terminal amides. The identities of three Aib-substituted Djeya1 analogues were confirmed by mass spectrometry, Agilent 6200 TOF: [Aib^{10}]Djeya1, calculated 2403.8 Da, observed 2403.4 Da; [Aib^{13}]Djeya1, calculated 2403.8 Da, observed 2403.4 Da; and [Aib^{16}]Djeya1, calculated 2429.9 Da, observed 2429.4 Da. Spectra are presented in Supplementary Materials (Figures S3–S5).

3.2. CD Spectral Analyses of Peptide α-Helicity

CD conformational analyses indicated that in aqueous solution (PBS, pH 7.4), Djeya1, [Aib^{10}]Djeya1, [Aib^{13}]Djeya1, and [Aib^{16}]Djeya1 displayed spectra typical of α-helical and random coil structure (Figure 1a). Figure 1a shows two minima at 220 nm and 207 nm, which is typical of α-helical peptides, and a negative band at 195 nm, which is characteristic of a random coil structure. The low intensities of these bands indicate a mixture of these structures, which is confirmed by further analysis in Table 3a. The estimated α-helical content of these peptides was 43.00 ± 1.73% for Djeya1, 41.33 ± 1.15% for [Aib^{10}]Djeya1, 40.67 ± 1.52% for [Aib^{13}]Djeya1, and 39.67 ± 4.04% for [Aib^{16}]Djeya1.

Table 3. (a) Calculated secondary structures of Djeyal analogues in PBS. (b) Calculated secondary structures of Djeyal analogues in 50% (*v*/*v*) TFE. (c) Calculated secondary structures of Djeyal analogues in DMPC vesicles. (d) Calculated secondary structures of Djeyal analogues in DMPS vesicles.

		(a)		
Peptide	**α-helix**	**ß-Strand**	**ß-Turns**	**Unordered**
Djeyal	43.00 ± 1.73	30.33 ± 0.58	6.67 ± 0.58	20.00 ± 1.73
[Aib[10]]Djeyal	41.33 ± 1.15	31.33 ± 0.58	7.00 ± 0.00	20.67 ± 0.58
[Aib[13]]Djeyal	40.67 ± 1.52	30.00 ± 1.00	7.00 ± 0.00	22.33 ± 1.52
[Aib[16]]Djeyal	39.67 ± 4.04	31.67 ± 2.87	7.33 ± 0.58	21.00 ± 1.00
		(b)		
Peptide	**α-helix**	**ß-Strand**	**ß-Turns**	**Unordered**
Djeyal	45.50 ± 4.94	27.55 ± 3.53	7.00 ± 0.00	20.00 ± 1.41
[Aib[10]]Djeyal	51.50 ± 0.07	21.00 ± 0.02	7.50 ± 0.71	20.00 ± 0.00
[Aib[13]]Djeyal	40.00 ± 2.64	31.00 ± 2.08	7.33 ± 0.05	22.33 ± 1.50
[Aib[16]]Djeyal	46.00 ± 1.41	27.50 ± 0.71	7.00 ± 0.00	19.50 ± 0.71
		(c)		
Peptide	**α-helix**	**ß-Strand**	**ß-Turns**	**Unordered**
Djeyal	44.00 ± 1.53	34.00 ± 2.12	8.00 ± 0.0	22.00 ± 0.00
[Aib[10]]Djeyal	45.00 ± 1.58	35.00 ± 0.70	7.00 ± 1.41	22.00 ± 1.41
[Aib[13]]Djeyal	42.00 ± 0.71	30.00 ± 0.71	8.00 ± 0.71	22.00 ± 0.71
[Aib[16]]Djeyal	42.00 ± 0.71	30.00 ± 0.71	7.00 ± 0.71	22.00 ± 0.71
		(d)		
Peptide	**α-helix**	**ß-Strand**	**ß-Turns**	**Unordered**
Djeyal	47.00 ± 1.24	26.00 ± 1.28	8.00 ± 0.00	20.00 ± 0.00
[Aib[10]]Djeyal	63.00 ± 2.82	15.00 ± 0.00	5.00 ± 0.00	16.00 ± 0.00
[Aib[13]]Djeyal	42.00 ± 0.00	30.00 ± 0.71	7.00 ± 0.71	21.00 ± 0.71
[Aib[16]]Djeyal	42.00 ± 0.00	30.00 ± 0.71	8.00 ± 0.71	22.00 ± 0.71

A key determinant in the membrane interaction of these peptides likely involves the adoption of secondary structures in the anisotropic environment of the interface [35]. This phenomenon is often investigated using TFE, a membrane-mimicking solvent with an α-helix-enhancing effect [36,37]. Figure 1b indicates that in a 50% (*v*/*v*) TFE/PBS (pH 7.4) mixture, all Djeyal analogues adopted conformations characterized by two minima near 205 and 225 nm, respectively, and a maximum at 193 nm, which is typical of α-helical peptides [38]. Further analysis of these CD spectra showed that enhanced α-helicity was observed for Djeyal, [Aib[10]]Djeyal, and [Aib[16]]Djeyal (45.50 ± 4.94%, 51.50 ± 0.07 and 46.00 ± 1.41%; Table 3b) with the remaining structural contributions to the peptide provided by random coil and β-type architectures. However, the presence of 50% TFE/PBS (*v*/*v*) mixture did not enhance the α-helical structure of [Aib[13]]Djeyal.

It is widely accepted that peptides, including CPPs, adopt and enhance an α-helical structure in the presence of a lipid. Therefore, the structure of the peptides used here (Table 3c,d) were analysed in the presence of lipid vesicles at a lipid:peptide ratio of 100 to 1. Figure 1c shows that in the presence of the zwitterionic lipid DMPC, peptides underwent conformational change and adopted an α-helical structure, which ranged from 42% to 45% (Table 3c). In the presence of the anionic lipid DMPS, Figure 1d shows that Djeyal and [Aib[10]]Djeyal displayed 47% and 63% α-helicity, respectively (Table 3d), indicating that the presence of an anionic lipid can strongly initiate α-helix formation by these peptides at the lipid interface. No difference in percentage α-helicity was observed for [Aib[13]]Djeyal and [Aib[16]]Djeyal in the presence of these membranes.

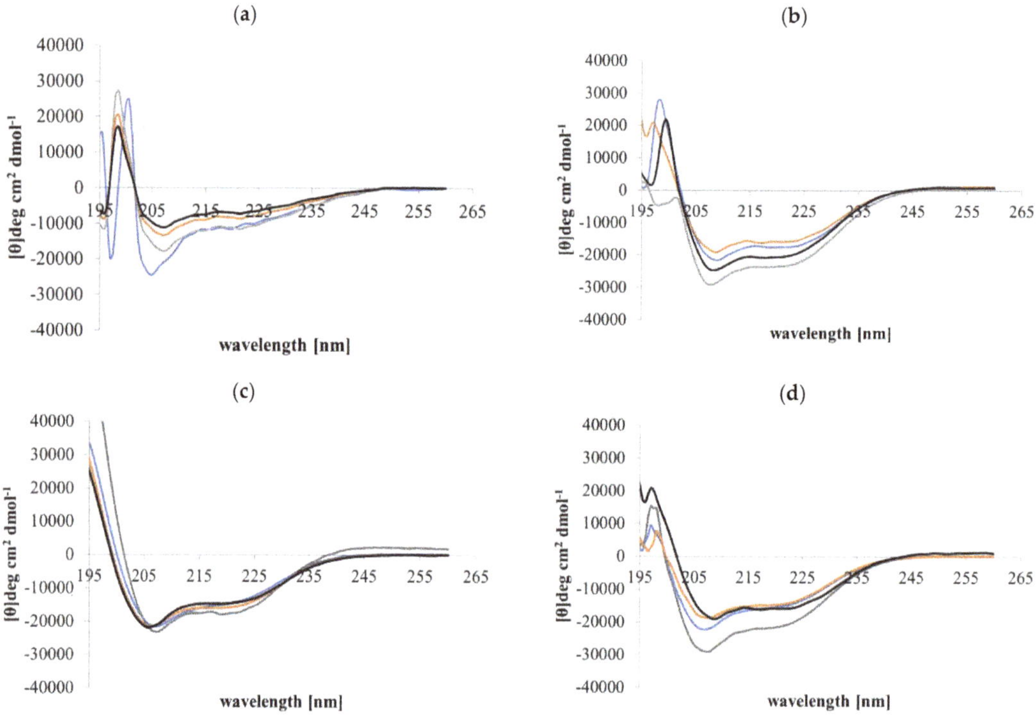

Figure 1. Conformational behavior of Djeya1 analogues. (**a**) Spectra obtained in PBS. (**b**) Spectra obtained in 50% (v/v) TFE. (**c**) Spectra obtained in the presence of DMPC vesicles. (**d**) Spectra obtained in the presence of DMPS vesicles. Djeya1, blue; [Aib^{10}]Djeya1, grey; [Aib^{13}]Djeya1, orange; [Aib^{16}]Djeya1, black.

3.3. Exogenous Application of [Aib^{13}]Djeya1 and [Aib^{16}]Djeya1 Delays Anterior Pole Regeneration and Eye Development in S. mediterranea

Immediately following transverse amputation (Figure 2a), planaria were treated with 25 μM of Aib-substituted Djeya1 analogues and monitored for morphological changes over an 8-day period. Additional treatments at the same concentration were administered on day 4 to compensate for any premature proteolytic degradation of the peptides. On day 8, planaria treated with [Aib^{13}]Djeya1 and [Aib^{16}]Djeya1 were unresponsive to light and regeneration of the anterior pole, including the eye spots, containing photoreceptors and the pigment cup, were absent (Figure 2c,d), compared to control planaria treated with PAM alone (Figure 2b). On day 9, 67% (n = 9) of both the [Aib^{13}]Djeya1 and [Aib^{16}]Djeya1-treated animals had commenced the development of photoreceptors within the unpigmented blastema. A total of 30 days post amputation, in PAM, all of the [Aib^{16}]Djeya1-treated planaria had fully regenerated, though only 67% (n = 9) of the [Aib^{13}]Djeya1-treated had achieved full regeneration and 33% remained unregenerated and in stasis. Further observations included an absence of generalized necrosis without excessive mucous production as a general indicator of animal stress. In these assays both Djeya1 and [Aib^{10}]Djeya1 were inactive.

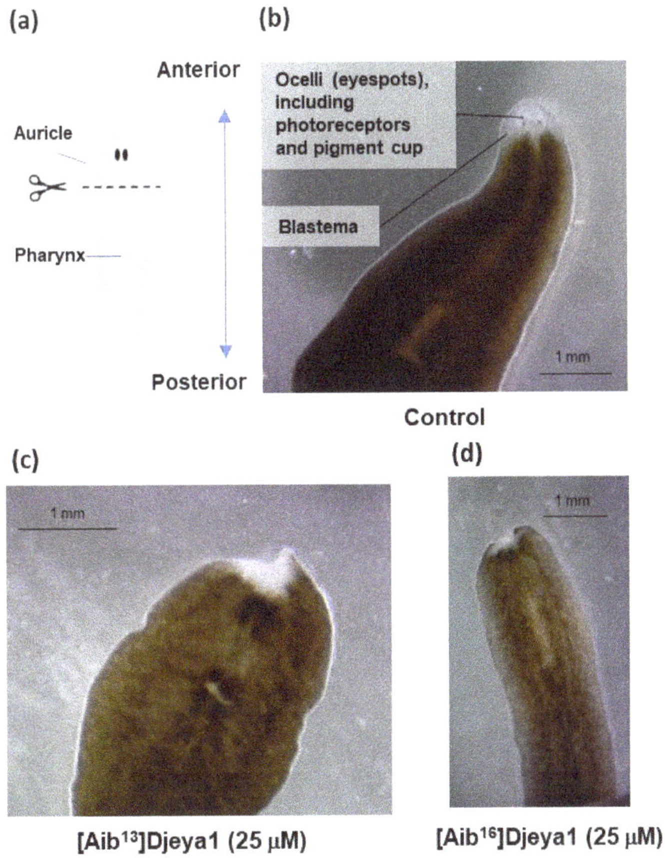

Figure 2. Influence of bioportides upon head remodelling. Transverse amputation at the post-auricle and pre-pharynx level induces the formation of the unpigmented blastema, eye regeneration, and head remodelling (**a**,**b**). (**b**) This shows the regeneration of the anterior pole and the eyespots in a representative planaria treated with PAM alone 8 days post amputation. (**c**,**d**) This demonstrates a noticeable absence of anterior pole regeneration and development of eyespots. Representative images were taken at day 8 post amputation.

3.4. Aib-Substituted Djeya1 Analogues Are Efficient CPPs

Red fluorescent TAMRA-conjugated peptides were utilised to compare the efficacy of cellular uptake. Comparative investigations utilised Tat and C105Y CPPs [11] as positive controls.

3.4.1. Qualitative Peptide Uptake Analyses

Confocal analyses compared the intracellular distribution of peptides in U373MG cells [11]. Observations after 1 h of exogenous peptide application (Figure 3a) revealed a similar distribution of Djeya1 analogues, which were, in part, perinuclear, and which were enhanced compared with both Tat and C105Y controls. Even after an extended period of 5 h (Figure 3b), Djeya1 analogues were absent from the cell nucleus.

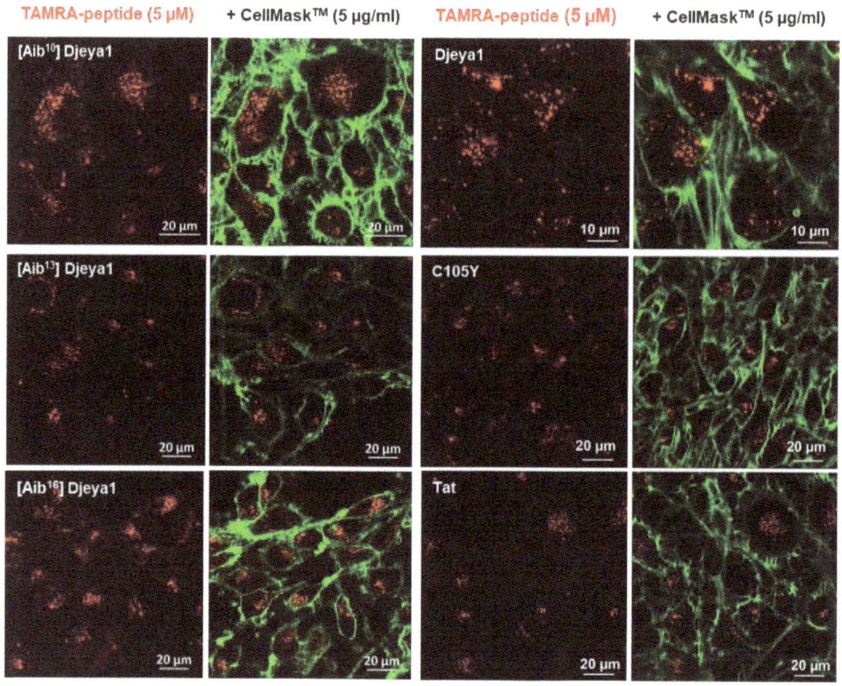

(a)

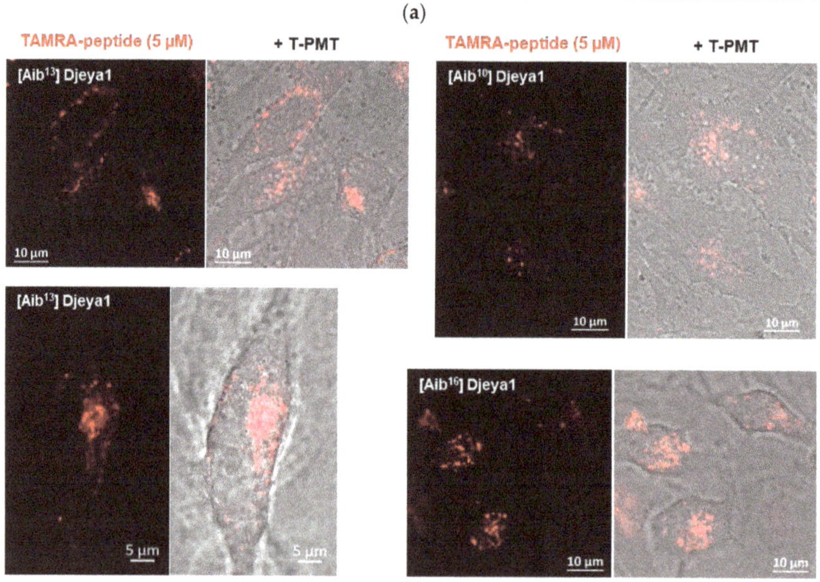

(b)

Figure 3. Qualitative uptake analyses using live confocal cell imaging. (**a**) U373MG cells were treated with 5 µM TAMRA-labelled peptides and 5 µg mL^{-1} CellMask^{TM} to label the plasma membrane for 1 h prior to visualization. Tat and C105Y were used as positive controls. (**b**) U373MG cells were treated for 5 h with 5 µM TAMRA-labelled Aib-substituted Djeya1 analogues. Confocal images are also presented with images using photomultiplier for transmitted light (T-PMT) to highlight subcellular distribution.

3.4.2. Quantitative Peptide Uptake Analyses

These studies (Figure 4) compared the intracellular uptake of Aib-substituted Djeya1 analogues into U373MG cells and U251 cells, the latter of which is reported to overexpress Eya proteins. All three Aib-substituted Djeya1 analogues are highly efficient CPPs, broadly comparable with Djeya1 and C105Y but superior to Tat, though they displayed a different rank order of uptake efficacy in the two cell lines: U373MG [Aib^{16}]Djeya1 = [Aib^{10}]Djeya1 > [Aib^{13}]Djeya1; U251 [Aib^{13}]Djeya1 > [Aib^{16}]Djeya1 > [Aib^{10}]Djeya1. Of particular note is the increased uptake efficacy of [Aib^{13}]Djeya1 into U251cells (Figure 4b) compared with U373MG cells (Figure 4a).

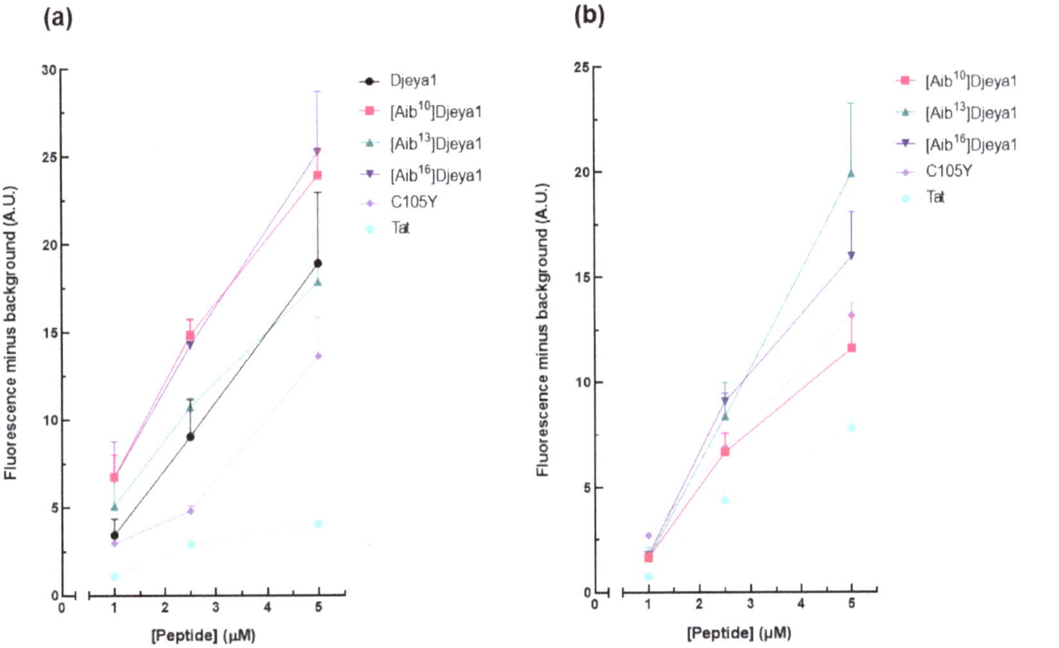

Figure 4. Quantitative peptide uptake analyses. Comparative analyses of peptide translocation efficacies of Aib-substituted Djeya1 analogues were performed using U373MG cells (**a**) and U251 cells (**b**) incubated with TAMRA-labelled peptides for 1 h at 37 °C at the concentrations indicated. Both Tat and C105Y were used as positive controls. Data are expressed as mean fluorescence minus background ± SEM. Data are from two experiments performed in triplicate.

3.5. Cytotoxicity of Djeya1 Analogues

A negative influence of peptides upon the viability of neoblasts could underlie their ability to inhibit head morphogenesis in *S. mediterranea* (Figure 2). As indicated in Figure 5, Djeya1 analogues had no influence upon the viability of U373MG cells following exposure to peptides at concentrations of 0.1–30 μM for 4 h. Treatment of PC-3 cells with Djeya1 analogues for a longer time of 24 h similarly induced no significant changes in cellular viability (Supplementary Materials; Figure S1).

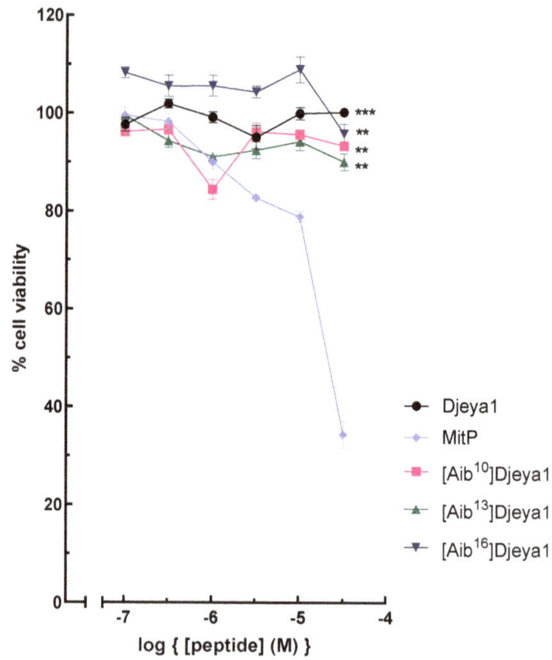

Figure 5. Cytotoxicity profiles of Aib-substituted Djeya1 analogues. U373MG cells were treated with Djeya1 and Aib-substituted analogues for 4 h at the concentrations indicated. Cell viability was measured by MTT conversion and expressed as a percentage of those cells treated with vehicle (medium) alone. The mitochondriotoxic peptide Mitoparan (MitP) was used as a positive control [11,31]. Data points are mean ± SEM from two experiments performed in triplicate. Statistical analyses employed a non-parametric Mann–Whitney test to compare changes in viability to that of MitP at 30 μM, (*** $p = 0.0001$, ** $p = 0.0022$), GraphPad Prism 9 software.

3.6. Impact of Djeya1 Analogues upon U251 Cellular Proliferation

Planarian neoblasts replenish lost organs and tissues by proliferation within hours of any injury [7,8,39,40]. Many wound-induced transcriptional changes necessary for these homeostatic events are regulated by the activity of extracellular signal-regulated kinase (ERK) [29,40]. U251 has been reported to have a high endogenous expression of Eya2, and siRNA knockdown of this protein decreases the proliferation and invasion of these cells, whilst Eya2 positively regulates p42/44 MAPK activity [29]. We have previously reported [11] that p42/44 inhibition prevents anterior pole regeneration in *S. mediterranea*. Thus, in these experiments, we compared the impact of Aib-substituted Djeya1 analogues upon the proliferation of U251 cells employing the p42/44 MAPK inhibitor U1026 as a positive control. As revealed in Figure 6, bioengineered Djeya1 analogues had minimal impact upon cellular proliferation and, whilst statistically significant data were recorded at some higher peptide concentrations, this influence was independent of concentration.

(a)

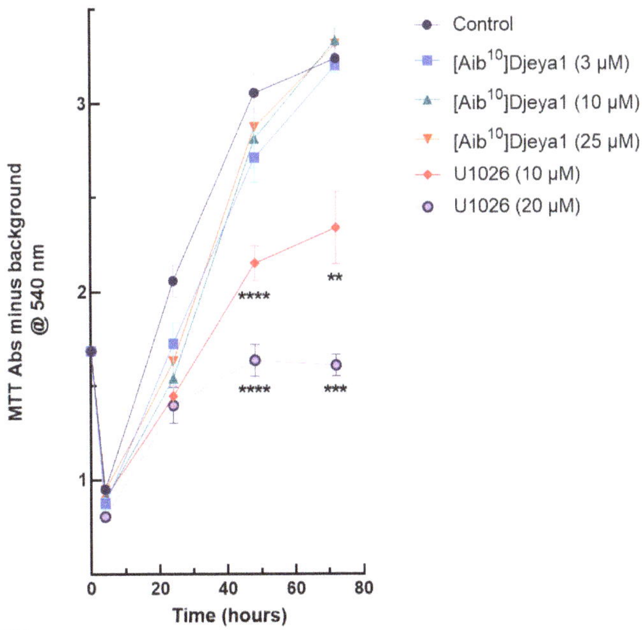

(b)

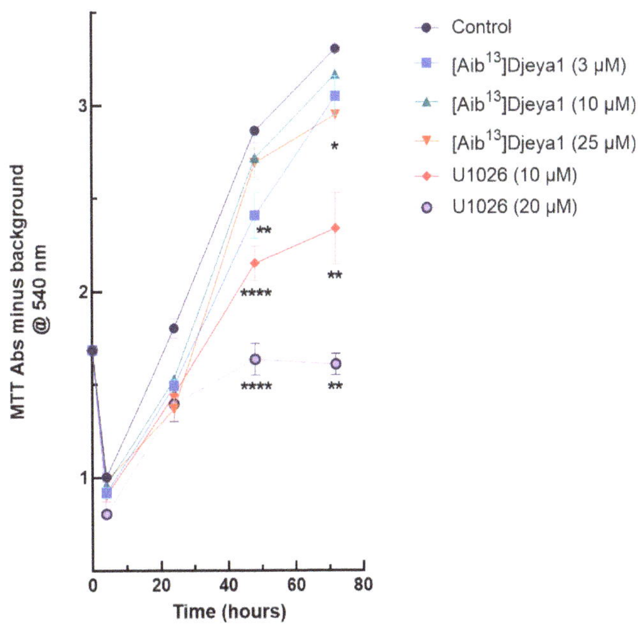

Figure 6. *Cont.*

(c)

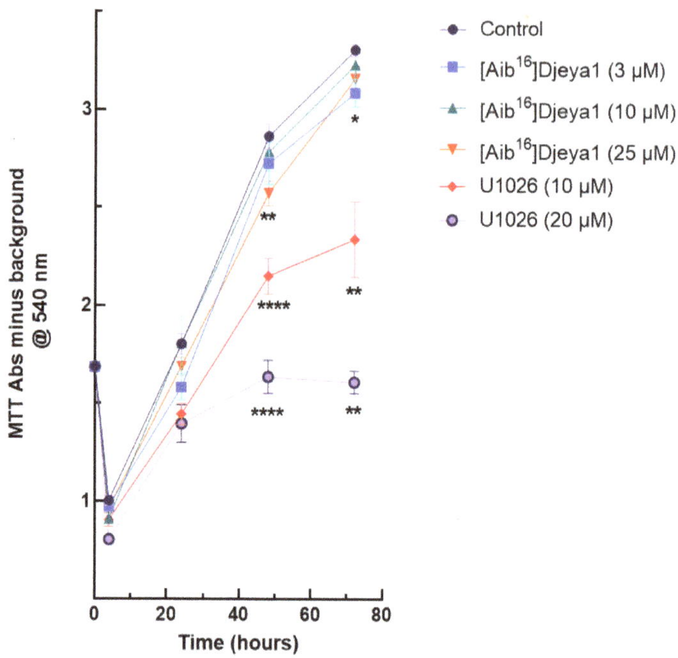

Figure 6. Growth curves of high Eya expressing U251 cells expressed as optical density. U251 cells were treated with Aib-substituted analogues or the p42/44 MAPK inhibitor U1026 at the concentrations indicated from 4–72 h (**a**–**c**). MTT conversion is expressed as Abs@540 nm minus background, and data points are mean ± SEM from two experiments performed in sextuplicate. A total of 4 h and 72 h, with one experiment performed in sextuplicate. Control denotes cells treated with medium alone. Statistical analyses comparing significant differences in cell viability to the untreated control were performed at 48 h and 72 h using the unpaired, 2-tailed, non-parametric Mann–Whitney test, (* $p < 0.05$, ** $p < 0.005$, *** $p = 0.0001$, **** $p < 0.0001$), using GraphPad Prism 9 software.

3.7. Aib-Substituted Djeya1 Analogues Enhance Cell Migration

Stem cell migration is considered a fundamental process of tissue maintenance in metazoans and of particular importance following tissue loss and wounding. Migrating neoblasts and their progeny adopt distinctive behaviours to selectively regenerate appropriate missing tissues in planaria and form the regeneration blastema where differentiation is completed [40,41]. Thus, we hypothesized that bioportides influencing head morphogenesis in *S. mediterranea* might adversely inhibit cellular migration in PC-3 cells that express high levels of Eya. Contrary to these expectations, all Aib-substituted Djeya analogues enhanced cellular migration, measured in wound closure assays, at concentrations of both 3 µM and 25 µM (Figure 7). Further details of these assays are provided in Supplementary Materials (Figure S2).

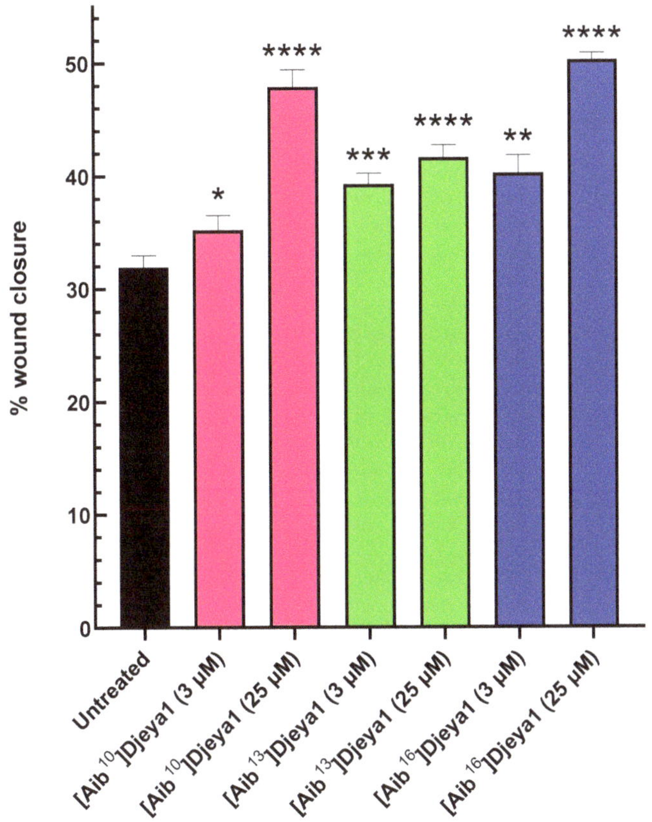

Figure 7. Aib-substituted Djeya1 analogues enhance PC-3 cell migration. Wound closure was measured after 48 h. All analogues demonstrated a significant enhancement in percentage wound closure compared to untreated cells, $p < 0.05$ using the paired, 2-tailed non-parametric Wilcoxon matched pairs signed rank test (* $p = 0.0353$, ** $p = 0.0015$, *** $p = 0.0002$, **** $p < 0.0001$), GraphPad Prism 9 software.

4. Discussion

The molecular bioengineering of CPPs and bioportides can enhance intracellular uptake and confer or increase biological activity [4–6,42]. As recently reviewed [43], biophysical methodologies, including circular dichroism, have impacted the understanding of the structural determinants of CPP trafficking and the molecular mechanisms of bioportides. Numerous studies of the secondary structures of CPPs have highlighted the possibility that a helical conformation, perhaps induced by contact with phospholipid membranes, may support the passage of CPPs into cells [43,44]. One or more cationic alpha helices, particularly those containing statistically enriched Arg residues, are commonly located at protein-protein interaction (PPI) sites. Cation-π interactions, which stabilize these PPIs [45], commonly involve interactions between arginine and tyrosine residues. Hence, we hypothesized that Aib-substitution might induce helicity within Djeya1 to further increase penetration efficacy [11]. We also anticipated that the structural constraints imposed by Aib-substitution would enable bioengineered Djeya1 analogues, by a dominant-negative mechanism, to modulate PPIs within the ED domain. Interference with EYA functions as a transcriptional regulator would likely manifest as inefficient head morphogenesis in decapitated *S. mediterranea* [15,33,34].

This study identified two bioportides, [Aib^{13}]Djeya1 and [Aib^{16}]Djeya1, which delay head regeneration in the *S. mediterranea* model; [Aib^{10}]Djeya1, in common with Djeya1 [11], is an effective CPP, but unable to influence planarian morphogenesis. Djeya1, in common with Aib-substituted analogues, adopts significant α-helical structure in aqueous solution that is only marginally influenced by a change to 50% (v/v) TFE and in the presence of a lipid environment. It is noteworthy that the mitochondriotoxic bioportide mitoparan, a peptide that accretes within mitochondria to promote intrinsic apoptosis, includes Aib to replace Ala at position-10 of mastoparan, a known α-helix adopting peptide [32]. Thus, the molecular bioengineering of CPPs by Aib-substitution can produce significant variations in bioactivities that are not necessarily the result of gross changes in peptide secondary structure.

The bioactivities of [Aib^{13}]Djeya1 and [Aib^{16}]Djeya1 cannot be readily explained by increased uptake into cells. Both qualitative and quantitative analyses show that the uptake of all Aib-substituted Djeya1 analogues are generally comparable to that of the parent CPP and better than other CPPs, including Tat and C105Y. It is, however, notable that a different rank order of penetrative efficacy was evident between U373MG and U251 cell lines, which in part may be attributable to the U251 cell line overexpressing the EYA protein [29]. We have previously proposed [22] that the propensity for cellular penetration and intracellular accumulation of CPPs involves two distinct processes: the first is translocation across the plasma membrane (hydrophobicity and cationic charge are significant factors), and the second is accretion at intracellular loci. Thus, the increased uptake of [Aib^{13}]Djeya1 within U251 cells could reflect, in part, enhanced accretion at intracellular loci, particularly since [Aib^{10}] Djeya1, also with a conservative aliphatic side chain substitution for Aib, showed a reduced uptake efficacy compared to [Aib^{13}]Djeya1.

It would be intriguing to establish the mechanism of uptake for these novel Aib-substituted Djeya1 analogues. Following live confocal cell imaging, all analogues showed a clear punctate intracellular distribution—-data which suggest an endocytotic mechanism of intracellular uptake. Further investigations, including the co-labelling with specific markers of endocytosis or micropinocytosis, in addition to quantitative uptake analyses performed at both 4 °C and 37 °C, could define a mechanism of cellular uptake. Additionally, all analogues demonstrated a distinct absence of cytotoxicity even at a peptide concentration of 30 μM. Thus, these peptides do not induce gross perturbations of the plasma membrane to induce Ca^{2+}-dependent necrosis.

Efforts to define a molecular mechanism of action of both [Aib^{13}]Djeya1 and [Aib^{16}]Djeya are hampered both by a lack of accessible planarian neoblast cultures and knowledge of a defined function for the α5 helix of ED that they mimic. However, we are confident that our investigations can exclude peptide-induced neoblast death as a biological explanation for the impact of [Aib^{13}]Djeya1 and [Aib^{16}]Djeya1 on anterior pole remodeling. Neither would it seem likely that these bioportides adversely influence cellular proliferation, a process fundamental to this regenerative process [16–18,39,40]. The observation that all three Aib-substituted Djeya1 analogues moderately enhance cell migration (wound healing) is intriguing and worthy of further investigation. However, since [Aib^{10}]Djeya1 does not inhibit heard morphogenesis in *S. mediterranea*, it would appear unlikely that this positive influence upon cell migration underlies the mechanism of action of [Aib^{13}]Djeya1 and [Aib^{16}]Djeya1, bioportides that delay head regeneration.

In common with the dominant negative action of many other bioportides [4–6,42], both sychnologic and rhegnylogic in molecular organization, we propose that [Aib^{13}]Djeya1 and [Aib^{16}]Djeya1 modulate head regeneration by interfering with PPIs mediated by the ED of Eya proteins. Support for this hypothesis is provided by a study of the G393S mutation of ED, identified from a patient with cataracts and both renal and optic abnormalities [46]. In the human ED, this missense mutation site is located close to R^{399}, the first residue of ED mimetic Djeya1 analogues. Further analyses of this mutation indicated this site to be critical for the interaction of ED with unknown proteins that bridge well characterized PPIs between Eya and SIX family proteins [46].

Considering that there are so many similarities between planarian proteins and those expressed in higher vertebrates [47], we anticipate that bioportides able to influence the distribution and function of Eya proteins may prove valuable for other studies of physiology and pathology. For example, mutations in the human *EYA1* gene cause branchio-oto-renal (BOR) syndrome [46]. The multifunctional nature of Eya proteins also influences tumour progression through multiple mechanisms [48]. Moreover, it is probable that [Aib^{13}]Djeya1 and [Aib^{16}]Djeya1 could be utilized to further understand and directly influence human stem cells, thus negating the requirement for genetic manipulation.

5. Conclusions

With the aim of modulating planarian stem cell biology through targeting PPIs integral to anterior pole and eye regeneration, rational design of helicogenic peptides, corresponding to the α5 helix of the highly conserved ED domain, generated the bioportides [Aib^{13}]Djeya1 and [Aib^{16}]Djeya1. Thus far, manipulations of planarian stem cell biology have predominantly utilized siRNA interference. To the best of our knowledge, this is the first instance in which bioportide technologies have given a functional response in this three-dimensional model, a response which is unlikely to be a detrimental consequence of cytotoxicity, inhibition of cellular proliferation, or migration.

Supplementary Materials: The following supporting information can be downloaded at: https://www.mdpi.com/article/10.3390/pharmaceutics15082018/s1. Figure S1: Viability of PC-3 cells treated with Aib-substituted Djeya1 analogues. Figure S2: Impact of Aib-substituted Djeya1 peptides (3 μM and 25 μM) on prostate cancer (PC-3) cell migration. Figure S3. Identification of [Aib^{10}]Djeya 1 by mass spectrometry, Agilent 6200 TOF. Figure S4. Identification of [Aib^{13}]Djeya 1 by mass spectrometry, Agilent 6200 TOF. Figure S5. Identification of [Aib^{16}]Djeya1 by mass spectrometry, Agilent 6200 TOF.

Author Contributions: Conceptualization, S.J. and J.H.; methodology, S.J. and J.H.; validation, S.J., B.M., S.D. and J.H.; formal analysis, S.J., B.M. and S.D.; investigation, S.J., B.M., S.D. and J.H.; writing—original draft preparation, S.J., B.M., S.D., M.F. and J.H.; writing—review and editing, S.J., B.M., S.D., M.F. and J.H.; visualization, S.J. and J.H.; supervision, M.F. and J.H.; project administration, M.F. and J.H.; funding acquisition, M.F. and J.H. All authors have read and agreed to the published version of the manuscript.

Funding: Bárbara Matos was supported by an individual grant from the Portuguese Foundation for Science and Technology (FCT), grant identifier SFRH/BD/146032/2019.

Institutional Review Board Statement: Not applicable.

Informed Consent Statement: Not applicable.

Data Availability Statement: The data presented in this study are available on request from the corresponding author.

Conflicts of Interest: The authors declare no conflict of interest.

References

1. Kurrikoff, K.; Vunk, B.; Langel, Ü. Status update in the use of cell-penetrating peptides for the delivery of macromolecular therapeutics. *Expert Opin. Biol. Ther.* **2021**, *21*, 361–370. [CrossRef]
2. Guidotti, G.; Brambilla, L.; Rossi, D. Cell-penetrating peptides: From basic research to clinics. *Trends Pharmacol. Sci.* **2017**, *38*, 406–424. [CrossRef] [PubMed]
3. Xie, J.; Zhang, H.; Dong, S.; Teng, L.; Lee, R.J.; Yang, Z. Cell-penetrating peptides in diagnosis and treatment of human diseases: From preclinical research to clinical application. *Front. Pharmacol.* **2020**, *11*, 697. [CrossRef] [PubMed]
4. Howl, J.; Matou-Nasri, S.; West, D.S.; Farquhar, M.; Slaninová, J.; Östenson, C.-G.; Zorko, M.; Östlund, P.; Kumar, S.; Langel, Ü.; et al. Bioportide: An emergent concept of bioactive cell penetrating peptides. *Cell. Mol. Life Sci.* **2012**, *69*, 2951–2966. [CrossRef]
5. Howl, J.; Jones, S. Insights into the molecular mechanisms of action of bioportides: A strategy to target protein-protein interactions. *Expert Rev. Mol. Med.* **2015**, *17*, e1. [CrossRef]
6. Silva, J.V.; Freitas, M.J.; Santiago, J.; Jones, S.; Guimarães, S.; Vijayaraghavan, S.; Publicover, S.; Colombo, G.; Howl, J.; Fardilha, M. Disruption of protein phosphatase 1 complexes with the use of bioportides as a novel approach to target sperm motility. *Fertil. Steril.* **2021**, *115*, 348–362. [CrossRef]

7. Aboobaker, A.A. Planarian stem cells: A simple paradigm for regeneration. *Trends Cell. Biol.* **2011**, *21*, 304–311. [CrossRef] [PubMed]
8. Owlarn, S.; Bartscherer, K. Go ahead, grow a head! A planarian's guide to anterior regeneration. *Regeneration* **2016**, *24*, 139–155. [CrossRef] [PubMed]
9. Robb, S.M.C.; Gotting, K.; Ross, E.; Sánchez Alvarado, A. SmedGD 2.0: The *Schmidtea mediterranea* genome database. *Genesis* **2015**, *53*, 535–546. [CrossRef]
10. Sánchez Alvarado, A.; Newmark, P.A.; Robb, S.M.; Juste, R. The *Schmidtea mediterranea* database as a molecular resource for studying platyhelminthes, stem cells and regeneration. *Development* **2002**, *129*, 5659–5665. [CrossRef]
11. Jones, S.; Osman, S.; Howl, J. The planarian *Schmidtea mediterranea* as a model system for the discovery and characterization of cell penetrating peptides and bioportides. *Chem. Biol. Drug Des.* **2019**, *93*, 1036–1049. [CrossRef]
12. Futaki, S.; Suzuki, T.; Ohashi, W.; Yagami, T.; Tanaka, S.; Ueda, K.; Sugiura, Y. Arginine-rich peptides. An abundant source of membrane-permeable peptides having potential as carriers for intracellular protein delivery. *J. Biol. Chem.* **2001**, *276*, 5836–5840. [CrossRef] [PubMed]
13. Mannini, L.; Rossi, L.; Deri, P.; Gremigni, V.; Salvetti, A.; Saló, E.; Batistoni, R. *Djeyes absent* (Djeya) controls prototypic planarian eye regeneration by cooperating with the transcription factor Djsix-1. *Dev. Biol.* **2004**, *269*, 346–359. [CrossRef] [PubMed]
14. Rayapureddi, J.P.; Kattamuri, C.; Steinmetz, B.D.; Frankfort, B.J.; Ostrin, E.J.; Mardon, G.; Hedge, R.S. Eyes absent represents a class of protein tyrosine phosphatases. *Nature* **2003**, *426*, 295–298. [CrossRef]
15. Hedge, R.S.; Roychoudhury, K.; Pandey, R.M. The multi-functional eyes absent proteins. *Crit. Rev. Biochem. Mol. Biol.* **2020**, *55*, 372–385. [CrossRef]
16. Eisenhoffer, G.T.; Kang, H.; Sánchez Alvarado, A. Molecular analysis of stem cells and their descendants during cell turnover and regeneration in the planarian Schmidtea mediterranea. *Cell Stem Cell.* **2008**, *3*, 327–339. [CrossRef]
17. Baguñà, J. The planarian neoblast: The rambling history of its origin and some current black boxes. *Int. J. Dev. Biol.* **2012**, *56*, 19–37. [CrossRef]
18. Sánchez Alvarado, A. Stem cells and the planarian *Schmidtea mediterranea*. *Comptes Rendus Biol.* **2007**, *330*, 498–503. [CrossRef] [PubMed]
19. Xu, P.X.; Zheng, W.; Luclef, C.; Maire, P.; Maas, R.L.; Peters, H.; Xu, X. Eya1 is required for the morphogenesis of mammalian thymus. *Development* **2002**, *129*, 3033–3034. [CrossRef]
20. Jones, S.; Osman, S.; Howl, J. A high-throughput synthetic platform enables the discovery of proteomimetic cell penetrating peptides and bioportides. *Int. J. Pept. Res. Ther.* **2018**, *25*, 1–8. [CrossRef]
21. Subirs-Fumosas, R.; Prohens, R.; Barbas, R.; El-Faham, A.; Albericio, F. Oxyma: An efficient additive for peptide synthesis to replace the benzotrialzole-based HOBt and HOAt with a lower risk of explosion. *Chemistry* **2009**, *15*, 9394–9403. [CrossRef] [PubMed]
22. Jones, S.; Uusna, J.; Langel, Ü.; Howl, J. Intracellular target-specific accretion of cell penetrating peptides and bioportides: Ultrastructural and biological correlates. *Bioconjugate Chem.* **2016**, *27*, 121–129. [CrossRef] [PubMed]
23. Kelly, S.M.; Jess, T.J.; Price, N.C. How to study proteins by circular dichroism. *Biochim. Biophys. Acta* **2005**, *1751*, 119–139. [CrossRef] [PubMed]
24. Tanaka, F.; Forster, L.S.; Pal, P.K.; Rupley, J.A. The circular dichroism of lysozyme. *J. Biol. Chem.* **1975**, *250*, 6977–6982. [CrossRef]
25. Johnson, M.P.; Ruban, A.V. Rethinking the existence of a steady-state $\Delta\psi$ component of the proton motive force across plant thylakoid membranes. *Photosynth. Res.* **2014**, *119*, 233–242. [CrossRef]
26. Cebriá, F.; Newmark, P.A. Planarian homologs of *netrin* and *netrin receptor* are required for proper regeneration of the central nervous system and the maintenance of nervous system architecture. *Development* **2005**, *132*, 3691–3703. [CrossRef]
27. Howl, J.; Howl, L.; Jones, S. The cationic tetradecapeptide mastoparan as a privileged structure for drug discovery: Enhanced antimicrobial properties of mitoparan analogues modified at position-14. *Peptides* **2018**, *101*, 95–105. [CrossRef]
28. Liu, Z.; Zhao, L.; Song, Y. Eya2 is overexpressed in human prostate cancer and regulates docetaxel sensitivity and mitochondrial membrane potential through AKT/Bcl-2 signaling. *BioMed. Res. Int.* **2019**, *2019*, 3808432. [CrossRef]
29. Wen, Z.; Liang, C.; Pan, Q.; Wang, Y. Eya2 overexpression promotes the invasion of human astrocytoma through the regulation of ERK/MMP9 signaling. *Int. J. Mol. Med.* **2017**, *40*, 1315–1322. [CrossRef]
30. Li, Z.; Qiu, R.; Qiu, X.; Tian, T. EYA4 promotes cell proliferation through downregulation of p27Kip1 in glioma. *Cell. Physiol. Biochem.* **2018**, *49*, 1856–1869. [CrossRef]
31. Jones, S.; Martel, C.; Belzacq-Casagrande, A.S.; Brenner, C.; Howl, J. Mitoparan and target-selective chimeric analogues: Membrane translocation and intracellular redistribution induces mitochondrial apoptosis. *Biochim. Biophys. Acta* **2008**, *1783*, 849–863. [CrossRef]
32. Carmichael, J.; DeGraff, W.G.; Gazdar, A.F.; Minna, J.D.; Mitchell, J.B. Evaluation of a tetrazolium-based semiautomated colorimetric assay: Assessment of chemosensitivity testing. *Cancer Res.* **1987**, *47*, 936–942.
33. Tadjuidje, E.; Hedge, R.S. The eyes absent proteins in development and disease. *Cell. Mol. Life Sci.* **2013**, *70*, 1897–1913. [CrossRef]
34. Jung, S.-K.; Jeong, D.G.; Chung, S.J.; Kim, J.H.; Park, B.C.; Tonks, N.K.; Ryu, S.E.; Kim, S.J. Crystal structure of ED-Eya2: Insight into dual roles as a protein tyrosine phosphatase and a transcription factor. *FASEB J.* **2010**, *24*, 560–569. [CrossRef]
35. Seelig, J. Thermodynamics of lipid–peptide interactions. *Biochim. Biophys. Acta* **2004**, *1666*, 40–50. [CrossRef]

36. Luo, P.Z.; Baldwin, R.L. Mechanism of helix induction by trifluoroethanol: A framework for extrapolating the helix-forming properties of peptides from trifluoroethanol/water mixtures back to water. *Biochemistry* **1997**, *36*, 8413–8421. [CrossRef]
37. Cammers-Goodwin, A.; Allen, T.J.; Oslick, S.L.; McClure, K.F.; Lee, J.H.; Kemp, D.S. Mechanism of stabilization of helical conformations of polypeptides by water containing trifluoroethanol. *J. Am. Chem. Soc.* **1996**, *118*, 3082–3090. [CrossRef]
38. Greenfield, N.J. Using circular dichroism spectra to estimate protein secondary structure. *Nat. Protoc.* **2006**, *1*, 2876–2890. [CrossRef]
39. Pelletieri, J.; Sánchez Alvarado, A. Cell turnover and adult tissue homeostasis: From humans to planarians. *Ann. Rev. Genet.* **2007**, *41*, 83–105. [CrossRef]
40. Bohr, T.E.; Shiroor, D.A.; Adler, C.E. Planarian stem cells sense the identity of the missing pharynx to launch its targeted regeneration. *eLife* **2021**, *10*, e68830. [CrossRef]
41. Abnave, P.; Aboukhatwa, E.; Kosaka, N.; Thompson, J.; Hill, M.A.; Aboobaker, A.A. Epithelial-mesenchymal transition transcription factors control pluripotent adult stem cell migration *in vivo* in planarians. *Development* **2017**, *144*, 3440–3453. [CrossRef]
42. Lukanowska, M.; Howl, J.; Jones, S. Bioportides: Bioactive cell-penetrating peptides that modulate cellular dynamics. *Biotechnol. J.* **2013**, *8*, 918–930. [CrossRef]
43. Zorko, M.; Langel, Ü. Studies of cell-penetrating peptides by biophysical methods. *Q. Rev. Biophys.* **2022**, *55*, e3. [CrossRef]
44. Hadjicharalambous, A.; Bournakas, N.; Newman, H.; Skynner, M.J.; Beswick, P. Antimicrobial and Cell-Penetrating Peptides: Understanding Penetration for the Design of Novel Conjugate Antibiotics. *Antibiotics* **2022**, *11*, 1636. [CrossRef]
45. Infield, D.T.; Rasouli, A.; Galles, G.D.; Chipot, C.; Tajkhorshid, E.; Ahern, C.A. Cation-π interactions and their functional roles in membrane proteins. *J. Mol. Biol.* **2021**, *433*, 167035. [CrossRef]
46. Buller, C.; Xu, X.; Marquis, V.; Schwanke, R.; Xu, P.-X. Molecular effects of Eya1 domain mutations causing organ defects in BOR syndrome. *Hum. Mol. Genet.* **2001**, *10*, 2775–2781. [CrossRef]
47. Lapan, S.W.; Reddien, P.W. Transcriptome analysis of the planarian eye identifies ovo as a specific regulator of eye regeneration. *Cell Rep.* **2012**, *2*, 294–307. [CrossRef]
48. Zhou, H.; Zhang, L.; Vartuli, R.L.; Ford, H.L.; Zhao, R. The Eya phosphatase: Its unique role in cancer. *Int. J. Biochem. Cell Biol.* **2018**, *96*, 165–170. [CrossRef]

Disclaimer/Publisher's Note: The statements, opinions and data contained in all publications are solely those of the individual author(s) and contributor(s) and not of MDPI and/or the editor(s). MDPI and/or the editor(s) disclaim responsibility for any injury to people or property resulting from any ideas, methods, instructions or products referred to in the content.

pharmaceutics

MDPI

Article

Stereoisomer-Dependent Membrane Association and Capacity for Insulin Delivery Facilitated by Penetratin

Ditlev Birch [1], Edward J. Sayers [2], Malene V. Christensen [3], Arwyn T. Jones [2], Henrik Franzyk [3] and Hanne M. Nielsen [1,*]

1 Center for Biopharmaceuticals and Biobarriers in Drug Delivery (BioDelivery), Department of Pharmacy, Faculty of Health and Medical Sciences, University of Copenhagen, Universitetsparken 2, 2100 Copenhagen, Denmark
2 Cardiff School of Pharmacy and Pharmaceutical Sciences, Cardiff University, Cardiff CF10 3NB, UK
3 Cancer and Infectious Diseases, Department of Drug Design and Pharmacology, Faculty of Health and Medical Sciences, University of Copenhagen, Universitetsparken 2, 2100 Copenhagen, Denmark
* Correspondence: hanne.morck@sund.ku.dk; Tel.: +45-3533-6346

Abstract: Cell-penetrating peptides (CPPs), such as penetratin, are often investigated as drug delivery vectors and incorporating D-amino acids, rather than the natural L-forms, to enhance proteolytic stability could improve their delivery efficiency. The present study aimed to compare membrane association, cellular uptake, and delivery capacity for all-L and all-D enantiomers of penetratin (PEN) by using different cell models and cargos. The enantiomers displayed widely different distribution patterns in the examined cell models, and in Caco-2 cells, quenchable membrane binding was evident for D-PEN in addition to vesicular intracellular localization for both enantiomers. The uptake of insulin in Caco-2 cells was equally mediated by the two enantiomers, and while L-PEN did not increase the transepithelial permeation of any of the investigated cargo peptides, D-PEN increased the transepithelial delivery of vancomycin five-fold and approximately four-fold for insulin at an extracellular apical pH of 6.5. Overall, while D-PEN was associated with the plasma membrane to a larger extent and was superior in mediating the transepithelial delivery of hydrophilic peptide cargoes compared to L-PEN across Caco-2 epithelium, no enhanced delivery of the hydrophobic cyclosporin was observed, and intracellular insulin uptake was induced to a similar degree by the two enantiomers.

Keywords: cell-penetrating peptide; stereoisomers; intracellular uptake; membrane binding; transepithelial permeation; peptide delivery

Citation: Birch, D.; Sayers, E.J.; Christensen, M.V.; Jones, A.T.; Franzyk, H.; Nielsen, H.M. Stereoisomer-Dependent Membrane Association and Capacity for Insulin Delivery Facilitated by Penetratin. *Pharmaceutics* **2023**, *15*, 1672. https://doi.org/10.3390/pharmaceutics15061672

Academic Editors: Prisca Boisguérin and Sébastien Deshayes

Received: 16 March 2023
Revised: 8 May 2023
Accepted: 26 May 2023
Published: 7 June 2023

Copyright: © 2023 by the authors. Licensee MDPI, Basel, Switzerland. This article is an open access article distributed under the terms and conditions of the Creative Commons Attribution (CC BY) license (https://creativecommons.org/licenses/by/4.0/).

1. Introduction

Cell-penetrating peptides (CPPs) can be efficient vectors for the delivery of associated cargos into the cell cytosol [1,2], as well as across the epithelia [3,4]. Typically, CPPs are employed in tandem with cargo as ionic complexes or as covalent conjugates [5,6]. Al-though the mechanism of CPP internalization is currently not fully understood, it has been reported to involve direct membrane translocation, endocytosis or a combination of both [2,7–9]. While there have been no specific structural preferences for either translocation mechanism, it is recognized that several factors affect these processes. These include the CPP structure (both primary [10–12] and secondary [13]), concentration [14], cell type [15], peptide-to-cell ratio [16], and cargo type [2]. Since direct translocation delivers the cargo to the cytosol, thereby avoiding the endocytic system, this pathway has the advantage of a rapid introduction of therapeutics to intracellular targets. Direct translocation may be advantageous, e.g., for the treatment of intracellular infections or for inducing necrosis in cancer cells [17]. By comparison, endocytosis may constitute a slower route for delivery, which is primarily caused by the need for endosomal escape following internalization [18,19]. The advantage of endocytosis is that it does not compromise the cell membrane's integrity, thereby avoiding concurrent cytotoxicity that is related to membrane damage [6,20] as opposed to direct translocation, where this is a risk [21]. In addition to aiming at cargo delivery to intracellular targets, CPPs may be applied as carriers for the delivery of cargo across the epithelia, either by transcytosis, cell membrane perturbation [22], or by affecting the epithelial integrity [3,23].

Penetratin is a very well-studied CPP, and both L- and D-enantiomers (L-PEN and D-PEN, respectively) of this peptide have been examined for their potential to deliver a variety of cargoes to intracellular [6,24–26] or transepithelial [5,23,27–31] targets in vitro, as well as in situ [4,32] and in vivo [3,33,34] Nevertheless, their detailed mode of action for enhancing deliveries remains unresolved. While initially believed to translocate across cell membranes by direct translocation [35,36], L-PEN was later, at lower concentrations, demonstrated to internalize by endocytosis [36,37] without compromising the membrane integrity of the examined cells [37]. In contrast, little is known regarding the translocation mechanism of D-PEN into cells, despite having been reported to interact differently in multiple in vitro cell models [25]. The detrimental effects of all-D CPPs have previously been observed at concentrations that were non-toxic for the corresponding L-form [36].

The aim of the present study was to elucidate the mode of action for D-PEN in comparison to L-PEN with respect to cell entry and delivery capacity. The cellular internalization of the carboxyfluorescein (CF)-labeled enantiomers was studied in relevant cell culture models of the gut (Caco-2, IEC-6) and liver (HepG2) as well as for their delivery potential in intracellular and transepithelial delivery. The intracellular delivery of tetramethylrhodamine (TAMRA)-insulin was also studied in Caco-2 cells by confocal laser-scanning microscopy (CLSM) and flow cytometry, while the transepithelial delivery propensity across a well-differentiated Caco-2 cell epithelium was examined for TAMRA-insulin as well as cyclosporin, vancomycin, and mannitol (as a paracellular marker).

2. Materials and Methods

2.1. Materials

Human colorectal Caco-2 and human liver HepG2 cells were derived from American Type Culture Collection (ATCC, Manassas, VA, USA), and rat small intestine IEC-6 cells were retrieved from the European Collection of Authenticated Cell Cultures (ECACC, Public Health England, Salisbury, UK). (3-(4,5-Dimethylthiazol-2-yl)-5-(3-carboxymethoxyphenyl)-2-(4-sulfophenyl)-2H-tetrazolium (MTS) and phenazine methosulfate (PMS) were from Promega (Madison, WI, USA). Hoechst33342 was from Fisher Scientific (Loughborough, UK). DRAQ7 from Biostatus (Leicestershire, UK), trifluoroacetic acid from Iris Biotech (Marktredwitz, Germany), fetal bovine serum (FBS) was from PAA laboratories (Brøndby, Denmark), Hank's Balanced Salt Solution (HBSS) from Gibco Life Technologies (Paisley, UK), and N-2-hydroxyethylpiperazine-N'-2-ethanesulfonic acid (HEPES, $\geq$99.5%) from PanReac AppliChem (Darmstadt, Germany). ^{14}C-D-mannitol (57.2 Ci/mmol) and Ultima Gold™ scintillation fluid were obtained from Perkin Elmer (Boston, MA, USA). ^{3}H-L-cyclosporin A (20 Ci/mmol) was from American Radiolabeled Chemicals (St. Louis, MO, USA) and ^{3}H-L-vancomycin (4.6 Ci/mmol) from ViTrax Company (Placentia, CA, USA). NHS-rhodamine (5/6-carboxy-tetramethyl-rhodamine succinimidyl ester) (TAMRA) was from Thermo Fisher Scientific (Waltham, MA, USA). All other materials were from Sigma-Aldrich (Buchs, Switzerland). Ultrapure water from a Barnstead NanoPure system (Thermo Scientific, Waltham, MA, USA) was used throughout the studies. Unless otherwise specified, all chemicals were of analytical grade or higher.

2.2. Methods

2.2.1. Peptide Synthesis, Purification and Labelling

L-Penetratin (L-PEN, RQIKIWFQNRRMKWKK) and D-Penetratin (D-PEN, rqikiwfqnrrmkwkk) as well as their carboxyfluorescein (CF)-labeled analogs (L-PEN$_{CF}$, D-PEN$_{CF}$, respectively) were synthesized by Fmoc-based solid-phase peptide synthesis on an automated CEM Liberty (Matthews, NC, USA) as previously described [38]. Briefly, N-terminal labeling was performed manually overnight in Teflon reactors. The labeling of insulin with TAMRA performed manually by dissolving insulin in water at an acidic pH (pH 1–2) and subsequently increasing the pH to >10 with 1 M Na_2CO_3. Hereafter, the solution was mixed with TAMRA (dissolved in DMSO) at a mass ratio of 2:1 insulin:TAMRA and was mixed overnight on a rotary mixer protected from light. The purification of the labeled CPPs, as well as TAMRA-insulin, were performed by RP-HPLC, and purities of >92% were obtained for all conjugates. Conjugate identity was confirmed by MALDI-ToF-MS (Bruker, Microflex LT/SH system). The compounds were freeze-dried, followed by storage at −18 °C.

2.2.2. Cell Culturing

Caco-2, HepG2, and IEC-6 cells were maintained in Corning Costar polystyrene culture flasks (75 cm^2 or 175 cm^2 surface area from Sigma-Aldrich) in either Dulbecco's mo-dified Eagles medium (Caco-2 and IEC-6 cells) or Eagle's minimum essential medium (HepG2 cells). The media were supplemented with penicillin (100 U/mL), streptomycin (100 μg/mL), L-glutamine (2 mM), non-essential amino acids (1% (*v*/*v*)), and FBS (10% (*v*/*v*) for Caco-2 and HepG2 cells, 5% (*v*/*v*) for IEC-6 cells). HepG2 cells were further supplemented with sodium pyruvate (1 mM) and IEC-6 cells with insulin (5 μg/mL). The cells were kept in a humidified incubator at 37 °C and 5% CO_2 and subcultured weekly at approximately 90% confluency by trypsinization. The cells used for flow cytometry were seeded in 24-well plates (1.90 cm^2 surface area, Corning Costar, Sigma-Aldrich, Brøndby, Denmark) or 12-well plates (3.80 cm^2 surface area) at densities of 181,800 cells/cm^2, 69,700 cells/cm^2, and 48,500 cells/cm^2 for Caco-2, HepG2 cells, and IEC-6 cells, respectively, and were kept for 24 h in a humidified incubator (37 °C and 5% CO_2). The cells used for microscopy were seeded in flat-bottomed, glass-bottomed microchambers (1 cm^2 surface area, Ibidi, Planegg/Martinsried, Germany) at densities of 89,300 cells/cm^2, 15,000 cells/cm^2, and 10,000 cells/cm^2 for Caco-2, HepG2, and IEC-6, respectively, and were kept for 72 h in a humidified incubator. Plastic-bottom plates for Caco-2 cells were coated using collagen (8.93 μg/cm^2) prior to seeding. The variations in cell densities and incubation length were necessary to ensure uniform confluence (approx. 90%). For permeation studies, Caco-2 cells were seeded at a density of 1×10^5 cells/filter insert (pore size: 0.4 μm, area: 1.12 cm^2; Corning) and were cultured for 18–20 days to obtain well-differentiated Caco-2 cell monolayers. All experiments were performed in at least two passages of cells, over 16, 6, and 9 passages of Caco-2, HepG2 and IEC-6 cells, respectively.

2.2.3. Intracellular Distribution and Uptake of CF-Labelled PEN Enantiomers and TAMRA-Insulin Evaluated by Confocal Microscopy

Cells seeded in microchambers were washed with hHBSS (10 mM HEPES, 0.05% (*v*/*v*) bovine serum albumin (BSA) in HBSS at pH 7.4) and preheated to 37 °C. The uptake and distribution of PEN-enantiomers were investigated in Caco-2 cells, HepG2 cells, and IEC-6 cells by the addition of 250 μL L-PEN_{CF} or D-PEN_{CF} at 20 μM in hHBSS (pH 7.4). Additionally, the uptake of TAMRA-insulin:PEN mixtures was investigated in Caco-2 cells at a 1:4 molar (5 μM TAMRA-insulin, 20 μM PEN enantiomer) using the same buffer as described above. The cells were kept for 1 h on a shaking table (37 °C, 50 rpm) and were washed with hHBSS and stained with Hoechst 33,342 and DRAQ7 (5 μg/mL and 0.3% (*v*/*v*), respectively) for 10 min. Hereafter, the cells were kept in hHBSS prior to imaging with an LSM 780 Zeiss inverted confocal microscope (Carl Zeiss, Oberkochen, Germany) or a Leica SP5 inverted laser scanning confocal microscope (Leica Microsystems, Milton Keynes, UK), equipped with 20× air or 63× oil-immersion objectives. Fluorescence was recorded for Hoechst 33342, CF-labelled PEN, TAMRA-insulin, and DRAQ7 using excitation lasers at 405 nm, 488 nm, 543 nm, and 647 nm, respectively. Pulse/chase experiments of L-PEN_{CF} and D-PEN_{CF} (20 μM in hHBSS, pH 7.4) were performed as above, with the addition that the cells were incubated (37 °C, 5% CO_2) in hHBSS (pH 7.4) after the first image acquisition (t = 0). Following 1 h and 4 h of the chase, the cells were imaged using the same settings. All images were analyzed using Fiji ImageJ [39].

2.2.4. Cellular Uptake of CF-Labelled PEN Enantiomers and TAMRA-Insulin Evaluated by Flow Cytometry

The cellular uptake of CF-labelled L-PEN and D-PEN into Caco-2 cells, HepG2 cells, and IEC-6 cells, and the PEN-mediated delivery of TAMRA-insulin were evaluated by flow cytometry. Cells were washed twice with hHBSS, followed by the addition of test solutions containing 1–20 μM L-PEN_{CF} or D-PEN_{CF}, TAMRA-insulin (5 μM), or TAMRA-insulin:PEN mixtures (5 μM:20 μM), prepared in hHBSS (pH 7.4). The controls were incubated with a buffer without CPP or insulin. The cells were kept on a shaking table (37 °C, 50 rpm) for 1 h. Hereafter, the test solutions were removed, and the cells were washed twice with cold phosphate-buffered saline (4 °C) and trypsinized for 10–15 min. Trypsinization was terminated by the addition of excess cold hHBSS (4 °C) containing 10% (*v*/*v*) FBS (hHBSS/FBS). The cells were then collected and centrifuged at 4000 rpm (1076× *g*, 4 °C). The supernatant was removed, and the cells were washed once with hHBSS/FBS and centrifuged again at the same conditions before suspension in hHBSS/FBS containing 0.3% (*v*/*v*) DRAQ7 as a marker of the plasma membrane's integrity. To quench extracellular plasma membrane-associated fluorescence, a parallel experiment was performed using the same test solutions, with the addition of trypan blue (TB) at a final concentration of 0.1% (*v*/*v*) just prior to analysis. The mean cellular fluorescence intensity of CF-labelled PEN, TAMRA-insulin, and DRAQ7 was determined from intensities in 10,000 viable cells by flow cytometry using a Gallios flow cytometer (Beckman Coulter, Fullerton, CA, USA) with gating for forward and sideways scattering.

2.2.5. Transepithelial Permeation of Cargoes Facilitated by L-PEN and D-PEN

The permeation of the different cargoes, TAMRA-insulin, vancomycin, and cyclosporin, was evaluated across well-differentiated Caco-2 epithelium in the presence of L-PEN and D-PEN. Cell monolayers were washed twice with hHBSS (37 °C) and equilibrated for 20 min prior to the measurement of the transepithelial electrical resistance (TEER) using a resistance chamber connected to a voltmeter (Endohm and EVOM, respectively, both from World Precision Instruments, Sarasota, FL, USA). Monolayers displaying an initial TEER below 200 $\Omega \times cm^2$ were discarded. Thereafter, the wash medium was replaced with 350 μL of mHBSS (pH 6.5, 37 °C) and 1000 μL of hHBSS (pH 7.4, 37 °C) on the apical and basolateral sides, respectively. The cell monolayers were kept on a shaking table (37 °C, 50 rpm) for 5 min before replacing the buffers on the apical side with 350 μL test solution (10 μM:40 μM cargo:PEN in mHBSS at pH 5 or 6.5). Monolayers exposed to 10 μM of cargo in the absence of PEN were used as controls. Cell monolayers exposed to vancomycin or cyclosporin were spiked with ^{3}H-vancomycin or ^{3}H-cyclosporin for a final activity of 1 μCi/mL on the apical side. In addition, apical volumes on all cell monolayers were spiked with ^{14}C-D-mannitol (final activity of 1 μCi/mL on the apical side) for the simultaneous measurement of the paracellular marker. These cells were kept on a shaking table (37 °C, 50 rpm) for 4 h with the collection of samples from the basolateral side at 15–30 min intervals. The collected sample volume was replaced with hHBSS (37 °C, pH 7.4). After the final sampling, TEER was measured using the previously described equilibration procedure. Afterward, the viability of the cell in the monolayers was determined by using 350 μL of the MTS/PMS buffer (240 μg/mL MTS and 2.4 μg/mL PMS in hHBSS pH 7.4) and 1000 μL hHBSS on the apical and basolateral side, respectively. After 1 h, the MTS/PMS buffer was collected from the apical side and analyzed at 492 nm using a plate reader (FLUOStar OPTIMA, BMG LABTECH, Ortenberg, Germany). The fluorescence of TAMRA-insulin samples was analyzed at a wavelength of 545/590 nm ($\lambda_{ex}/\lambda_{em}$) on the same plate reader. Scintillation fluid (2 mL) was mixed with each sample that was subsequently quantified by liquid scintillation (Perkin-Elmer Tri-Carb 2910 TR, Perkin Elmer, Boston, MA, USA).

2.2.6. Data Analysis

All statistical analysis was performed in GraphPad Prism 6 (GraphPad Software Inc., La Jolla, CA, USA) using a one-way ANOVA combined with Tukey's multiple comparison analysis. The calculation of P_{app} from the permeation experiments was calculated as described by Artursson et al. [40]. N represents the number of individual passages, and n represents the repeats within one passage. In the cases where we compared data from less than three passages, the technical replicates in each passage were at least three.

3. Results

3.1. Effect of Stereoisomerism on the Uptake and Cellular Distribution of L- and D-Penetratin

To investigate the translocation of both PEN enantiomers across the cell membranes, three cell culture models were incubated with L-PEN_{CF} and D-PEN_{CF} for 1 h while monitoring the uptake by CLSM. All cell models incubated with 20 μM L-PEN_{CF} displayed vesicular staining of the cytoplasm (Figure 1), which was consistent with endocytic uptake. Additionally, L-PEN_{CF} strongly stained the plasma membrane of Caco-2 cells, whilst IEC-6 cells demonstrated diffuse cytosolic staining. Caco-2 cells incubated with D-PEN_{CF} also displayed vesicular staining; however, it was prominent that clusters of fluorescence, which did not resemble any typical cytosolic pattern, were observed in this case. A similar phenomenon was seen in the other two cell lines, whereas D-PEN_{CF} labeling was not prominent on the plasma membrane in any cell model but appeared as aggregates in all cell lines. The white arrows depicting DRAQ7 clearly show that in HepG2 cells and, to a lesser extent, in IEC-6 cells, 20 μM D-PEN_{CF} caused plasma membrane porosity under these experimental conditions.

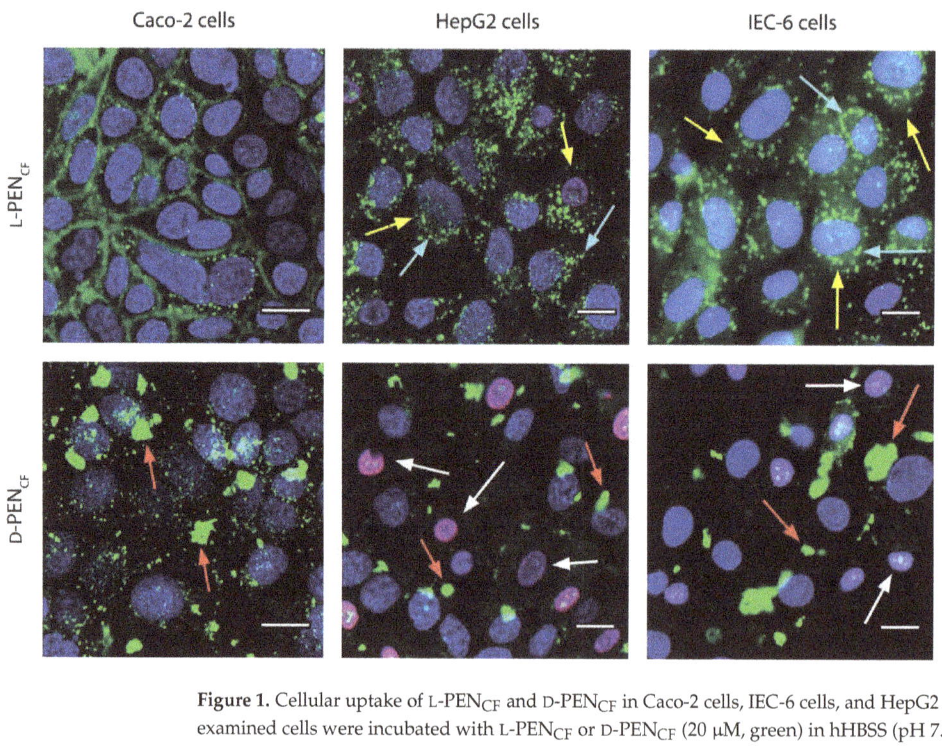

Figure 1. Cellular uptake of L-PEN_{CF} and D-PEN_{CF} in Caco-2 cells, IEC-6 cells, and HepG2 cells. The examined cells were incubated with L-PEN_{CF} or D-PEN_{CF} (20 μM, green) in hHBSS (pH 7.4) for 1 h, followed by the staining of nuclei with Hoechst 33,342 (blue) and DRAQ7 (magenta, white arrows). Vesicular staining (cyan arrows), cytosolic staining (yellow arrows), and CPP clusters (red arrows) are indicated by arrows. Micrographs are representative of two passages of cells. Scale bar = 20 μm.

To further investigate the internalization of L-PEN$_{CF}$ and D-PEN$_{CF}$, they were incubated with Caco-2 cells and subjected to a 0 h, 1 h, or 4 h endocytic chase study to follow the intracellular distribution of internalized CPP (Figure 2). The 0 h chase time point was identical to previous observations (Figure 1); however, after 1 h of the chase, it was observed that the level of plasma membrane-associated L-PEN$_{CF}$ decreased. An enrichment of fluorescence in the perinuclear region of cells was pronounced after 4 h of chase upon L-PEN$_{CF}$ incubation. A very different spatiotemporal fluorescence profile was shown for D-PEN$_{CF}$, showing the aggregation of fluorescence at all time points, which is consistent with the observations presented in Figure 1.

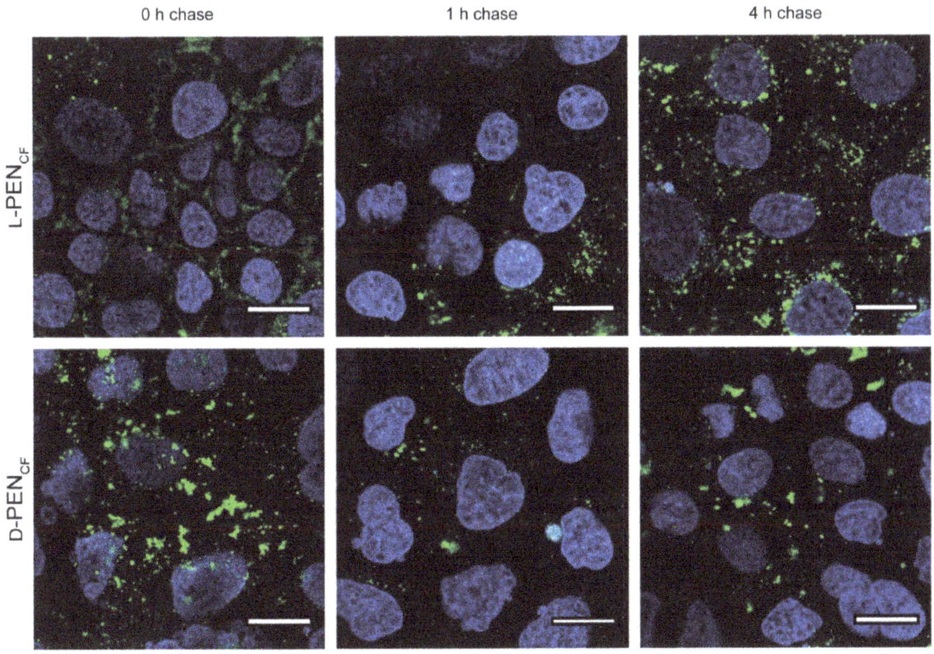

Figure 2. Cellular uptake of L-PEN$_{CF}$ and D-PEN$_{CF}$ in Caco-2 cells following 0 h, 1 h, and 4 h chase in hHBSS (pH 7.4). Caco-2 cells were incubated with 20 μM peptide (green) in hHBSS (pH 7.4) for 1 h at 37 °C. The cells were washed, and nuclei were stained with Hoechst 33,342 (blue) prior to imaging. The images are representative of the same passage of cells, and the experiment was repeated in two different passages of cells. Scale bar 20 μm.

Caco-2 cells and HepG2 cells were also subjected to analysis by flow cytometry to provide a semi-quantitative evaluation of the cellular uptake of L-PEN$_{CF}$ and D-PEN$_{CF}$ after incubation with the 1–20 μM peptide. The enantiomers of PEN$_{CF}$ were taken up at similar levels in the cells when incubated at concentrations in the range of 5–20 μM (Figure 3). A parallel experiment was performed under the same conditions as previously described, with the exception that Trypan Blue (0.1% (v/v) final concentration) was added as an extracellular quencher to the samples just prior to the analysis by flow cytometry. For data analysis, the relative mean fluorescence intensity (MFI) was normalized to the fluorescence of cells incubated with 20 μM L-PEN$_{CF}$ without Trypan Blue treatment. Under those conditions, the cells exposed to D-PEN$_{CF}$ exhibited an >95% reduction in fluore-scence at any concentration. By comparison, cells exposed to L-PEN$_{CF}$ displayed an approximately 50% decrease in the detected fluorescence at any concentration. HepG2 cells exposed to D-PEN$_{CF}$ at concentrations above 10 μM displayed morphological aberrations, and these were omitted from further analysis.

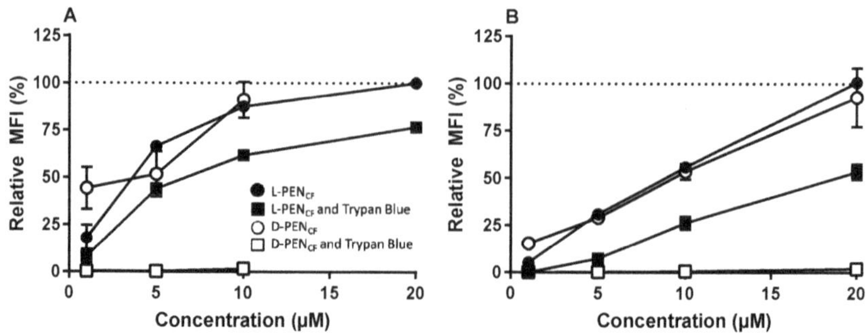

Figure 3. Relative uptake of L-PEN$_{CF}$ and D-PEN$_{CF}$ into HepG2 cells (**A**) or Caco-2 cells (**B**). 1–20 µM CPP in hHBSS (pH 7.4) was incubated with cells. L-PEN$_{CF}$ (black circle), L-PEN$_{CF}$ followed by 0.1% (v/v) Trypan Blue (black square), D-PEN$_{CF}$ (white circle), and D-PEN$_{CF}$ followed by 0.1% (v/v) Trypan Blue (white square). Data are relative to those of cells incubated with 20 µM L-PEN$_{CF}$ (indicated by dotted line). Mean ± SEM, N = 2–3, n = 3.

3.2. *L-PEN and D-PEN as Carrier Peptides for Transepithelial Peptide Delivery*

The delivery propensities of PEN enantiomers were investigated by performing permeation assays across well-differentiated Caco-2 cell monolayers. Cargoes displaying distinct physicochemical properties (Table 1) were employed to assess the PEN delivery efficiency for different structures. The delivery was assessed at different pH values resembling the variation that could be encountered in the small intestines, as the delivery efficiency of L-PEN was previously shown to depend on pH [27,31].

Table 1. Test compounds were included for investigating the delivery capacities of PEN enantiomers.

Compound	Mass (Da)	Log P	Note
Insulin	5808	−13.1	Hydrophilic two-chain peptide
Mannitol	182	−3.7	Sugar-derived polyol (paracellular marker)
Vancomycin	1449	−4.4	Hydrophilic tricyclic glycopeptide
Cyclosporin	1202	3.6	Hydrophobic monocyclic peptide

Physicochemical properties (mass and predicted log P) were obtained from Drugbank.com and pubchem.ncbi.nlm.nih.gov (accessed on 3 January 2023).

3.2.1. Intracellular Delivery of TAMRA-Insulin

The uptake and distribution of TAMRA-insulin mediated by PEN-enantiomers were examined in undifferentiated Caco-2 cells by using CLSM. These cells displayed the uptake of the TAMRA-insulin conjugate at both investigated pH values (pH 6.5 and 7.4), independent of the enantiomer present. In contrast, for the control cells incubated with TAMRA-insulin alone at pH 7.4, uptake was absent (Figure 4). At pH 6.5, there was a relatively strong localization of fluorescence at the plasma membrane with some punctate intracellular fluorescence, whereas cells exposed to TAMRA-insulin at pH 7.4 displayed only intracellular vesicular staining. The apparent cellular uptake was higher at pH 6.5 compared to that seen at pH 7.4, albeit it was not possible to clearly distinguish the membrane associated with internalized TAMRA-insulin at pH 6.5.

Flow cytometry analysis showed that L-PEN and D-PEN both increased the cellular uptake of TAMRA-insulin compared to the control cells at pH 6.5 and 7.4 (Figure 5), and the uptake of TAMRA-insulin was higher for the D-PEN-mediated delivery compared to that mediated by L-PEN at both pH 6.5 and 7.4. However, it was evident that quenching extracellular fluorescence with Trypan Blue only reduced the mean cell fluorescence intensity, indicative of TAMRA-insulin uptake, when this was co-incubated with D-PEN and not when it was co-incubated with L-PEN (Table 2).

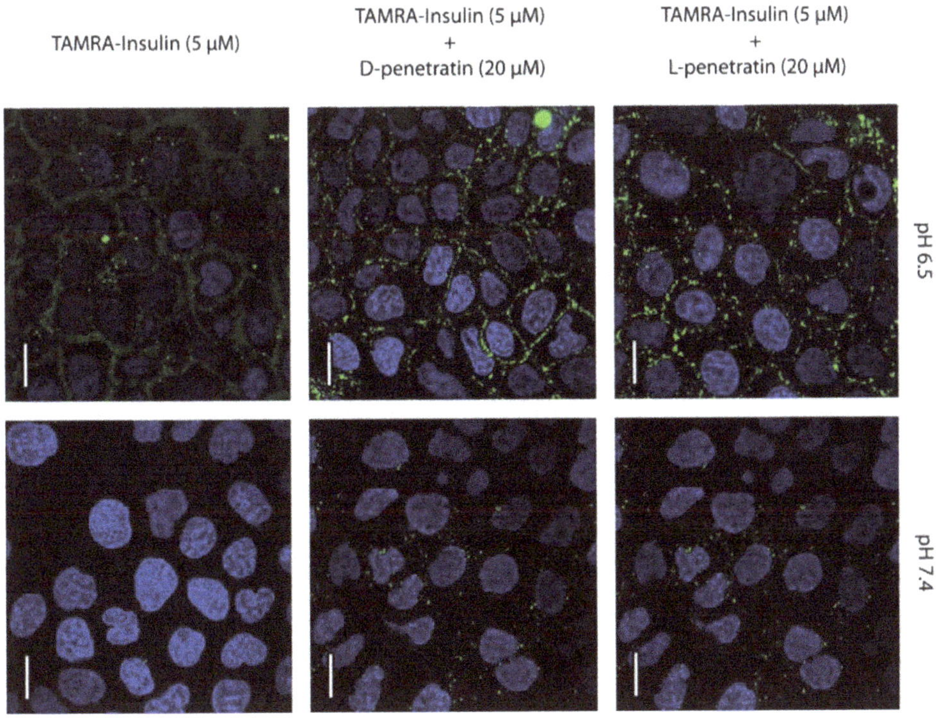

Figure 4. PEN-mediated delivery of TAMRA-insulin into Caco-2 cells at pH 6.5 and 7.4. Cells were incubated with TAMRA-insulin (5 μM, green) and L-PEN or D-PEN (20 μM) for 1 h, followed by staining of nuclei with Hoechst 33,342 (blue), while no membrane-compromised cells were found upon including staining with DRAQ7 as no magenta DRAQ7 fluorescence was detected. The images are representative of the same passage of cells, and the experiment was repeated in two different passages of cells. Scale bar 20 μm.

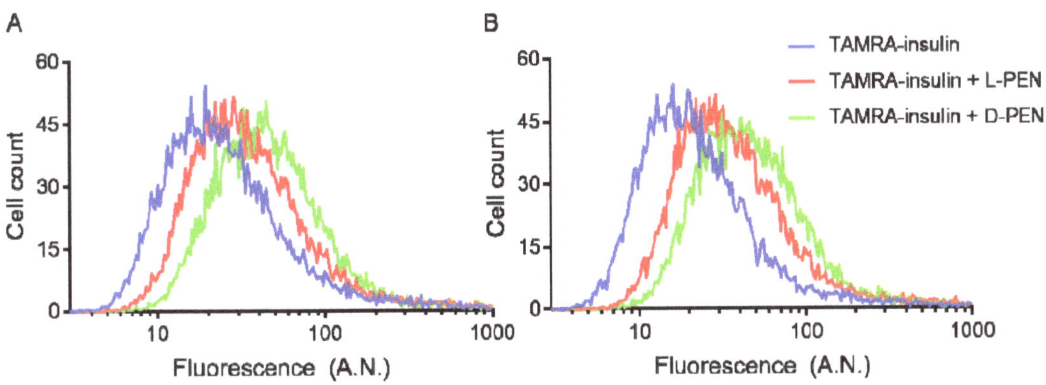

Figure 5. Uptake of TAMRA-insulin in Caco-2 cells displayed as histograms at pH 6.5 (**A**) and 7.4 (**B**). Cells were treated with test solution (5 μM TAMRA-insulin only (blue) as well as with 20 μM L-PEN (red) or D-PEN (green)) for 1 h, followed by trypsinization and analysis by flow cytometry. Mean fluorescence in arbitrary units (A.N.) from two independent experiments.

Table 2. Uptake of TAMRA-insulin (5 μM) in Caco-2 cells incubated at pH 6.5 and 7.4 alone or in the presence of 20 μM L-PEN or D-PEN for 1 h, followed by trypsinization, and analysis by flow cytometry in the presence or absence of Trypan Blue.

	pH 6.5		pH 7.4	
	Relative Uptake (%)	**MFI (A.N.)**	**Relative Uptake (%)**	**MFI (A.N.)**
TAMRA-Insulin				
Without Trypan Blue	100.0 ± 1.6	48.0 ± 1.2	100.0 ± 2.2	41.2 ± 2.9
With Trypan Blue	103.4 ± 5.6	49.6 ± 1.9	100.8 ± 4.3	40.8 ± 2.6
TAMRA-Insulin + L-PEN				
Without Trypan Blue	126.5 ± 7.7	60.4 ± 2.6	152.6 ± 6.9	61.9 ± 1.5 ***
With Trypan Blue	124.7 ± 7.3	62.3 ± 5.1	151.6 ± 2.8	62.1 ± 4.5 ***
TAMRA-Insulin + D-PEN				
Without Trypan Blue	148.2 ± 2.0	71.1 ± 1.4	181.1 ± 15.3	72.6 ± 1.6
With Trypan Blue	120.0 ± 7.1	59.1 ± 2.0 *	164.6 ± 10.5	65.9 ± 1.0

Mean fluorescence in arbitrary numbers (A.N.) from two independent experiments. Asterisks indicate significant differences (* $p < 0.05$; *** $p < 0.001$). Mean ± SD, N = 2, n = 3.

3.2.2. PEN-Mediated Permeation of Cargo across Caco-2 Cell Epithelium

The capacity for L-PEN and D-PEN to mediate the delivery across the Caco-2 cell epithelium was investigated for cargoes with different physicochemical properties and, thus, different permeation properties. It was observed that incubation with both enantiomers combined with any of the cargoes led to a reduction (by ~25%) in the metabolic activity of Caco-2 cells at pH 5, and a similar reduction was observed for the D-PEN and cargo at pH 6.5 compared to the level for the buffer control cells (Figure 6A). By contrast, L-PEN elicited no metabolic activity changes at pH 6.5. In addition, cells exposed to D-PEN displayed a reduction of ~50–75% in TEER (Figure 6B) and an increase in the P_{app} of mannitol (Figure 7) at both investigated pH values.

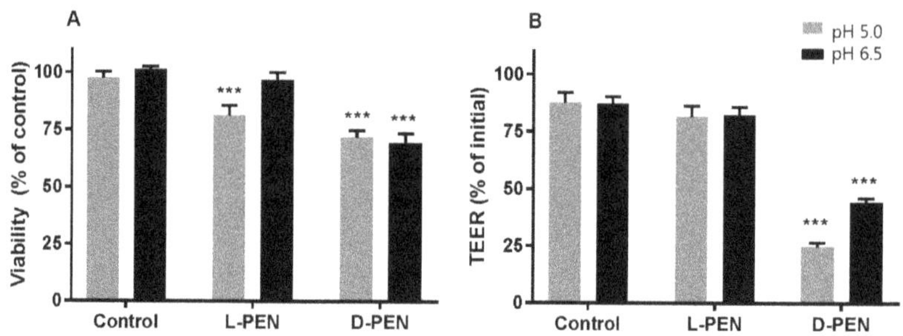

Figure 6. Viability (**A**) as indicated by metabolic activity using the MTS/PMS assay, and integrity (**B**) measured by TEER of Caco-2 epithelium upon exposure to L-PEN or D-PEN at 40 μM at pH 5 (grey) and pH 6.5 (black). Data were pooled from permeation experiments performed separately for each cargo. Metabolic activity was calculated relative to the activity of buffer-exposed control cell monolayers at pH 6.5. Asterisks indicate significant differences (*** $p < 0.001$). Mean ± SEM, N = 2–3, n = 12.

D-PEN mediated a 4-5-fold increased permeation of TAMRA-insulin at both pH 5.0 and 6.5 and for vancomycin at pH 5.0 compared to the permeation seen in the buffer alone. However, L-PEN elicited only a small increase in P_{app} for both cargoes at both pH values, and it did not differ from that observed in the control experiments without L-PEN. The permeation of cyclosporin was unaffected by the presence of either peptide enantiomer compared to controls at the respective pH values. The permeation of the paracellular marker mannitol increased significantly in the presence of D-PEN but not when combined with L-PEN.

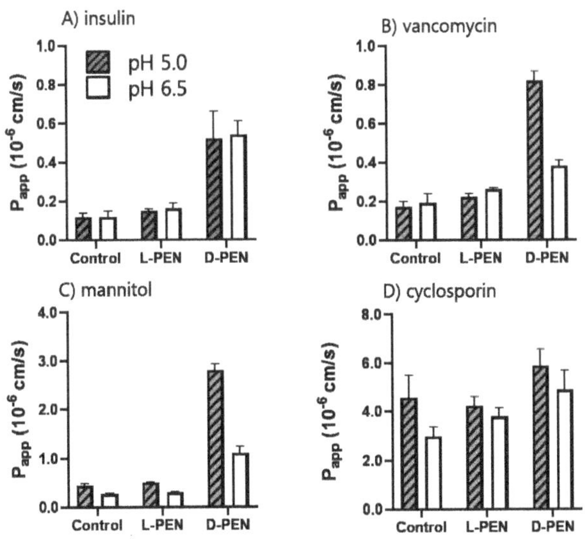

Figure 7. Apparent permeability coefficients (P_{app}) for different cargos across Caco-2 monolayers in the presence of 40 µM L-PEN or D-PEN at pH 5.0 (grey, dash) and 6.5 (white). TAMRA-insulin (**A**), vancomycin (**B**), mannitol, and (**C**) cyclosporin (**D**). Mean ± SEM, N = 2–3, *n* = 3.

4. Discussion

4.1. Enantiomers of PEN Interact Differently with Cell Membranes

The interactions between CPPs and synthetic and natural cell membranes have been extensively investigated [35]. The current consensus is that there is no universal cellular uptake mechanism for CPPs and that they internalize into cells by direct translocation, endocytosis, or via a combination of these processes [2,7,9]. The present study highlights that stereochemistry and cell type determine the internalization patterns, membrane binding, and cargo delivery. It was previously reported that different cell models exhibited different CPP internalization patterns, as shown by vesicular or cytosolic staining [6,20], corroborating our data.

In the current study, L-PEN$_{CF}$ displayed vesicular staining in all the investigated cell models, along with distinct membrane staining in the Caco-2 cell model (Figure 1). The vesicular staining was consistent with the uptake by endocytosis, as previously reported for both L-PEN [6] and fluorescently labeled L-PEN analogs [41]. Cytosolic labeling was very prominent in IEC-6 cells incubated with L-PEN$_{CF}$ and also had evidence of membrane permeabilization (DRAQ7). For the same experiment with D-PEN$_{CF,}$ there was some evidence of cytosolic labeling but also of the loss of plasma membrane integrity. The loss of membrane integrity was also noted in HepG2 cells incubated with the D-form. These observations are consistent with a study by Duchardt et al. [6], who observed a similar pattern in HeLa cells and also reported that direct translocation was only observed for CPP concentrations above a certain threshold and that it appeared to depend on both cell density and cell type. In the present study, the cell densities were comparable in terms of confluence at the time of use, and the concentrations of the enantiomers, when tested, were the same. Interestingly, the initial pronounced Caco-2 plasma membrane association of L-PEN$_{CF}$ was replaced by predominantly vesicular staining upon 1 h and 4 h chases (Figure 2), inferring endocytic trafficking of L-PEN$_{CF}$ from the plasma membrane. When applied in the same concentration as L-PEN$_{CF}$, D-PEN$_{CF}$ displayed clustering in all the examined cell models. These clusters were likely localized at the cell membrane as the fluorescence was fully quenched by the addition of extracellular Trypan Blue (Figure 3). Further, for HepG2 and IEC-6 cells, exposure to D-PEN$_{CF}$ resulted in pronounced membrane damage since a significant proportion of the cells displayed DRAQ7 in the staining of the nuclei (Figure 1). These cell models also displayed cytosolic staining and an absence of vesicular staining with D-PEN$_{CF}$, suggesting that this CPP was internalized by increasing the fluidity of the cell membrane, thereby bypassing the endocytic pathway. This may, in part, be explained by the increased stability of the all-D enantiomer compared to the all-L enantiomer when exposed to cells [29], thereby allowing for longer residence at the membrane. This is also consistent with a study by Watkins et al. [42], who reported cytosolic staining upon the exposure of HeLa cells to R$_8$-PAD at concentrations that also induced the uptake of DRAQ7.

In another study by Verdurmen et al. [26], D-enantiomers of CPPs, including PEN, were reported to display reduced uptake in HeLa cells and MC57 cells; these were identified as having high levels of plasma membrane heparan sulfate proteoglycans (HSPG). By comparison, Jurkat cells, lacking plasma membrane HSPG displayed no reduced uptake. However, HSPGs were previously reported not to interact with non-labeled L-PEN [43]. In addition, HSPGs were investigated together with CPPs for their potential role in peptide clustering [44]. In line with this, the observation that D-PEN$_{CF}$ exhibited significant aggregation suggests that there was a pronounced interaction between D-PEN$_{CF}$ and HSPGs on the cell membrane compared to that of L-PEN$_{CF}$. D-PEN$_{CF}$ also displayed aggregation when examined in Caco-2 cells and was accompanied by structures indicative of endocytic vesicles. Both Caco-2 and HepG2 cells were examined by flow cytometry and were found to display the pronounced quenching of fluorescence in response to the addition of extracellular Trypan Blue, suggesting that most of the detected L-PEN$_{CF}$ and especially D-PEN$_{CF}$ was membrane-bound rather than internalized. Further, membrane-bound D-PEN$_{CF}$ was not found to traffic intracellularly following 1 h and 4 h chases, as opposed to that seen for L-PEN$_{CF}$. This difference may have arisen from a combination of the increased proteolytic stability of all-D enantiomers [29,33] and the different binding of the PEN enantiomers to surface-bound HSPGs.

4.2. Intracellular Delivery and Distribution of Insulin Are Mediated by Both L-PEN and D-PEN

In the current study, TAMRA-insulin was used as a model cargo for delivery to and through Caco-2 cells. Insulin is a thoroughly studied peptide, which was previously found to permeate intestinal epithelium when co-administered with L-PEN [3,28]. At pH 6.5, TAMRA-insulin was associated with the plasma membrane in the presence of 20 μM L-PEN or D-PEN as well as in the control cells. However, this association appeared slightly more pronounced in cells exposed to D-PEN compared to cells treated with L-PEN (Figure 4). The flow cytometry studies demonstrated a reduction in the fluorescence of TAMRA-insulin in cells upon the quenching of extracellular fluorescence by Trypan Blue as well as in cells incubated with D-PEN but not with L-PEN (Table 2). While some vesicular staining was seen at pH 6.5, these observations inferred that a large part of the detected TAMRA-insulin was not internalized but was instead localized to the plasma membrane at pH 6.5. However, at pH 7.4, pronounced vesicular localization without membrane staining was observed when TAMRA-insulin was applied together with any of the PEN enantiomers, whereas no staining was evident in the insulin-only control. It appears likely that TAMRA-insulin was internalized by endocytosis, potentially as a result of complexation between insulin and PEN, which was especially prominent at pH 7.4 due to their opposite net charges at this pH [28]. No signs of negative effects, e.g., in terms of the decreased viability of the cells, were observed upon exposure to 20 μM L-PEN or D-PEN, which agrees with previously reported data [29].

4.3. Transepithelial Delivery of Hydrophilic Peptide Cargoes Is Mediated by Both Enantiomers, yet Preferentially by D-PEN

In the current study, it was observed that D-PEN interacted with Caco-2 cell monolayers through a mechanism, which at the investigated concentration affected the integrity of the epithelium as reflected in reduced TEER values and lowered metabolic activity after exposure for 4 h. Previously, we reported that the exposure of a Caco-2 cell monolayer to

50 μM D-PEN did not result in morphologies different from that seen for the buffer control as evaluated by transmission electron micrographs [29], while 60 μM L-PEN did not reduce the TEER of the monolayer [3]. An exact mechanism for the observed effects in our study of D-PEN in reducing the viability and TEER cannot be proposed, but the increased stability of D-PEN during incubation with the Caco-2 cell monolayers compared to that of L-PEN [29] may contribute. Further, the different membrane binding behaviors, as reported here for the CF-labelled versions of the enantiomers, may well be reflected in different effects on the TEER of the Caco-2 monolayer.

The widening of the paracellular space could allow for the increased paracellular diffusion of hydrophilic cargoes such as vancomycin, insulin, and mannitol. Recently, Maher et al. [30] reviewed several types of permeation enhancers and highlighted that many traditional permeation enhancers interact with the plasma membrane in a way that confers decreased TEER values. This indirect effect is for some enhancers hypothesized to be due to membrane perturbation, which in turn causes the instability of intracellular levels of Ca^{2+} or intracellular kinases, regulating the activity of tight junction proteins [30]. Additionally, studies on the interaction of the cationic AMP melittin with epithelium have reported that such interactions increased the paracellular diffusion of several compounds [22]. This effect was ascribed to the ability of melittin to elicit membrane perturbation, which in turn affected the tightness of the epithelium. In the present study, no signs of membrane perturbation were evident in undifferentiated Caco-2 cells (concluded from the lack of staining of nuclei with DRAQ7 included in the experiment); hence, membrane interaction without perturbation may likely account for the decreased TEER in the polarized monolayer. Membrane interactions have been proposed to be the primary mechanism for fatty acid-based permeation enhancers, which were found to induce similar effects on Caco-2 monolayers as observed for D-PEN in the current study. Here, Caco-2 cell epithelium, when exposed to 40 μM L-PEN elicited no increase in epithelial permeability as neither a decrease in the TEER (Figure 6B) nor increased permeation of the cargo (Figure 7) was observed. A previous report partly supported this by concluding that 20 μM L-PEN did not decrease TEER at either pH 5, 6.5, or 7.4, although it increased insulin permeation at pH 5 [28]. In contrast, a study by Kamei et al. [45] reported that both 60 μM L-PEN and D-PEN were found to increase the permeation of insulin across Caco-2 cell monolayers with an increasing pH from pH 5, 6, 7, to 8, while the TEER was unaffected upon exposure to any of the enantiomers and pH values. Neither L-PEN nor D-PEN increased the permeation of the hydrophobic peptide cyclosporin (Figure 7), supporting a mechanism that may include membrane interactions, while D-PEN had some effect on the TEER, but not to a level leading to the measurably increased permeation of a hydrophobic compound, which was expected to permeate primarily via the transcellular pathway. Overall, the increased proteolytic stability of the all-D stereoisomer, as demonstrated both on the Caco-2 cell monolayer [29] as well as in rat intestinal fluid [33], may indeed explain the improved delivery capacity of the CPP compared to the all-L stereoisomer. It is well-known [46] that one of the strategies to increase the stability of peptides in the gut is to introduce D-amino acids in the molecule, and this could naturally also be applied for cell-penetrating peptides when applied as carriers for oral drug delivery. Thus, the fact that D-PEN displayed improved delivery capacity for hydrophilic cargoes supports further studies on this stereoisomer as a carrier for therapeutic peptide delivery.

5. Conclusions

Overall, the stereochemistry of PEN influences the transepithelial delivery propensity but not the cellular uptake. The enantiomers were found to induce similar levels of cellular uptake for insulin in Caco-2 cells, and in particular, D-PEN induced noticeable membrane binding of insulin. The L-PEN enantiomer labelled with CF displayed a clear plasma membrane association prior to its uptake within vesicular structures, whereas the corresponding CF-labeled D-PEN at similar concentrations displayed aggregation at the plasma membrane. While L-PEN did not increase the transepithelial permeation of the studied cargo peptides at the investigated concentration, D-PEN increased the transepithelial delivery of the glycopeptide vancomycin five-fold at pH 5 and nearly four-fold for insulin irrespective of the pH. Moreover, the increased permeation of the paracellular marker pointed toward the increased paracellular permeation of the cargo peptides. By contrast, neither enantiomer was found to increase the transepithelial delivery of the hydrophobic cyclosporin. Thus, the mechanism of D-PEN may include a promoted cargo permeation via other pathways than those involved in L-PEN, which potentially is related to the significant membrane binding of D-PEN. Overall, these studies have implications for the use of this CPP and different CPP enantiomers for the oral delivery of macromolecular therapeutics.

Author Contributions: Conceptualization, D.B., A.T.J. and H.M.N.; methodology, D.B., E.J.S. and M.V.C.; validation, D.B. and E.J.S.; formal analysis, D.B.; investigation, D.B., E.J.S. and M.V.C.; resources, A.T.J., H.F. and H.M.N.; writing—original draft, D.B.; writing—review and editing, E.J.S., A.T.J., M.V.C., H.F. and H.M.N.; supervision, A.T.J., H.F. and H.M.N.; project administration, H.M.N.; funding acquisition, A.T.J., H.F. and H.M.N. All authors have read and agreed to the published version of the manuscript.

Funding: This research was funded by The Danish Council for Independent Research, Technology and Production Sciences (grant no. 4005-00455) (DB, MVC, HF, HMN), and the COMPACT Innovative Medicines Initiative Joint Initiative under grant agreement n° 115363 composed of financial contribution from the European Union's Seventh Framework Programme (FP7/2007-2013) and EFPIA companies' in kind contribution (HMN, ATJ, EJS), as well as by the Novo Nordisk Foundation (Grand Challenge Programme; NNF16OC0021948 (HMN). The Alfred Benzon Foundation and The Drug Research Academy are acknowledged for funding equipment used in this project.

Institutional Review Board Statement: Not applicable.

Informed Consent Statement: Not applicable.

Data Availability Statement: Data are available upon reasonable request.

Acknowledgments: Zishan Bazai is acknowledged for his assistance in synthesizing and purifying the test compounds. The Core Facility for Integrated Microscopy, Faculty of Health and Medical Sciences, University of Copenhagen is acknowledged for providing access to microscopes and help with CLSM.

Conflicts of Interest: The authors declare no conflict of interest.

References

1. Kristensen, M.; Birch, D.; Nielsen, H.M. Applications and challenges for use of cell-penetrating peptides as delivery vectors for peptide and protein cargos. *Int. J. Mol. Sci.* **2016**, *17*, 185. [CrossRef] [PubMed]
2. Jones, A.T.; Sayers, E.J. Cell entry of cell penetrating peptides: Tales of tails wagging dogs. *J. Control. Release* **2012**, *161*, 582–591. [CrossRef] [PubMed]
3. Diedrichsen, R.G.; Harloff-Helleberg, S.; Werner, U.; Besenius, M.; Kristensen, M.; Nielsen, H.M. Revealing the importance of carrier-cargo association in delivery of insulin and lipidated insulin. *J. Control. Release* **2021**, *338*, 8–21. [CrossRef] [PubMed]
4. Khafagy, E.S.; Morishita, M.; Kamei, N.; Eda, Y.; Ikeno, Y.; Takayama, K. Efficiency of cell-penetrating peptides on the nasal and intestinal absorption of therapeutic peptides and proteins. *Int. J. Pharm.* **2009**, *381*, 49–55. [CrossRef]
5. Kristensen, M.; de Groot, A.M.; Berthelsen, J.; Franzyk, H.; Sijts, A.; Nielsen, H.M. Conjugation of cell-penetrating peptides to parathyroid hormone affects its structure, potency, and transepithelial permeation. *Bioconjug. Chem.* **2015**, *26*, 477–488. [CrossRef]
6. Duchardt, F.; Fotin-Mleczek, M.; Schwarz, H.; Fischer, R.; Brock, R. A comprehensive model for the cellular uptake of cationic cell-penetrating peptides. *Traffic* **2007**, *8*, 848–866. [CrossRef]
7. Tréhin, R.; Krauss, U.; Beck-Sickinger, A.G.; Merkle, H.P.; Nielsen, H.M. Cellular uptake but low permeation of human calcitonin-derived cell penetrating peptides and Tat(47–57) through well-differentiated epithelial models. *Pharm. Res.* **2004**, *21*, 1248–1256. [CrossRef]
8. Jones, A.T. Macropinocytosis: Searching for an endocytic identity and role in the uptake of cell penetrating peptides. *J. Cell. Mol. Med.* **2007**, *11*, 670–684. [CrossRef]
9. Padari, K.; Säälik, P.; Hansen, M.; Koppel, K.; Raid, R.; Langel, Ü.; Pooga, M. Cell transduction pathways of transportans. *Bioconjug. Chem.* **2005**, *16*, 1399–1410. [CrossRef]

10. Kamei, N.; Kikuchi, S.; Takeda-Morishita, M.; Terasawa, Y.; Yasuda, A.; Yamamoto, S.; Ida, N.; Nishio, R.; Takayama, K. Determination of the optimal cell-penetrating peptide sequence for intestinal insulin delivery based on molecular orbital analysis with self-organizing maps. *J. Pharm. Sci.* **2013**, *102*, 469–479. [CrossRef]
11. Zhao, M.; Weissleder, R. Intracellular cargo delivery using tat peptide and derivatives. *Med. Res. Rev.* **2004**, *24*, 1–12. [CrossRef] [PubMed]
12. Vives, E.; Brodin, P.; Lebleu, B. A truncated HIV-1 Tat protein basic domain rapidly translocates through the plasma membrane and accumulates in the cell nucleus. *J. Biol. Chem.* **1997**, *272*, 16010–16017. [CrossRef] [PubMed]
13. Eiríksdóttir, E.; Konate, K.; Langel, Ü.; Divita, G.; Deshayes, S. Secondary structure of cell-penetrating peptides controls membrane interaction and insertion. *Biochim. Biophys. Acta Biomembr.* **2010**, *1798*, 1119–1128. [CrossRef] [PubMed]
14. Alves, I.D.; Goasdoué, N.; Correia, I.; Aubry, S.; Galanth, C.; Sagan, S.; Lavielle, S.; Chassaing, G. Membrane interaction and perturbation mechanisms induced by two cationic cell penetrating peptides with distinct charge distribution. *Biochim. Biophys. Acta* **2008**, *1780*, 948–959. [CrossRef]
15. Koren, E.; Torchilin, V.P. Cell-penetrating peptides: Breaking through to the other side. *Trends Mol. Med.* **2012**, *18*, 385–393. [CrossRef]
16. Hallbrink, M.; Oehlke, J.; Papsdorf, G.; Bienert, M. Uptake of cell-penetrating peptides is dependent on peptide-to-cell ratio rather than on peptide concentration. *Biochim. Biophys. Acta* **2004**, *1667*, 222–228. [CrossRef]
17. Madani, F.; Lindberg, S.; Langel, U.; Futaki, S.; Gräslund, A. Mechanisms of cellular uptake of cell-penetrating peptides. *J. Biophys.* **2011**, *2011*, 414729. [CrossRef]
18. Lönn, P.; Kacsinta, A.D.; Cui, X.S.; Hamil, A.S.; Kaulich, M.; Gogoi, K.; Dowdy, S.F. Enhancing Endosomal escape for intracellular delivery of macromolecular biologic therapeutics. *Sci. Rep.* **2016**, *6*, 32301. [CrossRef]
19. El-Sayed, A.; Futaki, S.; Harashima, H. Delivery of macromolecules using arginine-rich cell-penetrating peptides: Ways to overcome endosomal entrapment. *AAPS J.* **2009**, *11*, 13–22. [CrossRef]
20. Fischer, R.; Köhler, K.; Fotin-Mleczek, M.; Brock, R. A stepwise dissection of the intracellular fate of cationic cell-penetrating Peptides. *J. Biol. Chem.* **2004**, *279*, 12625–12635. [CrossRef]
21. Ziegler, A. Thermodynamic studies and binding mechanisms of cell-penetrating peptides with lipids and glycosaminoglycans. *Adv. Drug Deliv. Rev.* **2008**, *60*, 580–597. [CrossRef] [PubMed]
22. Maher, S.; Feighery, L.; Brayden, D.J.; McClean, S. Melittin as a permeability enhancer II: In vitro investigations in human mucus secreting intestinal monolayers and rat colonic mucosae. *Pharm. Res.* **2007**, *24*, 1346–1356. [CrossRef] [PubMed]
23. Diedrichsen, R.G.; Tuelung, P.S.; Foderà, V.; Nielsen, H.M. Stereochemistry and intermolecular interactions influence carrier peptide-mediated insulin delivery. *Mol. Pharm.* **2023**, *20*, 1202–1212. [CrossRef] [PubMed]
24. Letoha, T.; Somlai, C.; Takács, T.; Szabolcs, A.; Jármay, K.; Rakonczay, Z., Jr.; Hegyi, P.; Varga, I.; Kaszaki, J.; Krizbai, I.; et al. A nuclear import inhibitory peptide ameliorates the severity of cholecystokinin-induced acute pancreatitis. *World J. Gastroenterol.* **2005**, *11*, 990–999. [CrossRef]
25. Nakase, I.; Akita, H.; Kogure, K.; Graslund, A.; Langel, U.; Harashima, H.; Futaki, S. Efficient intracellular delivery of nucleic acid pharmaceuticals using cell-penetrating peptides. *Acc. Chem. Res.* **2012**, *45*, 1132–1139. [CrossRef]
26. Verdurmen, W.P.; Bovee-Geurts, P.H.; Wadhwani, P.; Ulrich, A.S.; Hällbrink, M.; van Kuppevelt, T.H.; Brock, R. Preferential uptake of L-versus D-amino acid cell-penetrating peptides in a cell type-dependent manner. *Chem. Biol.* **2011**, *18*, 1000–1010. [CrossRef]
27. Kristensen, M.; Franzyk, H.; Klausen, M.T.; Iversen, A.; Bahnsen, J.S.; Skyggebjerg, R.B.; Foderà, V.; Nielsen, H.M. Penetratin-mediated transepithelial insulin permeation: Importance of cationic residues and pH for complexation and permeation. *AAPS J.* **2015**, *17*, 1200–1209. [CrossRef]
28. Kamei, N.; Onuki, Y.; Takayama, K.; Takeda-Morishita, M. Mechanistic study of the uptake/permeation of cell-penetrating peptides across a caco-2 monolayer and their stimulatory effect on epithelial insulin transport. *J. Pharm. Sci.* **2013**, *102*, 3998–4008. [CrossRef]
29. Birch, D.; Christensen, M.V.; Staerk, D.; Franzyk, H.; Nielsen, H.M. Stereochemistry as a determining factor for the effect of a cell-penetrating peptide on cellular viability and epithelial integrity. *Biochem. J.* **2018**, *475*, 1773–1788. [CrossRef]
30. Maher, S.; Mrsny, R.J.; Brayden, D. Intestinal permeation enhancers for oral peptide delivery. *Adv. Drug Del Rev.* **2016**, *10*, 277–319. [CrossRef]
31. Kristensen, M.; Nielsen, L.H.; Zor, K.; Boisen, A.; Christensen, M.V.; Berthelsen, J.; Mørck Nielsen, H. Cellular effects and delivery propensity of penetratin is influenced by conjugation to parathyroid hormone fragment 1–34 in synergy with pH. *Bioconjug. Chem.* **2018**, *29*, 371–381. [CrossRef] [PubMed]
32. Khafagy, E.S.; Morishita, M.; Isowa, K.; Imai, J.; Takayama, K. Effect of cell-penetrating peptides on the nasal absorption of insulin. *J. Control. Release* **2009**, *133*, 103–108. [CrossRef]
33. Nielsen, E.J.; Yoshida, S.; Kamei, N.; Iwamae, R.S.; Khafagy, E.J.; Olsen, U.L.; Rahbek, B.L.; Pedersen, K.; Takayama, M. In vivo proof of concept of oral insulin delivery based on a co-administration strategy with the cell-penetrating peptide penetratin. *J. Control. Release* **2014**, *189*, 19–24. [CrossRef] [PubMed]
34. Iwase, Y.; Kamei, N.; Khafagy, E.S.; Miyamoto, M.; Takeda-Morishita, M. Use of a non-covalent cell-penetrating peptide strategy to enhance the nasal delivery of interferon beta and its PEGylated form. *Int. J. Pharm.* **2016**, *510*, 304–310. [CrossRef] [PubMed]

35. Derossi, D.; Joliot, A.H.; Chassaing, G.; Prochiantz, A. The third helix of the Antennapedia homeodomain translocates through biological membranes. *J. Biol. Chem.* **1994**, *269*, 10444–10450. [CrossRef]
36. Derossi, D.; Calvet, S.; Trembleau, A.; Brunissen, A.; Chassaing, G.; Prochiantz, A. Cell internalization of the third helix of the antennapedia homeodomain is receptor-independent. *J. Biol. Chem.* **1996**, *271*, 18188–18193. [CrossRef] [PubMed]
37. Illien, F.; Rodriguez, N.; Amoura, M.; Joliot, A.; Pallerla, M.; Cribier, S.; Burlina, F.; Sagan, S. Quantitative fluorescence spectroscopy and flow cytometry analyses of cell-penetrating peptides internalization pathways: Optimization, pitfalls, comparison with mass spectrometry quantification. *Sci. Rep.* **2016**, *6*, 36938. [CrossRef]
38. Birch, D.; Christensen, M.V.; Staerk, D.; Franzyk, H.; Nielsen, H.M. Fluorophore labelling of a cell-penetrating peptide induces differential effects on its cellular distribution and affects cell viability. *Biochim. Biophys. Acta—Biomembr.* **2017**, *1859*, 2483–2494. [CrossRef]
39. Schindelin, J.; Arganda-Carreras, I.; Frise, E.; Kaynig, V.; Longair, M.; Pietzsch, T.; Preibisch, S.; Rueden, C.; Saalfeld, S.; Schmid, B.; et al. Fiji: An open-source platform for biological-image analysis. *Nat. Meth* **2012**, *9*, 676–682. Available online: http://www.nature.com/nmeth/journal/v9/n7/abs/nmeth.2019.html#supplementary-information (accessed on 26 January 2022). [CrossRef]
40. Artursson, P.; Karlsson, J. Correlation between oral drug absorption in humans and apparent drug permeability coefficients in human intestinal epithelial (Caco-2) cells. *Biochem. Biophys. Res. Commun.* **1991**, *175*, 880–885. [CrossRef]
41. Sayers, E.J.; Cleal, K.; Eissa, N.G.; Watson, P.; Jones, A.T. Distal phenylalanine modification for enhancing cellular delivery of fluorophores, proteins and quantum dots by cell penetrating peptides. *J. Control. Release* **2014**, *195*, 55–62. [CrossRef] [PubMed]
42. Watkins, C.L.; Brennan, P.; Fegan, C.; Takayama, K.; Nakase, I.; Futaki, S.; Jones, A.T. Cellular uptake, distribution and cytotoxicity of the hydrophobic cell penetrating peptide sequence PFVYLI linked to the proapoptotic domain peptide PAD. *J. Control. Release* **2009**, *140*, 237–244. [CrossRef] [PubMed]
43. Ghibaudi, E.; Boscolo, B.; Inserra, G.; Laurenti, E.; Traversa, S.; Barbero, L.; Ferrari, R.P. The interaction of the cell-penetrating peptide penetratin with heparin, heparansulfates and phospholipid vesicles investigated by ESR spectroscopy. *J. Pept. Sci.* **2005**, *11*, 401–409. [CrossRef] [PubMed]
44. Ziegler, A.; Seelig, J. Contributions of glycosaminoglycan binding and clustering to the biological uptake of the nonamphipathic cell-penetrating peptide WR9. *Biochemistry* **2011**, *50*, 4650–4664. [CrossRef] [PubMed]
45. Kamei, N.; Aoyama, Y.; Khafagy, E.S.; Henmi, M.; Takeda-Morishita, M. Effect of different intestinal conditions on the intermolecular interaction between insulin and cell-penetrating peptide penetratin and on its contribution to stimulation of permeation through intestinal epithelium. *Eur. J. Pharm. Biopharm.* **2015**, *94*, 42–51. [CrossRef] [PubMed]
46. Kremsmayr, T.; Aljnabi, A.; Blanco-Canosa, J.B.; Tran, H.N.T.; Emidio, N.B.; Muttenthaler, M. On the utility of chemical strategies to improve peptide gut stability. *J. Med. Chem.* **2022**, *65*, 6191–6206. [CrossRef]

Disclaimer/Publisher's Note: The statements, opinions and data contained in all publications are solely those of the individual author(s) and contributor(s) and not of MDPI and/or the editor(s). MDPI and/or the editor(s) disclaim responsibility for any injury to people or property resulting from any ideas, methods, instructions or products referred to in the content.

pharmaceutics

MDPI

Article

The Balance between Hydrophobicity/Aromaticity and Positively Charged Residues May Influence the Cell Penetration Ability

Dóra Soltész [1], Ildikó Szabó [2] and Zoltán Bánóczi [1,*]

[1] Department of Organic Chemistry, Eötvös L. University, 1117 Budapest, Hungary
[2] ELKH-ELTE Research Group of Peptide Chemistry, 1117 Budapest, Hungary
* Correspondence: zoltan.banoczi@ttk.elte.hu

Abstract: Cell-penetrating peptides (CPPs) are commonly modified to increase their cellular uptake, alter the mechanism of penetration or enhance their endosomal release. Earlier, we described the internalization enhancement ability of the 4-((4-(dimethylamino)phenyl)azo)benzoyl (Dabcyl) group. We proved that this modification on the N-terminus of tetra- and hexaarginine enhanced their cellular uptake. The introduction of an aromatic ring 4-(aminomethyl) benzoic acid, AMBA) into the peptide backbone has a synergistic effect with Dabcyl, and the tetraarginine derivatives had outstanding cellular uptake. Based on these results, the effect of Dabcyl or Dabcyl-AMBA modification on the internalization of oligoarginines was studied. Oligoarginines were modified with these groups and their internalization was measured using flow cytometry. The concentration dependence of the cellular uptake of selected constructs was compared too. Their internalization mechanism was also examined by using different endocytosis inhibitors. While the effect of the Dabcyl group was optimal for hexaarginine, the Dabcyl-AMBA group increased the cellular uptake in the case of all oligoarginines. All derivatives, with the exception of only tetraarginine, were more effective than the octaarginine control. The internalization mechanism was dependent on the size of the oligoarginine and was independent of the modification. Our findings suggest that these modifications enhanced the internalization of oligoarginines and resulted in novel, very effective CPPs.

Keywords: cell-penetrating peptide; oligoarginines; aromatic modification

Citation: Soltész, D.; Szabó, I.; Bánóczi, Z. The Balance between Hydrophobicity/Aromaticity and Positively Charged Residues May Influence the Cell Penetration Ability. *Pharmaceutics* **2023**, *15*, 1267. https://doi.org/10.3390/pharmaceutics15041267

Academic Editors: Prisca Boisguérin and Sébastien Deshayes

Received: 15 March 2023
Revised: 5 April 2023
Accepted: 13 April 2023
Published: 18 April 2023

Copyright: © 2023 by the authors. Licensee MDPI, Basel, Switzerland. This article is an open access article distributed under the terms and conditions of the Creative Commons Attribution (CC BY) license (https://creativecommons.org/licenses/by/4.0/).

1. Introduction

Cell-penetrating peptides (CPPs) are promising tools for the delivery of biologically active compounds [1] e.g., small drug molecules [2–5], peptides, [6–8], proteins [9–11] and oligonucleotides [12,13]. Although they can transport many kinds of molecules across the cellular membrane, they have some critical drawbacks, such as endosomal entrapment. This results in a very low cytosolic delivery efficiency [14]. Thus, there is a focus on efforts to improve the direct penetration of CPPs and/or increase their endosomal escape. The direct penetration can be increased e.g., by the modification of well-known CPPs with fatty acid acylation [15] or using the 4-((4-(dimethylamino)phenyl)azo)benzoyl group (Dabcyl) as an internalization enhancer [16,17], or Dabcyl and aromatic residues [18,19]. The endosomal escape may be enhanced by inserting lytic peptides into the sequence [20].

We and Mandal et al. have described how the Dabcyl group enhances the internalization of cyclodecaarginine as a well-known CPP [17] and short non-penetrating oligoarginines [16]. Although the cellular uptake of tetraarginine was increased dramatically, it was far less efficient than octaarginine, a commonly used CPP. However, it was better than unmodified hexaarginine and its internalization showed a high extent of direct penetration at a low concentration. The efficiency of oligoarginines may be improved by further modifications [21]. The addition of tryptophan may enhance the internalization. In the case of Tat48-60, the Trp substitution of Pro resulted in an increased uptake [22], while replacing Trp with Leu diminished the uptake [23]. As Trp has a positive effect on the uptake of CPPs, its role has been extensively studied [24–26].

To increase the cellular uptake of the Dabcyl-Arg$_4$ peptide and retain its direct penetration ability we modified its sequence by the insertion of Trp residue. [18] The insertion of Trp into the N-terminus or middle of the tetraarginine resulted in a higher cellular uptake, while the direct penetration was more pronounced. As the aromatic side chain of the tryptophan had a high impact on the cellular uptake, we replaced it with some other aromatic unnatural amino acids, such as 4-aminobenzoic acid (PABA), 4-(aminomethyl) benzoic acid (AMBA) and 6-amino-2-naphthoic acid (NAPH) [19]. These residues were applied at the N-terminus or in the middle of the peptides instead of the Trp. The NAPH modification only retained the activity of the Trp-containing derivatives, but the PABA and AMBA modification at the N-terminus significantly increased the internalization in comparison with the Trp derivatives. These derivatives were 2–5 times better than the octaarginine.

In this study we synthesized oligoarginines with AMBA, Dabcyl and Dabcyl-AMBA modifications at their N-terminus and examined their cellular uptake on EBC-1 cells. We also examined the influence of the position of Dabcyl on cell penetration. We found that the modification effect was dependent on the number of arginine residues, and thus it seems to be determined by the hydrophobicity–hydrophilicity of the peptides. These modifications not only altered the extent of the internalization, but they had an effect on the mechanism of the internalization too.

2. Materials and Methods

2.1. Synthesis of Peptides

The peptides were synthesized manually by solid phase peptide synthesis (SPPS) on a Rink amide MBHA resin, using the Fmoc/tBu strategy. The arginine side chain was protected with 2,2,4,6,7-pentamethyldihydrobenzofuran-5-sulfonyl (Pbf), while in the case of lysine, the tert-butyloxycarbonyl (Boc) group was used. The cleavage solution used for the removal of the Fmoc-protecting group contained 2% piperidine and 2% 1,8-diazabicyclo [5.4.0]undec-7-ene (DBU) in N,N-dimethylformamide (DMF); the cleavage lasted 2 + 2 + 5 + 10 min according to the standard protocol. Extensive washing (8 × 1 min) was performed after the Fmoc-removal step to wash out any remaining cleavage solution. The coupling of the amino acid derivatives was carried out with N,N′-diisopropylcarbodiimide (DIC) and ethyl (hydroxyimino)cyanoacetate (OxymaPure) coupling reagents (three equimolar excess of each and the amino acid also) in DMF at room temperature (RT) for 60 min. Following the coupling, the resin was washed with DMF (2 × 1 min) and dichloromethane (DCM) (3 × 1 min). To ascertain the success of the coupling, the Kaiser test was conducted, and in the case of a negative test result (successful coupling), the procedure was repeated until the given peptide sequence was finished. The removal of the last Fmoc group was followed by the attachment of Fmoc-AMBA, Dabcyl or 5(6)-carboxyfluorescein (Cf) to the N-terminal amino acid on the resin using DIC/OxymaPure coupling reagents. In the case of the Dabcyl-AMBA peptides, the Fmoc-AMBA was deprotected and Dabcyl was coupled to the free amino group of AMBA. Peptides without one of these modifications were acylated on their N-terminal on the resin with a DMF solution containing acetic acid anhydride and DIEA. The peptides were cleaved from the resin with 5 mL TFA containing 0.365 g phenol, 0.25 mL distilled water, 0.25 mL thioanisole and 0.125 mL 1,2-ethanedithiol as scavengers. The obtained crude products were precipitated by dry diethyl-ether, dissolved in 10% acetic acid, lyophilized and subsequently purified by semi-preparative RP-HPLC. The purified compounds were characterized by analytical RP-HPLC and ESI-MS.

The fluorescent dye Cf or the Dabcyl molecule (2 eq relative to the peptide) was coupled to the ε-amino group of the C-terminal Lys residue of the peptides in a DMF solution containing 1 eq OxymaPure, 1 eq DIC and 5 eq DIEA, and the reaction took place overnight. Eluent A and a small amount of eluent B, if needed, was added to the reaction mixture and purified by RP-HPLC.

2.2. Determination of the Cellular Uptake by Flow Cytometry

For the examination of the internalization of the peptides, 10^5 cultured EBC-1 cells per well were plated on 24-well plates. The cells were incubated for 24 h at 37 °C then they were treated with peptides at different concentrations for 90 min in a serum-free medium. In the negative control the cells were treated with a serum-free medium. After the incubation, the peptide solutions were removed and 100 μL trypsin (0.25%) was added for 5–10 min to remove membrane bound peptides and detach the adherent cells from the plates. The activity of trypsin was terminated by the addition of 900 μL HPMI buffer (glucose, $NaHCO_3$, NaCl, HEPES, KCl, $MgCl_2$, $CaCl_2$, $Na_2HPO_4{\cdot}2H_2O$) containing 10% fetal bovine serum, and the cells were transferred from the plates to FACS-tubes for measurement. The cells were centrifuged at 216× *g* at 4 °C for 5 min and the supernatant was removed. The cells were resuspended in 250 μL HPMI, and the fluorescence intensity of the cells was quantified using flow cytometry (BD LSR II, BD Bioscience, San Jose, CA, USA). The data were analyzed with FACSDiva 5.0 software (BD Bioscience, San Jose, CA, USA).

The effect of the inhibitors was studied following the above-mentioned protocol, but with the pretreatment of cells with an inhibitor for 30 min, following treatment with a peptide at 5 μM. The cells were then incubated for 90 min at 37 °C. Macropinocytosis was inhibited using 5-(N-ethyl-N-isopropyl)amiloride (EIPA) [27], the clathrin-mediated endocytosis was prevented with chlorpromazine (CPZ) [28], methyl-β-cyclodextrin (CyD) was applied for the inhibition of caveolae/lipid raft-mediated endocytosis [29] and colchicine (Col) was used to ascertain the role of microtubules and thus the importance of pinocytosis [30].

2.3. Statistical Analysis

The results of the uptake studies are indicated as the mean value ± standard deviation. The Student's *t* test was used for the statistical analysis. Results with p values < 0.05 were regarded as statistically significant.

3. Results

3.1. Synthesis of Peptides

Our earlier results showed that both the Dabcyl and the aromatic residue have an exceptional cellular uptake enhancing effect on oligoarginines [8,16,17,31]. In these studies, hexa- and tetraarginines were modified to increase their internalization, thus we decided to investigate the effect of the number of arginine residues on the cell penetration of these kinds of constructs. In one of our previous peptide studies both groups were used, thus we planned to compare the effect of AMBA- and Dabcyl-AMBA modification based on the promising findings of our research group concerning the Dabcyl-AMBA-$(Arg)_4$-Lys(Cf) peptide, among others [19]. Therefore, peptides with different numbers of arginines in their sequence were prepared, and their N-termini were modified with Dabcyl, AMBA or Dabcyl-AMBA (Figure 1). As the synthetic aspect may cause a change in the positions of modifications in some cases, we also studied the impact of this change and studied peptides with the Dabcyl group at their C-terminal lysine side chain where the fluorescent molecule is located in general. To the best of our knowledge, relocating the Dabcyl molecule from the N-terminus is unique and the impact of this change in position on the internalization of peptides has not previously been investigated. The peptides were all synthesized manually by solid-phase peptide synthesis on a Rink-amide MBHA resin using the Fmoc/tBu strategy. The coupling reagents were DIC and OxymaPure, and the modification of the N-terminus (Dabcyl, AMBA or 5(6)-carboxyfluorescein (Cf)) was performed on the resin using the same coupling reagents. The commercially available AMBA was Fmoc-protected in advance according to a previously established method [19]. For the uptake studies the fluorescent molecule (Cf) was coupled either to the ε-amino group of an inserted Lys at the C-terminus in a solution, or to the N-terminus on a solid phase, such as the Dabcyl group.

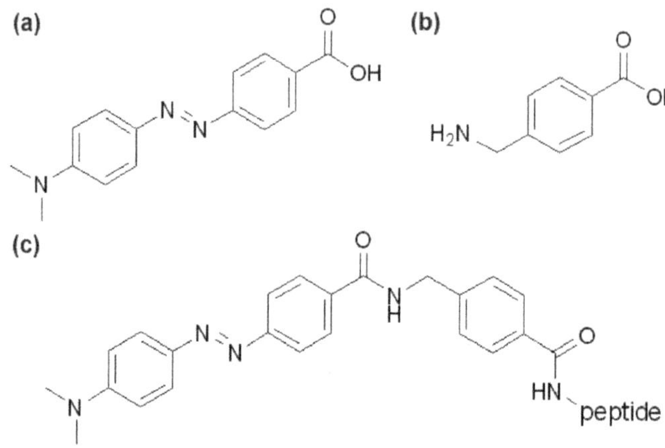

Figure 1. Structures of (**a**) Dabcyl, (**b**) AMBA, and (**c**) the Dabcyl-AMBA-peptides.

The characterization of the purified peptide conjugates was performed by analytical RP-HPLC and ESI-MS (Table 1; the analytical RP-HPLC chromatograms and MS spectra can be found in the Supplementary Materials, Figures S1–S38).

Table 1. Chemical characterization of the peptides.

Sequence	R_t	M_{calc}	M_{meas}
Cf-Arg_8	12.0	1623.9	1623.9
Ac-AMBA-$(Arg)_4$-Lys(Cf)	12.0	1302.6	1302.6
Ac-AMBA-$(Arg)_5$-Lys(Cf)	12.0	1458.7	1458.7
Ac-AMBA-$(Arg)_6$-Lys(Cf)	11.9	1614.8	1614.8
Ac-AMBA-$(Arg)_7$-Lys(Cf)	11.8	1770.9	1770.9
Ac-AMBA-$(Arg)_8$-Lys(Cf)	11.8	1927.0	1927.0
Dabcyl-$(Arg)_4$-Lys(Cf) [a]	13.9	1378.7	1378.3
Dabcyl-$(Arg)_5$-Lys(Cf) [a]	13.5	1534.8	1534.9
Dabcyl-$(Arg)_6$-Lys(Cf) [a]	13.0	1690.9	1690.9
Dabcyl-$(Arg)_7$-Lys(Cf) [a]	12.9	1847.0	1847.0
Dabcyl-$(Arg)_8$-Lys(Cf) [a]	13.3	2003.0	2003.0
Dabcyl-AMBA-$(Arg)_4$-Lys(Cf)	15.1	1511.7	1511.7
Dabcyl-AMBA-$(Arg)_5$-Lys(Cf)	14.9	1667.8	1667.8
Dabcyl-AMBA-$(Arg)_6$-Lys(Cf)	14.6	1823.9	1823.9
Dabcyl-AMBA-$(Arg)_7$-Lys(Cf)	14.3	1980.0	1980.0
Dabcyl-AMBA-$(Arg)_8$-Lys(Cf)	14.3	2136.1	2136.1
Cf-$(Arg)_4$-Lys(Dabcyl) [a]	14.2	1378.7	1378.7
Cf-$(Arg)_5$-Lys(Dabcyl) [a]	13.9	1534.9	1534.8
Cf-$(Arg)_6$-Lys(Dabcyl) [a]	13.7	1690.9	1690.9

Analytical RP-HPLC was performed on the Jupiter C18 column (4.6 mm × 150 mm, 3 μm, 300 Å). The applied linear gradient elution was 0 min 0% B, 2 min 0% B and 22 min 90% B at a 1 mL/min flow rate. The detection was conducted at λ = 220 nm. [a] analytical RP-HPLC was performed on the Hypersil Hypurity C18 column (4.6 mm × 150 mm, 5 μm, 190 Å). The applied linear gradient elution was 0 min 0% B, 2 min 0% B and 22 min 90% B at a 1 mL/min flow rate. The detection was conducted at λ = 220 nm. The mass spectrometric analysis was performed on a Bruker Amazon SL (Bremen, Germany). The samples were dissolved in acetonitrile-water (50:50, *v*/*v*), containing 0.1% formic acid.

3.2. Cellular Uptake

The cellular uptake of the peptides was determined using flow cytometry on EBC-1 human lung squamous carcinoma cells. The peptides were added to the cells at a concentration of 5 μM for 90 min at 37 °C. Selected constructs were used to study the concentration dependence at 0.125, 1.25 and 2.5 μM with 90 min treatment times. When evaluating the measurements data, the fluorescence of cells treated with peptides was corrected with the autofluorescence of cells without peptide treatment. The toxicity of the peptide conjugates was determined by comparing the live and dead cells ratio during the flow cytometry analysis. Based on this, none of the conjugates showed cytotoxicity at the highest concentration used (Supplementary Materials Figure S39).

In order to examine the effect of the hydrophobic AMBA group on the cell penetration of oligoarginines, we synthesized AMBA-modified peptides (Table 1). Our preliminary measurement showed that the AMBA group alone is unable to enhance the cellular uptake of oligoarginines (Supplementary Materials Figure S40).

In the next step, the effect of Dabcyl or the Dabcyl-AMBA modification was compared. All the constructs had a similar or higher cellular uptake than the octaarginine, except for the Dabcyl-Arg_4-Lys(Cf). In the case of the shortest oligoarginine (tetraarginine), the Dabcyl-AMBA modification increased the internalization better than the Dabcyl alone. Adding an extra arginine (pentaarginine) abolished the difference between the influence of the two modifications. Increasing the number of arginines (hexa-and heptaarginine) resulted in a higher cellular uptake of the Dabcyl-modified peptides. Interestingly, in the case of the longest oligoarginine (octaarginine) studied, the order of the internalization of the two modified peptides was reversed; the Dabcyl-AMBA was more efficient. Nonetheless, the Dabcyl-AMBA-$(Arg)_8$-Lys(Cf), Dabcyl-$(Arg)_6$-Lys(Cf) and Dabcyl-$(Arg)_7$-Lys(Cf) were prominent among the peptides, and their cellular uptake was higher, ~8 times, 6.5 times and 6 times, respectively, than the uptake of the control octaarginine (Figure 2).

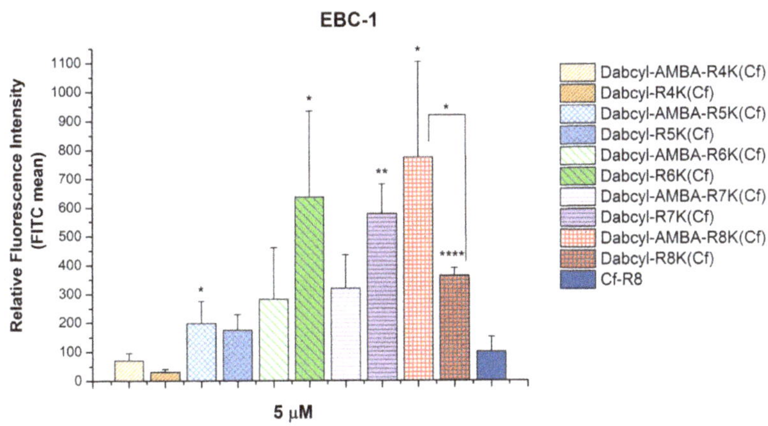

Figure 2. Comparing the effect of Dabcyl and the Dabcyl-AMBA group on the internalization of short oligoarginines into EBC-1 cells. The cells were treated with the peptides at a concentration of 5 μM at 37 °C for 90 min. After trypsinization the fluorescence intensity of the cells was studied using flow cytometry. The fluorescence intensities were normalized to the fluorescence intensity of cells that were treated with Cf-Arg_8 (100%). Any significant difference between the control Cf-Arg_8 and peptides with the same amount of arginine was measured using the Student's *t* test. The asterisks show a significant difference between the control octaarginine and the two modified octaarginines (* $p < 0.05$, ** $p < 0.01$, **** $p < 0.0001$).

We studied whether the arrangement of the internalization enhancer (Dabcyl) and the fluorescence labelling (5(6)-carboxyfluorescein) in tetra-, penta- and hexaarginine derivatives has an influence on the cellular uptake. There was an unexpected and marked difference only in the internalization of the hexaarginine derivatives: the Cf-$(Arg)_6$-Lys(Dabcyl) peptide showed a surprisingly poor cellular uptake (Figure 3), while the tetra- and pentaarginine derivatives were not sensitive to the position of Dabcyl.

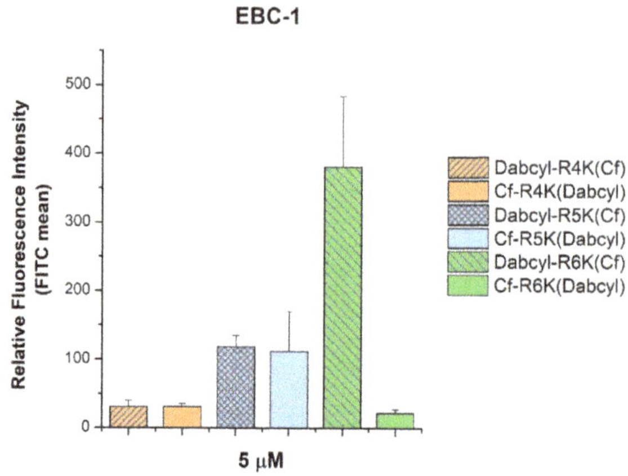

Figure 3. Effect of the position of Dabcyl and Cf on the cellular uptake of short oligoarginines into EBC-1 cells. The cells were treated with peptides at a concentration of 5 μM at 37 °C for 90 min. After trypsinization the fluorescence intensity of cells was determined using flow cytometry. The fluorescence intensities were normalized to the fluorescence intensity of cells that were treated with Cf-Arg_8 (100%). Any significant difference between the control Cf-Arg_8 and peptides with the same amount of arginine was measured using the Student's t test. No significant difference was found.

The concentration dependence of the cellular uptake was studied in the cases of the following selected peptides: Dabcyl-$(Arg)_6$-Lys(Cf), Cf-$(Arg)_6$-Lys(Dabcyl), Dabcyl-AMBA-$(Arg)_5$-Lys(Cf) and Dabcyl-AMBA-$(Arg)_8$-Lys(Cf). EBC-1 cells were treated with the peptides at concentrations of 0.125, 1.25 and 2.5 μM (Figure 4). The Dabcyl-AMBA-$(Arg)_8$-Lys(Cf) peptide demonstrated outstanding cell penetration at 2.5 μM: it was 30 times more effective in the cellular uptake than octaarginine, and, even at a concentration of 1.25 μM, it exceeded the internalization of Cf-Arg_8 (~6 times higher uptake). The two other peptides—Dabcyl-AMBA-$(Arg)_5$-Lys(Cf) and Dabcyl-$(Arg)_6$-Lys(Cf)—showed a similar or better uptake intensity than octaarginine. The Cf-$(Arg)_6$-Lys(Dabcyl) peptide was the only one which showed a significantly decreased uptake compared to Cf-$(Arg)_8$.

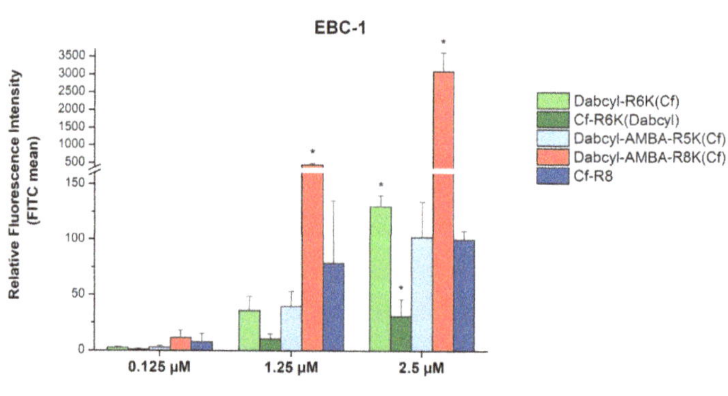

Figure 4. Cellular uptake of selected peptides by EBC-1 cells. The cells were treated with 0.125, 1.25 and 2.5 μM peptide solutions at 37 °C for 90 min. After trypsinization the fluorescence intensity of the cells was studied using flow cytometry. The fluorescence intensities were normalized to the fluorescence intensity of cells that were treated with Cf-Arg_8 at a concentration of 2.5 μM (100%). Any significant difference from octaarginine control of a given concentration was measured using the Student's *t* test (* $p < 0.05$).

3.3. Studying the Mechaanism of Internalization

The above-mentioned selected peptides were further examined concerning their preferred cellular uptake mechanisms. The EBC-1 cells were preincubated with various endocytic inhibitors, thereby investigating their effects on the internalization of peptides (Figure 5). In these experiments the indirect macropinocytosis inhibitor 5-(N-ethyl-N-isopropyl)amiloride (EIPA) [27], the clathrin-mediated endocytosis inhibitor chlorpromazine (CPZ) [28], the caveolae-mediated endocytosis inhibitor methyl-beta-cyclodextrin (mBCD) [29] and the pinocytosis inhibitor colchicine [30] were used.

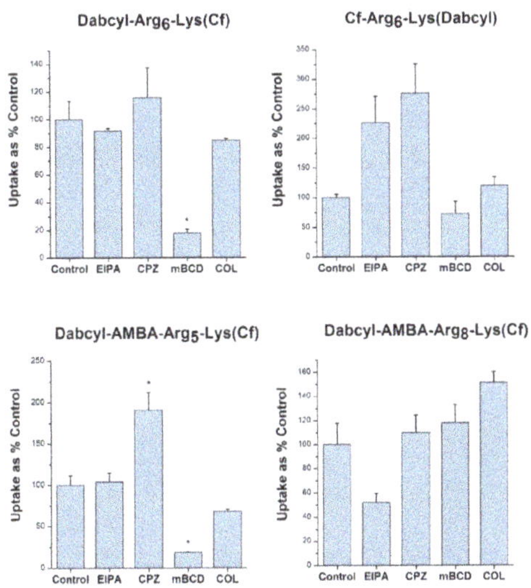

Figure 5. The role of different endocytic routes in the cellular uptake of peptides. EBC-1 cells were pretreated with the inhibitors EIPA (50 μM), CPZ (30 μM), mBCD (5 mM) and COL (20 mM) for 30 min, then the cells were treated with the peptide conjugates (5 μM) for 90 min. Any significant difference from the control was measured using the Student's *t* test (* $p < 0.05$).

The data demonstrated that EIPA remarkably decreased the cellular uptake only of Dabcy-AMBA-$(Arg)_8$-Lys(Cf) (to 51.7% of the untreated control). Chlorpromazine did not negatively affect the internalization of either of the peptides. The most remarkable effect was caused by methyl-beta-cyclodextrin (mBCD), which considerably reduced the uptake of Dabcyl-$(Arg)_6$-Lys(Cf) and Dabcyl-AMBA-$(Arg)_5$-Lys(Cf) (to 18.1% and to 18.7% of the control, respectively), and only moderately hampered the internalization of Cf-$(Arg)_6$-Lys(Dabcyl) (to 72.6%). The microtubule inhibitor colchicine decreased the cellular entry of Dabcyl-AMBA-$(Arg)_5$-Lys(Cf) and Dabcyl-$(Arg)_6$-Lys(Cf) to a smaller degree (to 68.3% and to 85.1% of the control, respectively).

4. Discussion

The interesting cellular uptake-enhancing effect of the Dabcyl group was further investigated in this study. Based on our previous results concerning the possible superior effect of the Dabcyl-AMBA tandem group compared to Dabcyl alone [19], we planned to analyze the effect of this construct on oligoarginines with different lengths. In the case of shorter oligoarginines, one can assume that the aromaticity/hydrophobicity has an enhanced role in driving the internalization compared to the positive character of the peptide, while the uptake of longer Dabcyl-AMBA-oligoarginines has a more pronounced charge-driven factor (ion-pair formation with glycosaminoglycans and with phosphate groups of membrane lipids). Our constructs with AMBA at their N-terminus did not show increased internalization; however, their cellular uptake was increased by the increasing number of arginine residues. When Dabcyl was introduced into the oligoarginines or AMBA-oligoarginines, its effect on the cellular uptake showed a dependence on the length of oligoarginines at a concentration of 5 μM (Figure 2). The effect of the Dabcyl group alone had a maximum value at the hexaarginine. The addition of more arginine residues decreased its effectiveness. It seems that the ratio of positive charges and hydrophobicity caused by aromatic moiety has an optimum in this arrangement (the N-terminus is hydrophobic, the C-terminus is positively charged). Earlier oligoarginines were acylated by a fatty acid and those results indicated that increasing the arginine residues enhances the cellular uptake up to Arg_{11} [32,33]. While these results correlated well with the internalization efficiency of oligoarginines, which reaches its maximum value at Arg_8-Arg_{12} [34,35], our findings suggest that there is an optimal balance between the aromatic moieties and the positively charged arginine residues in the case of less than eight arginines. The insertion of AMBA into the Dabcyl-oligoarginine peptides resulted in enhanced cell penetration compared to the unmodified peptides, independently of the length of the oligoarginine. The rise in internalization was linear until heptaarginine, while in the case of octaarginine there was an exponential increase in the cellular uptake.

In the case of tetraarginine, the double modification was two times better than Dabcyl alone and resulted in a derivative with the same efficiency as octaarginine. When the positive charge of the peptide was increased with the addition of an extra arginine (pentaarginine derivative), the Dabcyl alone was enough to achieve the same internalization. In this case, the addition of the extra aromatic group did not result in any synergistic effect. However, both derivatives showed a two times higher cellular uptake than the octaarginine. By increasing the amount of arginine residues (hexa- and heptaarginine) both the Dabcyl and Dabcyl-AMBA modifications increased the cellular uptake and resulted in better CPPs than the octaarginine, but Dabcyl alone was more efficient (the internalization of these derivatives was two times higher than those of the Dabcyl-AMBA peptides). Although the insertion of the AMBA increased the hydrophobicity of the Dabcyl-peptides based on the retention times, it did not enhance the cellular uptake. The effect of the Dabcyl-AMBA group compared to Dabcyl alone suggests that in the case of Dabcyl, not excluding its hydrophobic aromatic nature is the determining factor concerning the internalization enhancement, as was previously hypothesized [17], and as is the case in fatty acid-conjugated CPPs [36]. This corresponds to the observations of our group: when comparing the Dabcyl-PABA and Dabcyl-NAPH modified tetraarginines, the latter was inferior in uptake efficiency, despite its more hydrophobic character and the more extended electron delocalization [19]. This concurs with the findings of Mandal et. al. [17], who found that the Black Hole Quencher 2 (BHQ2), which contains an extra benzene ring connected with azo bonds, has no significant effect, similar to Dabcyl when conjugated to a cyclic CPP and its cargo. Recently, a novel Dabcyl derivative, azido-Dabcyl conjugated to cyclic decaarginine, was found to be three times more efficient in delivering ubiquitin into U2OS cells than the Dabcyl-conjugated CPP or a more hydrophobic Dabcyl derivative-conjugated construct [10]. Therefore, there might be other factors besides the hydrophobicity giving Dabcyl its special effect, and the possibility that extended electron delocalization and electronic or other effects together are essential for the uptake enhancement. It is worth mentioning that the Dabcyl-$(Arg)_4$-Lys(Cf) peptide showed lower cellular uptake than octaarginine (Figure 2), such as in the case of the HL-60 cells [16], although on HL-60 cells it was less efficient than on EBC-1 cells, which suggests an important cell-line dependence of the cellular uptake.

In the case of peptide conjugates, the position of the attached molecules (N- or C-terminus) may have a high impact on the biological activity. Therefore, we studied the cellular uptake of peptides containing the Dabcyl group at the N- or C-terminus (Figure 3). We noticed that its position has a strong influence only on the internalization of hexaarginine, the longest peptide studied. The alteration of the position increased the HPLC retention times of the derivatives. The increment was bigger in the case of longer peptides. The retention times of Dabcyl-containing derivatives correlates very well with their cellular uptake. A higher retention time means lower internalization at a concentration of 5 μM. The reverse arrangement of Dabcyl and Cf in the hexaarginine derivatives resulted in an increased retention time and a very low cellular uptake. Its retention time became similar to the retention time of Dabcyl-Arg_4-Lys(Cf). Based on the change in the retention time, it can be suggested that the position of the two groups has an influence on the structure of the peptides and thus on their retention time and cellular uptake.

Although the Dabcyl-AMBA-$(Arg)_8$-Lys(Cf) and Dabcyl-$(Arg)_6$-Lys(Cf) peptides had very similar internalization at a concentration of 5 μM, they had a very altered cellular uptake at low concentrations. It was highlighted that the Dabcyl-AMBA-$(Arg)_8$-Lys(Cf) peptide had exceptional efficiency and intense uptake at relatively low concentrations (30 times and ~6 times more effective than octaarginine at 2.5 μM and 1.25 μM concentrations, respectively). While the cellular uptake of the Dabcyl-AMBA-$(Arg)_5$-Lys(Cf) and Dabcyl-$(Arg)_6$-Lys(Cf) peptides were different at 5 μM (the hexaarginine derivative was three times better), their internalization was not significantly different at these low concentrations and was similar to those of octaarginine. However, Cf-$(Arg)_6$-Lys(Dabcyl) showed a significantly lower cellular uptake than Cf-Arg_8 at 2.5 μM, which reinforces previous findings concerning the inferiority of this peptide compared to its alternative (Dabcyl-$(Arg)_6$-Lys(Cf)) and octaarginine.

The investigation of the preferred endocytic routes of selected peptides revealed important internalization properties. In the case of the outstandingly effective octaarginine derivative, Dabcyl-AMBA-$(Arg)_8$-Lys(Cf), the results obtained with the utilization of various endocytic inhibitors suggest that this peptide mainly enters cells by macropinocytosis, as EIPA, the inhibitor of this pathway, decreased its uptake by half, while the other inhibitors had no deleterious effect on the internalization. The octaarginine part of the peptide can establish bidentate hydrogen bonds with proteoglycans on the cell surface, thereby accumulating on the membrane and inducing actin polymerization and macropinocytosis. The role of the Dabcyl-AMBA part of the construct is currently unknown. This effect of micropinocytosis inhibition was noticed in the case of octaarginine too [37]. This showed that the Dabcyl-AMBA part enhances the internalization but does not alter the route of cellular uptake. The Dabcyl-$(Arg)_6$-Lys(Cf) peptide was significantly affected by mBCD, revealing caveolae/lipid raft-mediated endocytosis as the main endocytic entry mode, similar to the peptide Dabcyl-AMBA-$(Arg)_5$-Lys(Cf). This is in harmony with previous findings concerning the Dabcyl-AMBA-$(Arg)_4$-Lys(Cf) peptide on the MDA-MB-231 cell line [19]. Caveolae/lipid raft-mediated endocytosis is the preferred type of endocytosis due to the possibility of avoiding lysosomal entrapment, thereby the peptide or peptide-cargo can reach the cytosol in an intact form [38]. The different effect of EIPA on the internalization of Dabcyl-$(Arg)_6$-Lys(Cf) in the case of EBC-1 and HeLa cells [16] may show cell type-dependent internalization. However, on HeLa cells only the first 10 min of internalization was examined, whereas here the cells were studied after 90 min of treatment. The uptake of Dabcyl-AMBA-$(Arg)_5$-Lys(Cf) was also inhibited by COL, but not as significantly as by mBCD, suggesting only a minor role of pinocytosis in the endocytic internalization. The difference between the preferred endocytic routes of the Dabcyl-AMBA oligoarginines denotes that the amount of arginine residue (thus the more hydrophobic or hydrophilic the character of the peptide) is more influential concerning the mechanism of uptake than the presence of the Dabcyl or Dabcyl-AMBA group. This finding is somewhat contrary to the findings of Zhang et al. who examined the mechanistic properties of liposome-mediated dermal delivery [39]. Their work showed that deoxycholate-mediated liposomes (DOC-LS) used the same endocytosis pathways as standard liposomes, despite their different surface charge and particle size. Here, oligoarginines with the same modification but a different charge and length (Dabcyl-AMBA-$(Arg)_5$-Lys(Cf) and Dabcyl-AMBA-$(Arg)_8$-Lys(Cf)) entered cells through diverse endocytic pathways. It is worth mentioning that in DOC-LS the bile salts provide a negative surface charge to the liposome, while in our case the positive charge of the peptide increased with more arginine residues (and simultaneously the hydrophilicity of the peptide increased as well), and this has an effect on the internalization mechanism, presumably caused by interactions with anionic structures on the membrane surface. Interestingly, the cellular entry of peptides can also be markedly enhanced when co-incubated with certain endocytosis inhibitors. This phenomenon can be explained by the theory that if one endocytic route is blocked, other pathways which lead to more effective internalization are facilitated.

Future studies should analyze the contribution of energy-independent internalization on the net cellular uptake of peptides, as well as the cellular distribution; thereby, we can obtain a more detailed picture of the cellular entry modes that the peptides can utilize. Another thing worth examining is the effect of Dabcyl on the membrane dipole potential. It was shown that certain CPPs can be taken up by some cell lines more intensely when the membrane dipole potential is reduced [19,40]. It is also known that the intercalation of dipolar molecules in the membrane affects its dipole potential [41], and the membrane dipole potential can influence molecule–membrane interactions [42]. Based on these findings, it is reasonable to assume that Dabcyl might be able to reduce the membrane dipole potential and thereby enhance the cellular uptake of peptides. However, in the case of the shorter Dabcyl-conjugated oligoarginines, which enter cells mainly via lipid raft-mediated endocytosis, there is an apparent contradiction: treating cells with mBCD lowers the membrane dipole potential [43], yet the uptake of peptides is reduced. Therefore, it is an interesting question as to what is the mechanism by which Dabcyl exerts its special effect.

Supplementary Materials: The following supporting information can be downloaded at: https://www.mdpi.com/article/10.3390/pharmaceutics15041267/s1, Figures S1–S19: HPLC chromatogram of compounds. Figures S20–S38: MS Spectrum of compounds.; Figure S39: Determination of the cytotoxicity of peptide conjugates by examining the percentage of live cells by flow cytometry.; Figure S40: Cellular uptake of AMBA-modified oligoarginines.

Author Contributions: Conceptualization, Z.B. and I.Sz.; methodology, D.S., Z.B. and I.Sz.; investigation, I.Sz. and D.S.; writing—original draft preparation, D.S; writing—review and editing, Z.B. and I.Sz.; visualization, D.S.; supervision, Z.B. All authors have read and agreed to the published version of the manuscript.

Funding: We kindly appreciate the support from the Hevesy György PhD school of Chemistry, Eötvös Loránd University and the support from the ELTE Thematic Excellence Programme supported by the Hungarian Ministry for Innovation and Technology.

Institutional Review Board Statement: Not applicable.

Informed Consent Statement: Not applicable.

Data Availability Statement: Not applicable.

Conflicts of Interest: The authors declare no conflict of interest.

References

1. Hudecz, F.; Bánóczi, Z.; Csík, G. Medium-Sized Peptides as Built in Carriers for Biologically Active Compounds. *Med. Res. Rev.* **2005**, *25*, 679–736. [CrossRef] [PubMed]
2. Szabó, I.; Orbán, E.; Schlosser, G.; Hudecz, F.; Bánóczi, Z. Cell-Penetrating Conjugates of Pentaglutamylated Methotrexate as Potential Anticancer Drugs against Resistant Tumor Cells. *Eur. J. Med. Chem.* **2016**, *115*, 361–368. [CrossRef] [PubMed]
3. Bánóczi, Z.; Gorka-Kereskényi, Á.; Reményi, J.; Orbán, E.; Hazai, L.; Tokési, N.; Oláh, J.; Ovádi, J.; Béni, Z.; Háda, V.; et al. Synthesis and In Vitro Antitumor Effect of Vinblastine Derivative- Oligoarginine Conjugates. *Bioconjug. Chem.* **2010**, *21*, 1948–1955. [CrossRef] [PubMed]
4. Bánóczi, Z.; Peregi, B.; Orbán, E.; Szabó, R.; Hudecz, F. Synthesis of Daunomycin-Oligoarginine Conjugates and Their Effect on Human Leukemia Cells (HL-60). *Arkivoc* **2008**, *3*, 140–153. [CrossRef]
5. Bánóczi, Z.; Keglevich, A.; Szabó, I.; Ranđelović, I.; Hegedüs, Z.; Regenbach, F.L.; Keglevich, P.; Lengyel, Z.; Gorka-Kereskényi, Á.; Dubrovay, Z.; et al. The Effect of Conjugation on Antitumor Activity of Vindoline Derivatives with Octaarginine, a Cell-Penetrating Peptide. *J. Pept. Sci.* **2018**, *24*, e3118. [CrossRef]
6. Bánóczi, Z.; Alexa, A.; Farkas, A.; Friedrich, P.; Hudecz, F. Novel Cell-Penetrating Calpain Substrate. *Bioconjug. Chem.* **2008**, *19*, 1375–1381. [CrossRef]
7. Bánoczi, Z.; Tantos, Á.; Farkas, A.; Tompa, P.; Friedrich, P.; Hudecz, F. Synthesis of Cell-Penetrating Conjugates of Calpain Activator Peptides. *Bioconjug. Chem.* **2007**, *18*, 130–137. [CrossRef]

8. Alexa, A.; Ember, O.; Szabó, I.; Mo'ath, Y.; Póti, Á.L.; Reményi, A.; Bánóczi, Z. Peptide Based Inhibitors of Protein Binding to the Mitogen-Activated Protein Kinase Docking Groove. *Front. Mol. Biosci.* **2021**, *8*, 629. [CrossRef]
9. Sauter, M.; Strieker, M.; Kleist, C.; Wischnjow, A.; Daniel, V.; Altmann, A.; Haberkorn, U.; Mier, W. Improving Antibody-Based Therapies by Chemical Engineering of Antibodies with Multimeric Cell-Penetrating Peptides for Elevated Intracellular Delivery. *J. Control. Release* **2020**, *322*, 200–208. [CrossRef]
10. Saha, A.; Mandal, S.; Arafiles, J.V.V.; Gómez-González, J.; Hackenberger, C.P.R.; Brik, A. Structure–Uptake Relationship Study of DABCYL Derivatives Linked to Cyclic Cell-Penetrating Peptides for Live-Cell Delivery of Synthetic Proteins. *Angew. Chem. Int. Ed.* **2022**, *61*, e202207551. [CrossRef]
11. Patel, S.G.; Sayers, E.J.; He, L.; Narayan, R.; Williams, T.L.; Mills, E.M.; Allemann, R.K.; Luk, L.Y.P.; Jones, A.T.; Tsai, Y.H. Cell-Penetrating Peptide Sequence and Modification Dependent Uptake and Subcellular Distribution of Green Florescent Protein in Different Cell Lines. *Sci. Rep.* **2019**, *9*, 6298. [CrossRef]
12. Klein, A.F.; Varela, M.A.; Arandel, L.; Holland, A.; Naouar, N.; Arzumanov, A.; Seoane, D.; Revillod, L.; Bassez, G.; Ferry, A.; et al. Peptide-Conjugated Oligonucleotides Evoke Long-Lasting Myotonic Dystrophy Correction in Patient-Derived Cells and Mice. *J. Clin. Investig.* **2019**, *129*, 4739. [CrossRef] [PubMed]
13. McClorey, G.; Banerjee, S. Cell-Penetrating Peptides to Enhance Delivery of Oligonucleotide-Based Therapeutics. *Biomedicines* **2018**, *6*, 51. [CrossRef] [PubMed]
14. LeCher, J.C.; Nowak, S.J.; McMurry, J.L. Breaking in and Busting out: Cell-Penetrating Peptides and the Endosomal Escape Problem. *Biomol. Concepts* **2017**, *8*, 131–141. [CrossRef] [PubMed]
15. Swiecicki, J.M.; Di Pisa, M.; Lippi, F.; Chwetzoff, S.; Mansuy, C.; Trugnan, G.; Chassaing, G.; Lavielle, S.; Burlina, F. Unsaturated Acyl Chains Dramatically Enhanced Cellular Uptake by Direct Translocation of a Minimalist Oligo-Arginine Lipopeptide. *Chem. Commun.* **2015**, *51*, 14656–14659. [CrossRef] [PubMed]
16. Szabó, I.; Illien, F.; Dókus, L.E.; Yousef, M.; Baranyai, Z.; Bősze, S.; Ise, S.; Kawano, K.; Sagan, S.; Futaki, S.; et al. Influence of the Dabcyl Group on the Cellular Uptake of Cationic Peptides: Short Oligoarginines as Efficient Cell-Penetrating Peptides. *Amino Acids* **2021**, *53*, 1033–1049. [CrossRef] [PubMed]
17. Mandal, S.; Mann, G.; Satish, G.; Brik, A. Enhanced Live-Cell Delivery of Synthetic Proteins Assisted by Cell-Penetrating Peptides Fused to DABCYL. *Angew. Chem. Int. Ed.* **2021**, *60*, 7333–7343. [CrossRef]
18. Yousef, M.; Szabó, I.; Biri-Kovács, B.; Szeder, B.; Illien, F.; Sagan, S.; Bánóczi, Z. Modification of Short Non-Permeable Peptides to Increase Cellular Uptake and Cytostatic Activity of Their Conjugates. *ChemistrySelect* **2021**, *6*, 10111–10120. [CrossRef]
19. Yousef, M.; Szabó, I.; Murányi, J.; Illien, F.; Soltész, D.; Bató, C.; Tóth, G.; Batta, G.; Nagy, P.; Sagan, S.; et al. Cell-Penetrating Dabcyl-Containing Tetraarginines with Backbone Aromatics as Uptake Enhancers. *Pharmaceutics* **2022**, *15*, 141. [CrossRef]
20. Azuma, Y.; Imai, H.; Kawaguchi, Y.; Nakase, I.; Kimura, H.; Futaki, S. Modular Redesign of a Cationic Lytic Peptide to Promote the Endosomal Escape of Biomacromolecules. *Angew. Chem. Int. Ed.* **2018**, *57*, 12771–12774. [CrossRef]
21. Szabó, I.; Yousef, M.; Soltész, D.; Bató, C.; Mező, G.; Bánóczi, Z. Redesigning of Cell-Penetrating Peptides to Improve Their Efficacy as a Drug Delivery System. *Pharmaceutics* **2022**, *14*, 907. [CrossRef] [PubMed]
22. Caesar, C.E.B.; Esbjörner, E.K.; Lincoln, P.; Nordén, B. Membrane Interactions of Cell-Penetrating Peptides Probed by Tryptophan Fluorescence and Dichroism Techniques: Correlations of Structure to Cellular Uptake. *Biochemistry* **2006**, *45*, 7682–7692. [CrossRef] [PubMed]
23. Walrant, A.; Correia, I.; Jiao, C.Y.; Lequin, O.; Bent, E.H.; Goasdoué, N.; Lacombe, C.; Chassaing, G.; Sagan, S.; Alves, I.D. Different Membrane Behaviour and Cellular Uptake of Three Basic Arginine-Rich Peptides. *Biochim. Biophys. Acta Biomembr.* **2011**, *1808*, 382–393. [CrossRef]
24. Walrant, A.; Bauzá, A.; Girardet, C.; Alves, I.D.; Lecomte, S.; Illien, F.; Cardon, S.; Chaianantakul, N.; Pallerla, M.; Burlina, F.; et al. Ionpair-π Interactions Favor Cell Penetration of Arginine/Tryptophan-Rich Cell-Penetrating Peptides. *Biochim. Biophys. Acta Biomembr.* **2020**, *1862*, 183098. [CrossRef]
25. Bechara, C.; Pallerla, M.; Burlina, F.; Illien, F.; Cribier, S.; Sagan, S. Massive Glycosaminoglycan-Dependent Entry of Trp-Containing Cell-Penetrating Peptides Induced by Exogenous Sphingomyelinase or Cholesterol Depletion. *Cell. Mol. Life Sci.* **2015**, *72*, 809–820. [CrossRef]
26. Khemaissa, S.; Walrant, A.; Sagan, S. Tryptophan, more than Just an Interfacial Amino Acid in the Membrane Activity of Cationic Cell-Penetrating and Antimicrobial Peptides. *Q. Rev. Biophys.* **2022**, *55*, e10. [CrossRef] [PubMed]
27. Koivusalo, M.; Welch, C.; Hayashi, H.; Scott, C.C.; Kim, M.; Alexander, T.; Touret, N.; Hahn, K.M.; Grinstein, S. Amiloride Inhibits Macropinocytosis by Lowering Submembranous PH and Preventing Rac1 and Cdc42 Signaling. *J. Cell. Biol.* **2010**, *188*, 547–563. [CrossRef]
28. Gomes dos Reis, L.; Lee, W.H.; Svolos, M.; Moir, L.M.; Jaber, R.; Engel, A.; Windhab, N.; Young, P.M.; Traini, D. Delivery of PDNA to Lung Epithelial Cells Using PLGA Nanoparticles Formulated with a Cell-Penetrating Peptide: Understanding the Intracellular Fate. *Drug. Dev. Ind. Pharm.* **2020**, *46*, 427–442. [CrossRef] [PubMed]
29. Fittipaldi, A.; Ferrari, A.; Zoppé, M.; Arcangeli, C.; Pellegrini, V.; Beltram, F.; Giacca, M. Cell Membrane Lipid Rafts Mediate Caveolar Endocytosis of HIV-1 Tat Fusion Proteins. *J. Biol. Chem.* **2003**, *278*, 34141–34149. [CrossRef]
30. Starling, D.; Duncan, R.; Lloyd, J.B. The Role of Microtubules in Pinocytosis. Inhibition of Fluid-Phase Pinocytosis in the Rat Visceral Yolk Sac by Mitoclasic and Related Agents. *Cell. Biol. Int. Rep.* **1983**, *7*, 593–602. [CrossRef]

31. Roloff, A.; Nelles, D.A.; Thompson, M.P.; Yeo, G.W.; Gianneschi, N.C. Self-Transfecting Micellar RNA: Modulating Nanoparticle Cell Interactions via High Density Display of Small Molecule Ligands on Micelle Coronas. *Bioconjug. Chem.* **2018**, *29*, 126–135. [CrossRef] [PubMed]
32. Pham, W.; Kircher, M.F.; Weissleder, R.; Tung, C.H. Enhancing Membrane Permeability by Fatty Acylation of Oligoarginine Peptides. *ChemBioChem* **2004**, *5*, 1148–1151. [CrossRef] [PubMed]
33. Lee, J.S.; Tung, C.H. Lipo-Oligoarginines as Effective Delivery Vectors to Promote Cellular Uptake. *Mol. Biosyst.* **2010**, *6*, 2049–2055. [CrossRef] [PubMed]
34. Futaki, S.; Suzuki, T.; Ohashi, W.; Yagami, T.; Tanaka, S.; Ueda, K.; Sugiura, Y. Arginine-Rich Peptides. An Abundant Source of Membrane-Permeable Peptides Having Potential as Carriers for Intracellular Protein Delivery. *J. Biol. Chem.* **2001**, *276*, 5836–5840. [CrossRef] [PubMed]
35. Mitchell, D.J.; Kim, D.T.; Steinman, L.; Fathman, C.G.; Rothbard, J.B. Polyarginine Enters Cells More Efficiently than Other Polycationic Homopolymers. *J. Pept. Res.* **2000**, *56*, 318–325. [CrossRef]
36. Zhang, P.; Lock, L.L.; Cheetham, A.G.; Cui, H. Enhanced Cellular Entry and Efficacy of Tat Conjugates by Rational Design of the Auxiliary Segment. *Mol. Pharm.* **2014**, *11*, 964–973. [CrossRef] [PubMed]
37. Nakase, I.; Niwa, M.; Takeuchi, T.; Sonomura, K.; Kawabata, N.; Koike, Y.; Takehashi, M.; Tanaka, S.; Ueda, K.; Simpson, J.C.; et al. Cellular Uptake of Arginine-Rich Peptides: Roles for Macropinocytosis and Actin Rearrangement. *Mol. Ther.* **2004**, *10*, 1011–1022. [CrossRef]
38. Cleal, K.; He, L.; Watson, D.P.; Jones, T.A. Endocytosis, Intracellular Traffic and Fate of Cell Penetrating Peptide Based Conjugates and Nanoparticles. *Curr. Pharm. Des.* **2013**, *19*, 2878–2894. [CrossRef]
39. Zhang, Y.T.; Zhang, K.; Li, Z.; Zhang, H.Y.; Guo, T.; Li, Y.Y.; Zhao, J.H.; Feng, N.P. DOC-LS, a New Liposome for Dermal Delivery, and Its Endocytosis by HaCaT and CCC-ESF-1 Cells. *IET Nanobiotechnol.* **2018**, *12*, 1037–1041. [CrossRef]
40. Batta, G.; Kárpáti, L.; Henrique, G.F.; Tóth, G.; Tarapcsák, S.; Kovacs, T.; Zakany, F.; Mándity, I.M.; Nagy, P. Statin-Boosted Cellular Uptake and Endosomal Escape of Penetratin Due to Reduced Membrane Dipole Potential. *Br. J. Pharmacol.* **2021**, *178*, 3667–3681. [CrossRef]
41. Wang, L. Measurements and Implications of the Membrane Dipole Potential. *Annu. Rev. Biochem.* **2012**, *81*, 615–635. [CrossRef] [PubMed]
42. Asawakarn, T.; Cladera, J.; O'Shea, P. Effects of the Membrane Dipole Potential on the Interaction of Saquinavir with Phospholipid Membranes and Plasma Membrane Receptors of Caco-2 Cells. *J. Biol. Chem.* **2001**, *276*, 38457–38463. [CrossRef] [PubMed]
43. Sarkar, P.; Chattopadhyay, A. Membrane Dipole Potential: An Emerging Approach to Explore Membrane Organization and Function. *J. Phys. Chem. B* **2022**, *126*, 4415–4430. [CrossRef] [PubMed]

Disclaimer/Publisher's Note: The statements, opinions and data contained in all publications are solely those of the individual author(s) and contributor(s) and not of MDPI and/or the editor(s). MDPI and/or the editor(s) disclaim responsibility for any injury to people or property resulting from any ideas, methods, instructions or products referred to in the content.

pharmaceutics

MDPI

Article

Inhibition of Hepatitis B Virus (HBV) by Tachyplesin, a Marine Antimicrobial Cell-Penetrating Peptide

Pankhuri Narula, Sankar Kiruthika, Shruti Chowdhari, Perumal Vivekanandan * and Archana Chugh *

Kusuma School of Biological Sciences, Indian Institute of Technology, Hauz Khas, New Delhi 110016, India
* Correspondence: vperumal@bioschool.iitd.ac.in (P.V.); achugh@bioschool.iitd.ac.in (A.C.)

Abstract: We investigate the role of Tachyplesin (Tpl), a marine antimicrobial cell-penetrating peptide, as an anti-HBV agent. Our findings, using confocal microscopy and flow cytometry, demonstrate the internalization of FITC-Tpl in both Huh7 and HepG2 cell lines. Further, our results show that Tpl inhibits the expression of HBV proteins, including hepatitis B surface antigen (HBsAg) and hepatitis B 'e' antigen (HBeAg) in cell supernatants of human liver cell lines transfected with 1.3× pHBV. Interestingly Tpl also reduces levels of HBV pre-core RNA and HBV pregenomic RNA, suggesting that Tpl-mediated inhibition occurs at the early stages of HBV replication, including viral transcription. In addition, Tpl led to a significant reduction in levels of hepatitis B virion secretion. In sum, here we demonstrate the potent anti-HBV activity of Tpl at non-cytotoxic concentrations indicating the potential of Tpl to emerge as an effective therapeutic peptide against HBV.

Keywords: antimicrobial peptide; antiviral peptide; cell-penetrating peptide; hepatitis B virus Tachyplesin (Tpl); virion secretion

Citation: Narula, P.; Kiruthika, S.; Chowdhari, S.; Vivekanandan, P.; Chugh, A. Inhibition of Hepatitis B Virus (HBV) by Tachyplesin, a Marine Antimicrobial Cell-Penetrating Peptide. *Pharmaceutics* **2023**, *15*, 672. https://doi.org/10.3390/pharmaceutics15020672

Academic Editors: Alyssa Panitch, Prisca Boisguérin, Sébastien Deshayes and Jun Dai

Received: 26 September 2022
Revised: 5 December 2022
Accepted: 14 February 2023
Published: 16 February 2023

Copyright: © 2023 by the authors. Licensee MDPI, Basel, Switzerland. This article is an open access article distributed under the terms and conditions of the Creative Commons Attribution (CC BY) license (https://creativecommons.org/licenses/by/4.0/).

1. Introduction

The Hepatitis B virus (HBV) is an enveloped DNA virus with a partially double-stranded relaxed circular DNA (rcDNA) genome of 3.2 kb. HBV can cause chronic infection and is the leading cause of liver cirrhosis and hepatocellular carcinoma (HCC) [1]. Partial double-stranded DNA of HBV consists of a minus(−) strand, which is complimentary to pregenomic RNA (pgRNA), and a shorter plus(+) strand. The minus strand consists of four overlapping open reading frames (ORFs), namely PreC/C, which codes for Hepatitis B 'e' Ag (HBeAg) and Hepatitis B core Ag (HBcAg), ORF P, which encodes the HBV DNA polymerase, PreS/S ORF encodes Hepatitis B surface antigen (HBsAg), and ORF X, the smallest ORF, encodes the HBx protein. HBV is hepatotropic [2], and after successful entry into the host hepatocytes and uncoating, the viral genome replicates in the nucleus. In the nucleus, viral rcDNA is repaired and converted into HBV covalently closed circular DNA (cccDNA/viral minichromosome) with the aid of host polymerases and DNA repair proteins. HBV cccDNA acts as a template for transcription for (a) the viral mRNAs, which are translated into viral proteins [3], and (b) the HBV pgRNA, which is packaged into the HBV core protein along with the viral polymerase. The HBV pgRNA serves as the template for reverse transcription, resulting in the generation of HBV DNA. This is followed by the maturation of viral nucleocapsid in the endoplasmic reticulum and viral budding. A significant proportion of HBV core particles with the virus genome may return to the nucleus to maintain HBV cccDNA levels [4].

Currently, interferons (IFNs) and nucleoside analogs (NAs) are used for the management of HBV. NAs act by inhibiting the activity of reverse transcriptase, leading to reduced viral replication. IFNs are known to act through various mechanisms and are found to be effective against a subset of patients. The toxicity and efficacy of antiviral agents, as well as the emergence of antiviral resistance, have been key issues with the existing anti-HBV therapeutic regime [5]. Moreover, since the persistence of cccDNA acts as a reservoir of viral relapse, eradication of HBV cccDNA is the ultimate goal. However, none of the currently approved therapies target HBV cccDNA.

Antimicrobial peptides (AMPs) are small peptides, also known as host defense peptides (HDPs). They are short amino acid sequences ranging from 5–30 amino acids (aa) and display large structural diversity. In the last few years, AMPs have gained attention due to their diverse range of action as therapeutics. Few AMPs have been identified for their antiviral activity against viruses [6], and multiple mechanisms of action have been elucidated [7]. AMPs are known to block the attachment of the virus to the host cell membrane by interfering with the membrane envelope and viral glycoproteins, prevent fusion of the virus to host cell by interacting with the host cell receptors, and interfere with viral signaling pathways or inhibit enzymes involved in replication and transcription processes such as polymerases and reverse transcriptase [7]. Mucroporin-M1, a scorpion venom-derived antimicrobial peptide, exhibits anti-HBV activity by activating the MAPK pathway, resulting in the inhibition of HBV replication in vitro and in vivo [8]. In addition to AMPs, cell-penetrating peptides (CPPs) such as TAT, Penetratin, and Transportan have been extensively studied for their internalization into mammalian cells [9]. Apart from their ability to translocate themselves across cell membranes, CPPs are known for the efficient delivery of macromolecules [10]. Therefore, CPPs have been used for the delivery of antiviral agents into cells that are not easily permeable. Both AMPs and CPPs exhibit common characteristics such as cationicity and amphiphilicity and, in some instances, have been shown to play interchangeable roles, as observed in the case of Tachyplesin [11]. The antibacterial and antifungal activity of CPPs have been widely reported, but their antiviral role has not been well studied [12].

The antiviral activity of CPPs has been investigated for Herpes Simplex Virus (HSV) and Human Immunodeficiency Virus (HIV). In the case of HBV, CPPs have been primarily studied as delivery vehicles for either peptide nucleic acids (PNAs) or siRNAs [13]. CPP-assisted delivery of siRNA has been reported to show inhibitory effects on HBV DNA, HBV RNA, and other viral proteins in a mouse model of HBV infection [14]. In addition, a recent study has demonstrated the anti-HBV activity of a human telomerase reverse transcriptase-derived peptide [15]. Since HBV resides in the host hepatocytes, the antiviral drug should be able to exhibit cellular translocation without being cytotoxic to the host cell.

The present work aims at evaluating the anti-HBV activity of Tachyplesin (Tpl), an antimicrobial cell-penetrating peptide. Tpl is 19 aa long, stabilized by two disulfide bridges imparting stability. It is derived from hemocytes of horseshoe crab *Tachypleus tridentatus* (K_5-WCFRVCYRGICYRRCRG-K_{15}, Cys3-Cys16, and Cys7-Cys12) [16]. The peptide used in this study is a fragment of the natural peptide, which consists of 77 aa. Tpl is known to exhibit significant cellular uptake in both mammalian as well as plant cell culture systems [11]. Tpl is also known to act as an efficient delivery vehicle for various biomolecules [17]. Further, Tachyplesin has shown inhibition against Gram-positive and Gram-negative bacteria, as well as fungi [18]. The anti-parasitic activity of Tpl against leishmania was reported in our lab recently [19]. Further, it has been shown that Tachyplesin or peptides derived from Tachyplesin exhibit anti-HIV-1 activity by inhibiting HIV-cell fusion. Furthermore, it is also known to obstruct the attachment of viral gp120 to CXCR4 and CCR5 co-receptors [20]. Tachyplesin I was also found to significantly inhibit the infection and replication of Singapore grouper iridovirus (SGIV) and Nervous Necrosis Virus (NNV) by interacting with the viral particles and also by inducing Type-1 interferon response [21].

In this study, the anti-HBV activity of Tpl in a transient HBV cell culture model has been evaluated. The cellular permeability of Tpl has been assessed both qualitatively and quantitatively in Huh7 and HepG2 cells. Further, the cytotoxicity of Tpl in both hepatocyte cell lines has been investigated. The effect of Tpl on the secretion of HBV proteins has also been evaluated. In addition, the ability of Tpl to regulate key HBV transcripts has also been investigated by quantitating HBV pcRNA and HBV pgRNA and hepatitis B virion secretion in Huh7 and HepG2 cells. Overall, the ability of Tpl to act as an anti-HBV agent has been determined by assessing its action at various life cycle stages of HBV at non-cytotoxic concentrations.

2. Methods

2.1. Peptide Synthesis

Fluorescein isothiocyanate (FITC)-labeled Tachyplesin (Tpl) and non-labeled Tpl were custom synthesized using solid phase peptide synthesis with more than 95% purity from GenPro Biotech, Noida, India. The mass spectroscopy and HPLC data of Tachyplesin are given as supplementary information (Figure S3).

2.2. Cell Culture

Human hepatoma cell lines, Huh7 and HepG2, were maintained in Dulbecco's Modified Eagle's Medium (DMEM, high glucose 4500 mg/L, L-glutamine, phenol red, without sodium pyruvate and HEPES, Thermo Fischer, Gibco™ 11965092 USA) supplemented with 10% Fetal Bovine Serum (FBS, Thermo Fischer Gibco, Miami, FL, USA) and 1% PenStrep (Thermo Fischer Gibco, USA) and incubated at 37 °C with 5% CO_2.

2.3. HBV Plasmid and Transfection

Huh7 and HepG2 cells were transfected with 1.3× pHBV plasmid using transfection reagent lipofectamine2000 (Invitrogen, Waltham, MA, USA), as described previously [22–24]. After 6 h of transfection, the media was replaced by fresh media, and cells were treated with different concentrations of peptide (Tpl). Supernatants were collected, and cells were harvested for RNA extraction after 48 h.

2.4. Cellular Uptake Studies

Confocal Laser Scanning Microscopy (CLSM): Huh7 and HepG2 cells were seeded in 24-well plates at a density of 10^5 cells per well on an autoclaved glass coverslip and incubated at 37 °C and 5% CO_2. After achieving 80% confluency, media was removed, and cells were washed with 1X phosphate buffer saline (PBS). Thereafter, cells were treated with varying concentrations of FITC-Tpl for 1 h. Following this, media was removed, and cells were treated with 0.05% trypan blue for 10 min to remove any membrane-bound peptide. Cells were then washed twice with PBS and observed under a confocal microscope (Olympus Fluoview FV1200, Olympus Corporation, Tokyo, Japan). The uptake of FITC-labeled mutated Tpl (M-Tpl3) where all arginines are replaced with alanines was also investigated. The uptake of Mut-Tpl3 was also investigated in our previous studies [11,17].

Flow Cytometry: Quantitative assessment of FITC-Tpl internalization into Huh7 and HepG2 cells was carried out using flow cytometry. About 80,000 cells were seeded in 24-well plates and incubated overnight at 37 °C and 5% CO_2. At 80% confluence, cells were treated with various concentrations of FITC-Tpl for 1 h. After the incubation period, cells were trypsinized using 0.1% trypsin-EDTA solution and washed twice with PBS to remove any unbound peptide. The cell pellet obtained was resuspended in sheath fluid and analyzed on BD FACS Aria III (BD Biosciences, San Jose, CA, USA). A total of 10,000 events were recorded, and data were analyzed using FACS Diva 6.0 software.

2.5. Cytotoxicity Analysis

MTT assay was used to identify the non-cytotoxic concentrations of Tpl. Cell viability was measured using MTT (3-(4,5_dimethylthaizol-2-yl)-2,5-diphenyltetrazolium bromide) assay. Approximately 10,000 cells were seeded per well in 96-well plates and kept overnight at 37 °C and 5% CO_2 to achieve more than 80% confluency. Thereafter, cells were treated with various concentrations of Tpl for 48 h. After 48 h, media was removed, and cells were washed with PBS. MTT reagent was added at a final concentration of 1 mg/mL and kept for 4 h at 37 °C and 5% CO_2. Thereafter, 100 μL of DMSO was added to dissolve the formazan crystals formed, and absorbance was determined at a wavelength of 570 nm with a background wavelength of 630 nm (Multiskan GO microplate spectrophotometer, Thermo Scientific, Waltham, MA, USA).

2.6. Dual-Luciferase Reporter Assay

Huh7 and HepG2 cells were seeded at a density of 2×10^4 in 96-well plates and transfected with psiCHECK-2 dual-luciferase (*Renilla*/firefly) reporter vector (Promega Corporation, Madison, WI, USA) using lipofectamine2000 (Invitrogen, USA) according to the manufacturer's protocol followed by addition of 12 μM of Tpl. Dual-luciferase assay was performed after 48 h of transfection according to the manufacturer's instructions, and luminescence was measured using the BioTek Citation 5 plate reader (Agilent, Santa Clara, CA, USA). *Renilla* luciferase (reporter) values were normalized to those of firefly (control) luciferase expression. The fold change in cellular transcription in Tpl-treated cells compared to untreated cells is represented as relative luciferase units (RLU) [24].

2.7. Estimation of HBV Proteins by ELISA

Huh7 and HepG2 cells were seeded at a density of 3×10^5 cells per well in 12-well plates and incubated at 37 °C and 5% CO_2 overnight to achieve >80% confluency. Thereafter, cells were transfected with 1 μg of 1.3× pHBV using transfecting agent lipofectamine 2000, as mentioned above. After 6 h of transfection, media was replaced by fresh media, and cells were treated with various concentrations of Tpl for an incubation period of 48 h. After 48 h, cell supernatants were collected to analyze HBV proteins, and cells were harvested to extract RNA. Commercially available HBsAg (MONOLISA, BioRAD, Hercules, CA, USA) and HBeAg (Diapro, Sesto San Giovanni, Italy) ELISA kits were used for the estimation of HBsAg and HBeAg, respectively, from cell culture supernatants as per given instructions. Dilutions were made so that OD values fell within the linear range, as described previously [25,26].

2.8. qRT-PCR Studies

Total RNA was extracted using TRIzol reagent (Invitrogen, USA) as per the manufacturer's protocol. The RNA pellet obtained was dissolved in nuclease-free water, and RNA concentration was determined using Nanodrop. Purified RNA was treated with DNase, and cDNA was synthesized from 1 μg of RNA using an iScript cDNA synthesis kit (BioRad, USA). Primers targeting HBV pre-core RNA (pcRNA) and pregenomic RNA (pgRNA) were used for quantification of HBV pcRNA and HBV pgRNA (listed in Table 1), and for quantification of IFN-1 pathway genes, the primers targeted for IFN-α and IFN-β were used. Gene expression was measured using iTaq Universal SYBR Green Supermix (BioRad, USA). GAPDH (Glyceraldehyde 3-phosphate dehydrogenase) was used as a control gene for the normalization of gene levels, respectively.

2.9. Estimation of Secreted Virion

Huh7 and HepG2 cells were seeded at a density of 3×10^5 cells per well in 12-well plates and transfected with 1.3× pHBV construct, as described earlier [27,28]. Briefly, 120 μL of supernatant collected after an incubation of 48 h was added to an ELISA microplate coated with anti-HBs antibodies (MONOLISA HBsAg ELISA kit, BioRad, USA) and incubated for 90 min at 37 °C. Thereafter, the microplates were washed with PBS and treated with Proteinase K, and DNA was then extracted using a QIAamp DNA mini kit (Qiagen, Germantown, MD, USA). DNA extracted from the captured virions was quantified with real-time PCR using virion DNA-specific primers.

2.10. Primers Used in the Study

The details of primers used in the study are given in Table 1.

Table 1. The following primers are used for the qPCR studies [23,29,30].

Primer	Sequence (5′ to 3′)
pgRNA	FP: CACCTCTGCCTAATCATC [23] RP: GGAAAGAAGTCAGAAGGCAA
pcRNA	FP: GGTCTGCGCACCAGCACC [23] RP: GGAAAGAAGTCAGAAGGCAA
Virion	FP: GGTCTGCGCACCAGCACC [23] RP: GAACTTTAGGCCCATATTAGTG
IFN-α	FP: GACTCCATCTTGGCTGTGA [29] RP: TGATTTCTGCTCTGACAACCT
IFN-β	FP: GCTTGGATTCCTACAAAGAAGCA [30] RP: ATAGATGGTCAATGCGGCGTC
GAPDH	FP: TGCACCACCAACTGCTTAGC [23] RP: GGCATGGACTGTGGTCATGAG

2.11. Statistical Analysis

The values obtained are from three independent experiments. The results are presented as means ± standard deviation (SD). Differences between control (untreated) and treated groups were determined using Student's *t*-test. *p*-values < 0.05 are represented by * and # and are considered statistically significant.

3. Results

3.1. Internalization of FITC-Tpl in Huh7 and HepG2 Cells

The cell-penetrating ability of Tpl in Huh7 and HepG2 cells was evaluated using confocal microscopy as well as flow cytometry. The data presented in Figure 1A show significant uptake of FITC-Tpl in Huh7 cells at all the concentrations tested (5, 7, 10, and 12 μM). To assess the role of arginine residues in the cellular penetration of Tpl in liver cells, mutated Tpl-3 (M-Tpl3) was employed where all arginines were replaced with alanines. No significant uptake of M-Tpl3 was observed in Huh7 cells as seen by confocal laser scanning microscopy (Figure 1A). Thus, M-Tpl3 was not employed for further antiviral studies.

Further, uptake of FITC-Tpl in Huh7 cells was found to be increasing with an increase in the Tpl concentration, as observed using flow cytometry (Figure 1C). In HepG2 cells, there was appreciable uptake of FITC-Tpl at all the concentrations as observed by confocal microscopy (Figure 1D). Notably, more than 90% of HepG2 cells showed uptake of FITC-Tpl at all concentrations (Figure 1F).

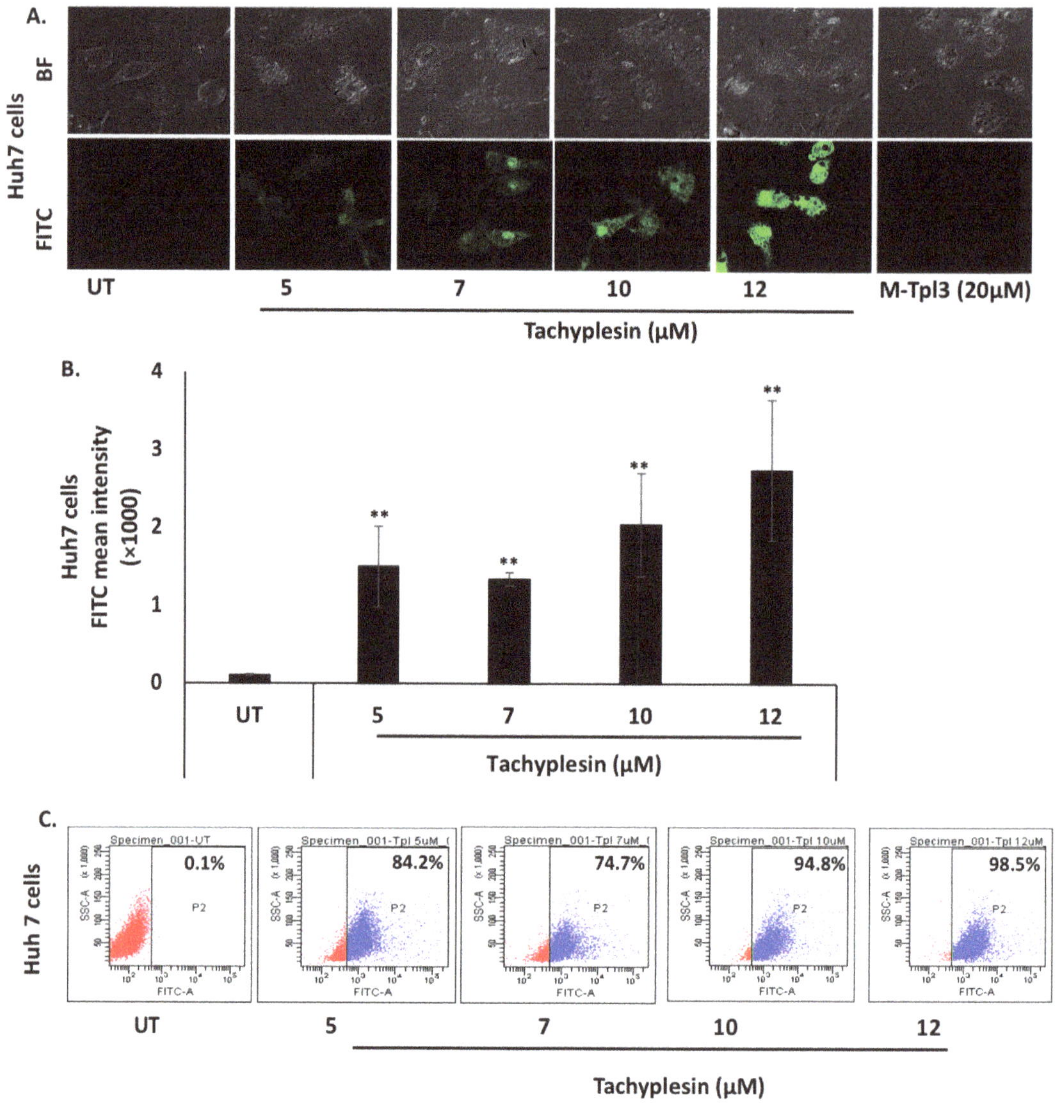

Figure 1. *Cont.*

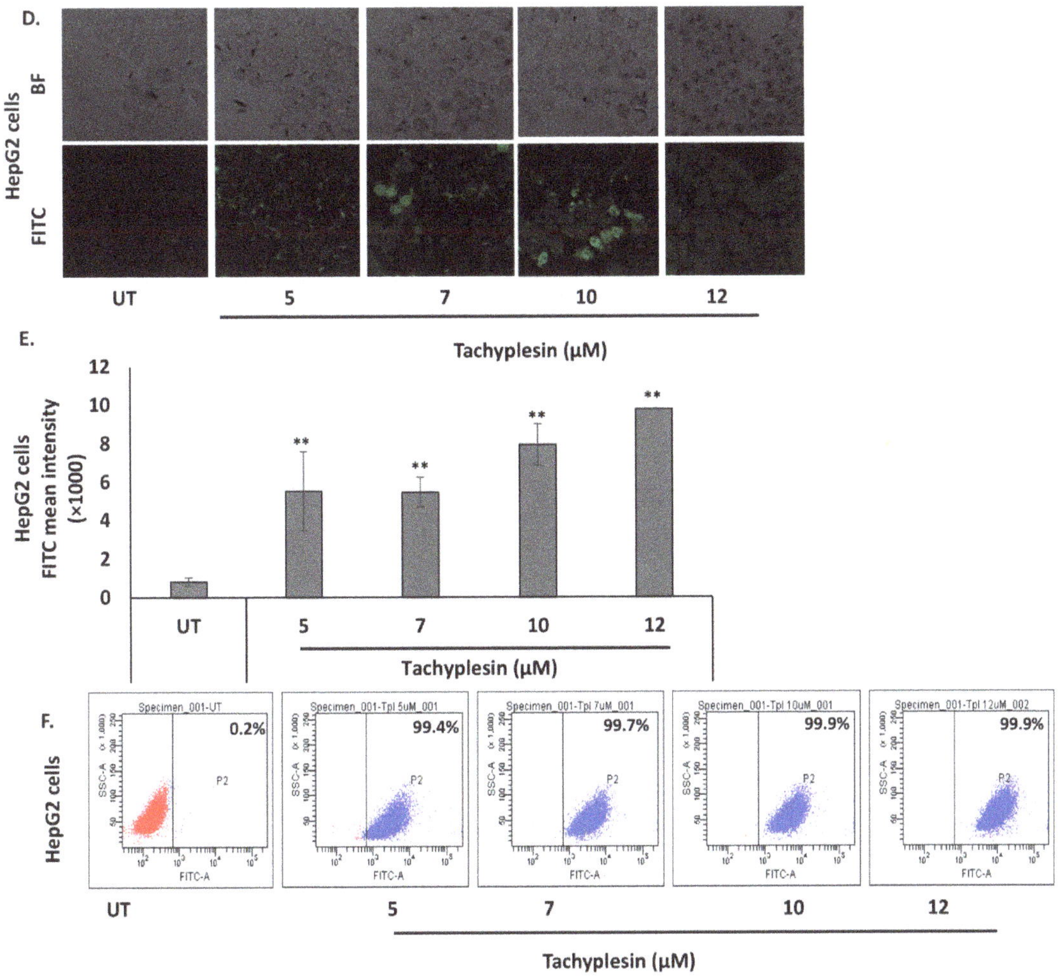

Figure 1. Significant uptake of FITC-Tpl was observed in hepatocyte cell lines at different concentrations of the peptide. (**A**) CLSM studies at magnification of 60× and (**B**,**C**) flow cytometry analysis suggests significant uptake of FITC-Tpl in Huh7 cells. (**D**) CLSM studies at magnification of 40× and (**E**,**F**) flow cytometry analysis show uptake of FITC-Tpl in HepG2 cells. ** represents significant difference with respect to untreated (p-value < 0.01, Student's t-test). UT = untreated.

3.2. Tpl Is Well Tolerated by Huh7 and HepG2 Cells

Cellular toxicity associated with Tpl in Huh7 and HepG2 cells was studied using an MTT assay. After 48 h of treatment with up to 12 µM Tpl, cell viability was found to be more than 85% in both cell lines (Figure 2A,B). Therefore, Tpl at 5, 7, 10, and 12 µM concentrations were further used to assess its anti-HBV activity.

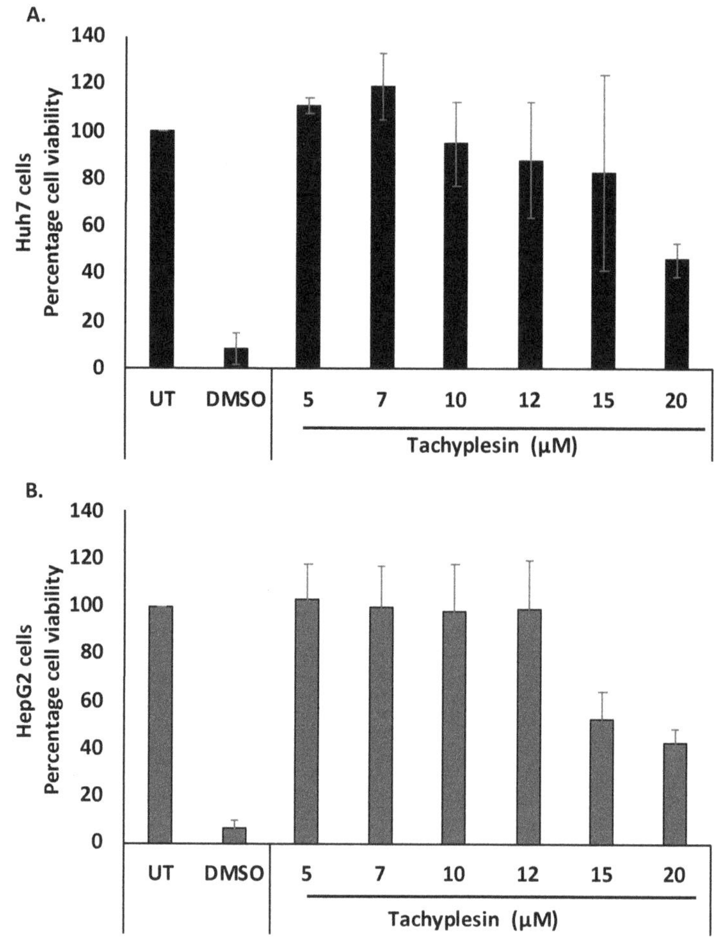

Figure 2. Cytotoxicity studies with Tpl treatment. Cell viability in (**A**) Huh7 cells and (**B**) HepG2 cells after 48 h.

3.3. Tpl Does Not Affect Transfection or the Cellular Transcription, Translation, and Secretion Machinery

To ascertain if the Tpl peptide has any effect on other proteins involved in normal cellular transcription or translation machinery, we transfected Huh7 and HepG2 cells with psiCHECK-2 dual-luciferase reporter vector followed by the addition of Tpl peptide. The relative luciferase expression was comparable in cells with or without Tpl, as shown in Figure 3A,B, thus, ruling out inhibition or activation of cellular transcription and translation machinery by Tpl. In addition to the luciferase assay, we also looked at the endogenous levels of two housekeeping genes (GAPDH and β-actin) in both Huh7 and HepG2 cell lines in the presence or absence of Tpl. We found no significant difference in the Ct values of GAPDH and β-actin genes between peptide-treated and untreated cells (Figure S1). In addition, qPCR data for IFNα/β gene expression further confirmed that Tpl (12 μM) treatment did not elicit an interferon response in both Huh7 and HepG2 cells, as shown in Figure 3C,D.

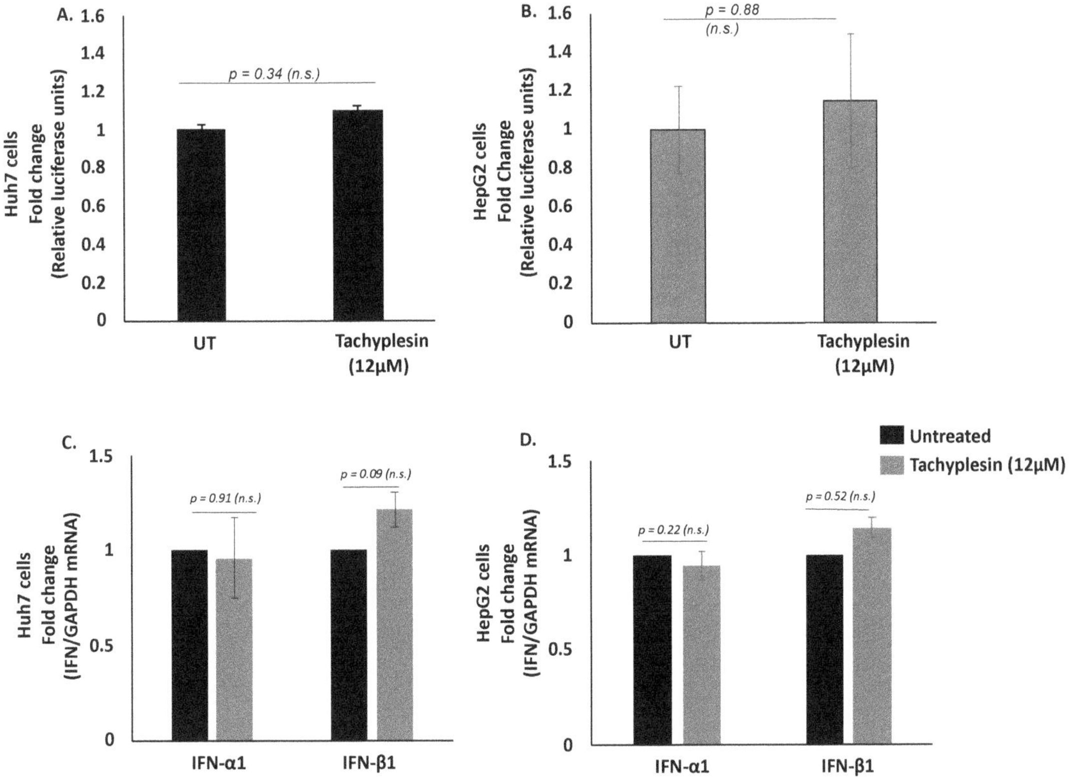

Figure 3. Dual-luciferase assay and IFN-1 expression level. Relative luciferase expression in (**A**) Huh7 and (**B**) HepG2 cells transfected with psiCHECK-2 dual-luciferase (*Renilla*/firefly) reporter vector following 48 h incubation with 12 μM of Tachyplesin. No significant change was observed in levels of IFN-α1 and IFN-β1 (**C**) Huh7 cells and (**D**) HepG2 cells treated with 12 μM of Tpl. *n.s*: non-significant, *p* represents *p*-value.

Additionally, to assess the binding affinity of Tpl with HBV DNA at different ratios, a gel retardation assay was performed. With the increase in the ratio of the complex, the intensity of the HBV band decreased, indicating the complexation of HBV DNA with peptide (Figure S2).

3.4. Tpl Inhibits Secreted HBV Proteins

The ability of Tpl to inhibit secreted HBV proteins (HBsAg and HBeAg) was estimated using ELISA quantification. HBsAg is secreted by infected hepatocytes, and its levels in the serum have been linked to disease outcome and response to antiviral drugs. Similarly, HBeAg, another secreted HBV protein, is used as an early serum marker of HBV infection, and an HBeAg-positive status has been linked to a high viral load.

Notably, treatment with Tpl at 5, 7, 10, and 12 μM in both Huh7 and HepG2 cells resulted in a significant reduction in HBsAg and HBeAg levels as compared to untreated cells (Figure 4). These results suggest that Tpl is a potent inhibitor of both HBsAg and HBeAg in hepatocyte cell lines.

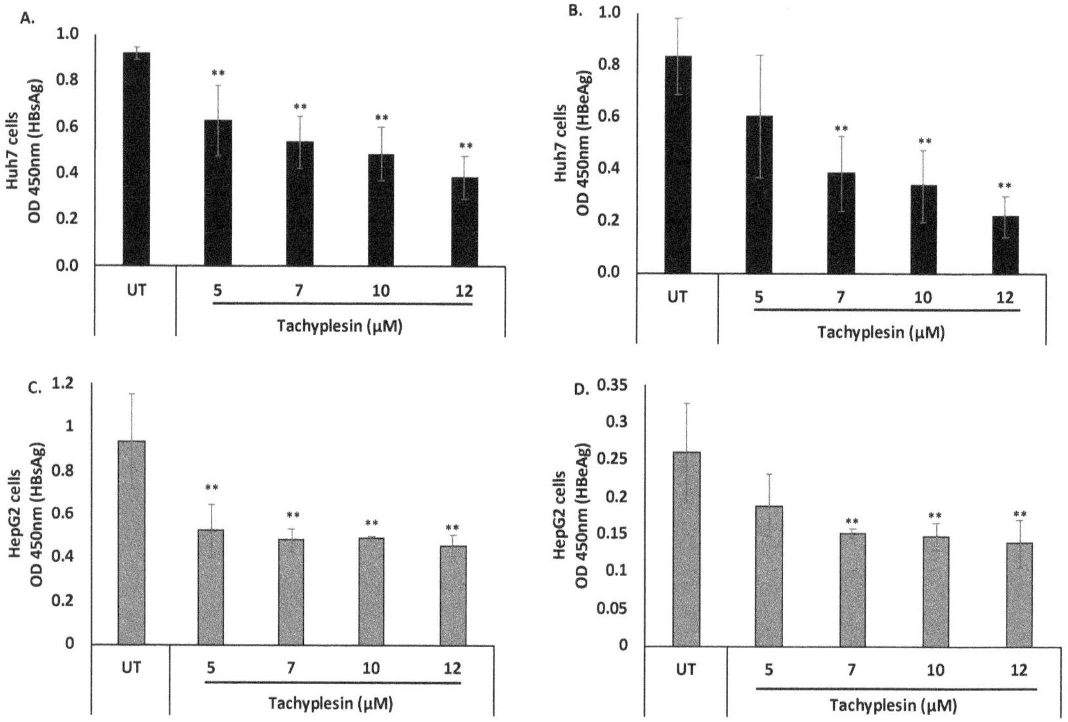

Figure 4. Inhibitory effect of Tpl on the secreted HBV proteins. Huh7 cells and HepG2 cells were transfected with HBV 1.3× plasmid, and supernatants were harvested at 48 h. The inhibitory effect of Tpl on the secretion of (**A**) HBsAg and (**B**) HBeAg was observed in Huh7 cells. (**C**) HBsAg and (**D**) HBeAg levels were also significantly reduced in HepG2 cells treated with Tpl at 7, 10, and 12 μM for 48 h. ** represents significant difference with respect to untreated (p-value < 0.01, Student's t-test). UT = untreated.

3.5. Tpl Inhibits HBV pcRNA and HBV pgRNA

HBV pgRNA and HBV pcRNA represent widely studied HBV replicative intermediates and transcripts, respectively. Therefore, the effect of Tpl on HBV pcRNA and HBV pgRNA was investigated. HBV pcRNA and HBV pgRNA levels were estimated quantitatively using real-time PCR. Translation of pcRNA leads to the formation of HBeAg. The levels of pcRNA were significantly decreased at all the concentrations of Tpl in both Huh7 (Figure 5A) and HepG2 cells (Figure 5C) as compared to the untreated (UT) control. This finding corroborates with Tpl-mediated reduction in HBeAg, a protein synthesized from HBV pcRNA. Tpl also led to a significant reduction in HBV pgRNA levels in both cell lines as compared to the untreated controls (Figure 5B,D). The HBV pgRNA is packaged into the HBV capsid, where it is converted to HBV DNA by the HBV polymerase. The HBV pgRNA represents a key replicative intermediate that is used as the template for the synthesis of HBV genomic DNA.

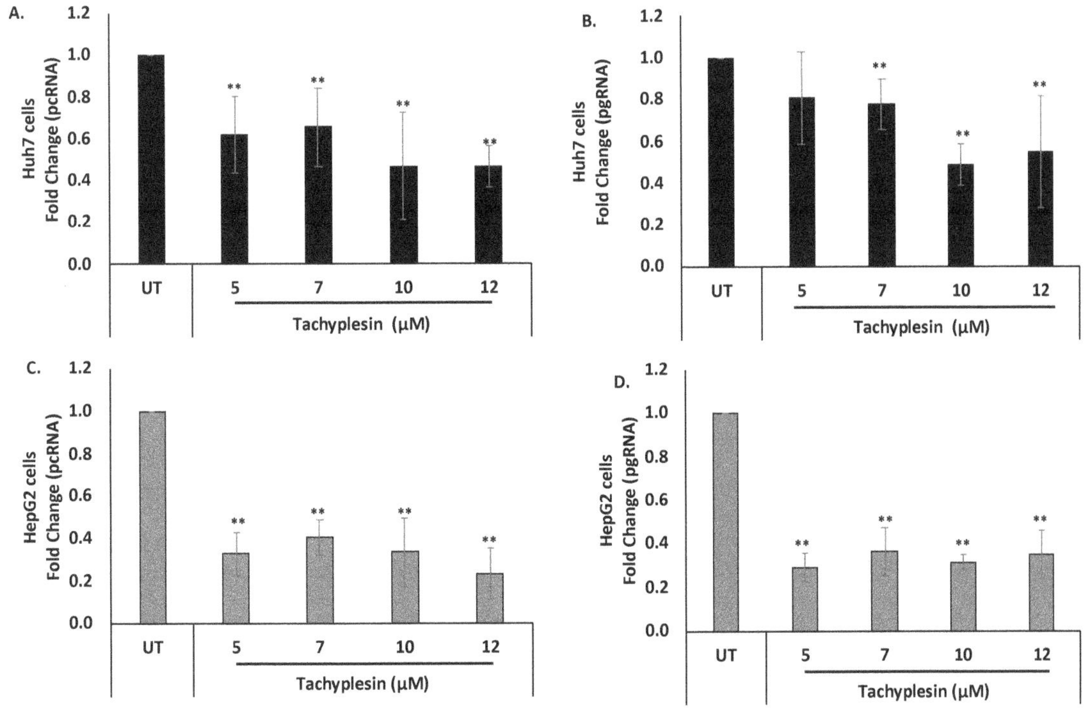

Figure 5. Reduction of HBV pcRNA and HBV pgRNA levels. Tpl led to the reduction of (**A**) HBV pcRNA and (**B**) HBV pgRNA in Huh7 cells. Significant reduction in levels of (**C**) HBV pcRNA and (**D**) HBV pgRNA was also observed in HepG2 cells treated with Tpl. GAPDH was used for normalization. ** represents significant difference with respect to untreated (p-value < 0.01, Student's t-test). UT = untreated control.

3.6. Tpl Inhibits Hepatitis B Virion Secretion

The action of Tpl as an anti-HBV agent was further explored by investigating virion secretion. Virions secreted in the supernatant were quantified using an immunocapture-based protocol followed by quantification of virion-associated DNA using RT-PCR as optimized previously [24,28]. A significant reduction in virion secretion was observed at the concentrations of Tpl employed (Figure 6A,B). The Tpl-mediated reduction in virion secretion was more pronounced than that by lamivudine (3TC) at most of the concentrations tested.

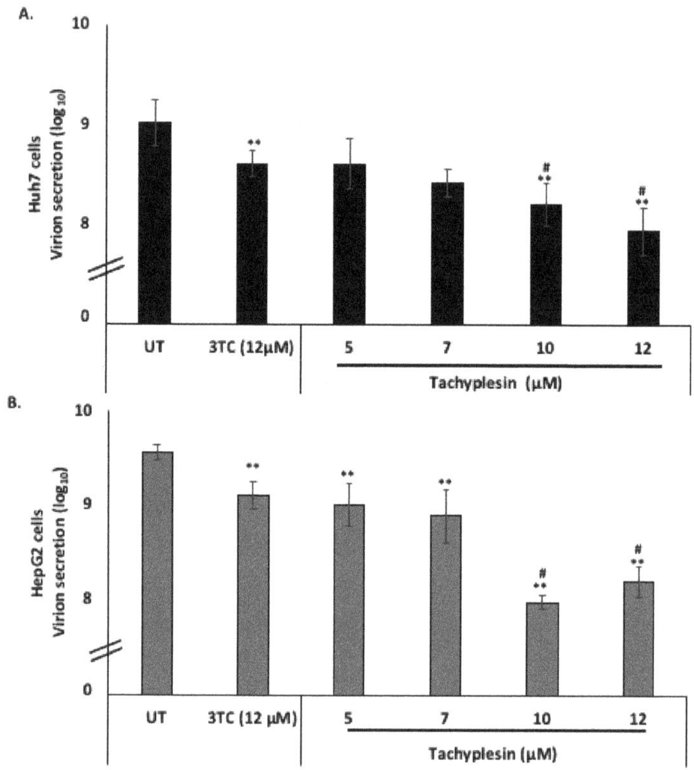

Figure 6. Tpl inhibits hepatitis B virus virion secretion in Huh7 cells and HepG2 cells. Tpl inhibited virion secretion in a concentration-dependent manner in both (**A**) Huh7 and (**B**) HepG2 cell lines. ** represents significant difference with respect to untreated (p-value < 0.01, Student's t-test). # represents significant difference with respect to 3TC (p-value < 0.05). UT = Untreated, 3TC = lamivudine.

4. Discussion

The current treatments available for HBV, such as interferons (IFNs) and nucleoside analogs (NAs), are known to be involved in the suppression of viral replication. However, the side effects, inability to suppress/inhibit cccDNA levels, and development of viral resistance are the major limitations associated with currently available anti-HBV therapies. Therefore, there is a need for alternative therapeutics that can overcome these challenges. Over the last few years, many efforts have been made toward the development of more promising antiviral agents, and the identification of antiviral peptides is one of them [31]. Peptide-based therapeutics have added a new dimension to the area of antiviral research owing to their high efficacy, high selectivity, low cytotoxicity, ease of modification, and small size [7]. Peptides have been known to show antiviral activity by either acting at the virus level, which involves destabilizing viral envelope and damaging virions, or by binding to host cell receptors which are involved in the initial attachment and entry of the virus. A study by Donkers et al. (2019) reported an interaction between Myrcludex B, a synthetic peptide mimicking the sodium taurocholate co-transporting polypeptide (NTCP)-binding domain of HBV and the NTCP-receptor which is a viral entry receptor for the hepatitis B and D virus (HBV/HDV). Myrcludex B was observed to bind non-covalently with the NTCP receptor, blocking HBV and HDV infection [32].

In this study, we identified the anti-HBV activity of antimicrobial cell-penetrating peptide Tpl and its effect on various stages of the HBV life cycle. Previous studies have shown wide spectrum antibacterial, antifungal, and anti-parasitic activities of Tachyplesin, but its efficiency as an anti-HBV agent has not been explored. To achieve anti-HBV activity, the drug molecule should possess efficient cell permeability properties without being cytotoxic to host cells. Taking into consideration these critical factors, such as cell penetration, low cytotoxicity, and antimicrobial activity, Tpl has been investigated in the present study for its antiviral activity against HBV. In both confocal microscopy and flow cytometry analysis, significant uptake of FITC-labeled Tpl was observed in Huh7 cells. However, at all the concentrations of FITC-Tpl, more than 90% of HepG2 cells exhibited fluorescence. Moreover, no significant change was observed in expression levels of luciferase, which suggests that Tpl does not affect cellular transcription or translation.

Current antiviral therapies target viral replication. The loss of serum HBsAg and HBeAg are used as serological indicators of anti-HBV therapy. Thus, the ability of anti-HBV therapies to inhibit the secretion of HBsAg and HBeAg is used for screening antiviral agents [33]. The ability of Tpl to impact the secretion of HBV proteins in both the liver cell lines was evaluated with ELISA. Tpl induced a reduction in HBsAg and HBeAg levels in Huh7 cells. The levels of both HBsAg and HBeAg were significantly reduced in the cell culture supernatants of HepG2 cells treated with Tpl.

Furthermore, we found a significant reduction in levels of HBV pcRNA on treatment with Tpl; this may explain the reduced HBeAg levels observed in the presence of Tpl. The formation of HBV pgRNA from HBV cccDNA in the nucleus is an essential step in the replication of HBV genomic DNA. Tpl was associated with a significant reduction in HBV pgRNA levels. Together, these findings indicate that Tpl may target the early stages of HBV replication, including virus transcription. Finally, Tpl significantly inhibited hepatitis B virion secretion from Huh7 and HepG2 cells. The inhibition of hepatitis B virion secretion in cell culture is an important property of potential anti-HBV agents. Importantly, at all concentrations, Tpl-mediated inhibition of virion secretion was comparable to or more pronounced than that by 3TC (lamivudine).

Overall, our work suggests significant internalization of Tpl into human hepatocyte cell lines, and it also highlights significant anti-HBV activity exhibited by Tpl at non-cytotoxic concentrations. Furthermore, we demonstrate Tpl-mediated inhibition of HBV proteins (HBsAg and HBeAg), HBV transcripts (pcRNA, pgRNA), and virion secretion. It is noteworthy that the ability of Tpl to inhibit hepatitis B virion secretion was comparable to or better than that of lamivudine. The precise mechanism of action of Tpl as an anti-HBV agent needs more elucidation in further studies. Our findings provide promising insights into the anti-HBV activity of Tpl. We believe that this work represents an important step toward the development of peptide-based anti-HBV therapies and their applications.

In sum, this work highlights the anti-HBV activity of Tpl. Importantly, our findings pave the way for further studies on developing efficient peptide-based anti-HBV strategies.

Supplementary Materials: The following supporting information can be downloaded at: https://www.mdpi.com/article/10.3390/pharmaceutics15020672/s1, Figure S1: Effect of Tpl on GAPDH and β-actin genes in both the cell lines. The bar graphs represent mean of Ct values from qRT-PCR assays for housekeeping genes: GAPDH and β-actin in Huh7 and HepG2 cell lines treated with or without Tpl. The error bars represent standard deviation (n = 3). Figure S2: Complexation of HBV DNA:Tpl. HBV DNA:Tpl were mixed at different w/w ratios and were then analysed for complexation by gel retardation assay. The band intensity was also quantified using Image lab 6.1 software. Figure S3: MS/HPLC data of Tachyplesin. The expected mass of Tpl was 2449.09 Daltons and the observed mass was found to be 2449.2 (M+3H+ = 817.4) respectively.

Author Contributions: P.N.: methodology, investigation, validation, and writing the original draft. S.K.: methodology, data analysis. S.C.: experimentation and analysis, P.V. and A.C.: conceptualization, supervision, formal analysis, resources, writing—review and editing. All authors have read and agreed to the published version of the manuscript.

Funding: This research did not receive any extramural funding.

Institutional Review Board Statement: Not applicable.

Informed Consent Statement: Not applicable.

Data Availability Statement: The data presented in this study are available in this article (and Supplementary Materials).

Acknowledgments: PN is thankful to the Council of Scientific & Industrial Research (CSIR), Govt. of India, for the award of Junior Research Fellowship and Senior Research Fellowship. SK is thankful to the IIT-D Institute fellowship for the Ph.D. program. SC was supported by the Indian Council of Medical Research-Research Associate (ICMR-RA) fellowship. PN is also thankful to Jasmine Samal for her help in the initialization of the investigations. The authors are thankful to Aditya Mittal Kusuma School of Biological Sciences, IIT-Delhi, for providing the confocal laser scanning microscope facility. We also thank Syed Naqui Kazim (Jamia Millia Islamia, New Delhi, India) for providing HBV 1.3× construct.

Conflicts of Interest: The authors declare no conflict of interest.

Abbreviations

AMP: antimicrobial peptide; AVP: antiviral peptide; cccDNA: covalently closed circular DNA; CHB: chronic hepatitis B; CLSM: confocal laser scanning microscope; CPP: cell-penetrating peptide; DMEM: Dulbecco's Modified Eagle's Medium; DNA: deoxyribonucleic acid; ELISA: enzyme-linked immunosorbent assay; FACS: fluorescence-activated cell sorter; FITC: Fluorescein isothiocyanate; HBV: hepatitis B virus; HBeAg: HBV 'e' antigen; HBsAg: HBV surface antigen; IFN: interferon; MTT: (3-(4,5_ dimethylthaizol-2-yl)-2,5-diphenyltetrazolium bromide); NA: nucleotide analog; ORF: open reading frame; PBS: phosphate buffer saline; pcRNA: pre-core RNA; pgRNA: pregenomic RNA; PNA: peptide nucleic acid; rcDNA: relaxed circular DNA; RNA: ribonucleic acid; TAT: transactivator of transcription; Tpl: Tachyplesin; 3TC: Lamivudine.

References

1. Tsukuda, S.; Watashi, K. Hepatitis B virus biology and life cycle. *Antiviral Res.* **2020**, *182*, 104925. [CrossRef] [PubMed]
2. Locarnini, S. Molecular virology of Hepatitis B Virus. *Semin. Liver Dis.* **2004**, *24*, 3–10. [CrossRef] [PubMed]
3. Wei, L.; PLoSs, A. Mechanism of Hepatitis B Virus cccDNA Formation. *Viruses* **2021**, *13*, 1463. [CrossRef]
4. Liang, T.J. Hepatitis B: The virus and disease. *Hepatology* **2009**, *49*, S13–S21. [CrossRef] [PubMed]
5. Tang, H. *Hepatitis Infection B Virus: Molecular Virology to Antiviral Drugs*; Springer Nature: Singapore, 2020.
6. Elnagdy, S.; AlKhazindar, M. The potential of antimicrobial peptides as an antiviral therapy against COVID-19. *ACS Pharmacol. Transl. Sci.* **2020**, *3*, 780–782. [CrossRef]
7. Vilas Boas, L.C.P.; Campos, M.L.; Berlanda, R.L.A.; de Carvalho Neves, N.; Franco, O.L. Antiviral peptides as promising therapeutic drugs. *Cell. Mol. Life Sci.* **2019**, *76*, 3525–3542. [CrossRef] [PubMed]
8. Zhao, Z.; Hong, W.; Zeng, Z.; Wu, Y.; Hu, K.; Tian, X.; Li, W.; Cao, Z. Mucroporin-M1 inhibits hepatitis B virus replication by activating the mitogen-activated protein kinase (MAPK) pathway and down-regulating HNF4α in vitro and in vivo. *J. Biol. Chem.* **2012**, *287*, 30181–30190. [CrossRef]
9. Järver, P.; Langel, Ü. Cell-penetrating peptides—A brief introduction. *Biochim. Biophys. Acta Biomembr.* **2006**, *1758*, 260–263. [CrossRef]
10. Desale, K.; Kuche, K.; Jain, S. Cell-penetrating peptides (CPPs): An overview of applications for improving the potential of nanotherapeutics. *Biomater. Sci.* **2021**, *9*, 1153–1188. [CrossRef]
11. Jain, A.; Yadav, B.K.; Chugh, A. Marine antimicrobial peptide Tachyplesin as an efficient nanocarrier for macromolecule delivery in plant and mammalian cells. *FEBS J.* **2015**, *282*, 732–745. [CrossRef]
12. Neundorf, I. Antimicrobial and Cell-Penetrating Peptides: How to understand two distinct functions despite similar physicochemical properties. *Adv. Exp. Med. Biol.* **2019**, *1117*, 93–109. [PubMed]
13. Ndeboko, B.; Hantz, O.; Lemamy, G.J.; Cova, L. Developments in cell-penetrating peptides as antiviral agents and as vehicles for delivery of peptide nucleic acid targeting hepadnaviral replication pathway. *Biomolecules* **2018**, *8*, 55. [CrossRef] [PubMed]
14. Ndeboko, B.; Ramamurthy, N.; Lemamy, G.J.; Jamard, C.; Nielsen, P.E.; Cova, L. Role of cell-penetrating peptides in intracellular delivery of peptide nucleic acids targeting hepadnaviral replication. *Mol. Ther. Nucleic Acids* **2017**, *9*, 162–169. [CrossRef]
15. Choi, Y.-M.; Kim, H.; Lee, S.-A.; Lee, S.-Y.; Kim, B.-J. A Telomerase-derived peptide exerts an anti-Hepatitis B Virus effect via mitochondrial DNA stress-dependent type I interferon production. *Front. Immunol.* **2020**, *11*, 652. [CrossRef] [PubMed]
16. Nakamura, T.; Furunaka, H.; Miyata, T.; Tokunaga, F.; Muta, T.; Iwanaga, S.; Niwa, M.; Takao, T.; Shimonishi, Y. Tachyplesin, a class of antimicrobial peptide from the hemocytes of the horseshoe crab (*Tachypleus tridentatus*). Isolation and chemical structure. *J. Biol. Chem.* **1988**, *263*, 16709–16713. [CrossRef] [PubMed]

17. Jana, A.; Narula, P.; Chugh, A.; Kulshreshtha, R. Efficient delivery of anti-miR-210 using Tachyplesin, a cell penetrating peptide, for glioblastoma treatment. *Int. J. Pharm.* **2019**, *572*, 118789. [CrossRef]
18. Vernen, F.; Harvey, K.J.; Dias, S.A.; Veiga, A.S.; Huang, Y.-H.; Craik, D.J.; Lawrence, N.; Henriques, S.T. Characterization of Tachyplesin peptides and their cyclized analogues to improve antimicrobial and anticancer properties. *Int. J. Mol. Sci.* **2019**, *20*, 4184. [CrossRef]
19. Kumar, V.; Chugh, A. Peptide-mediated leishmaniasis management strategy: Tachyplesin emerges as an effective anti-leishmanial peptide against Leishmania donovani. *Biochim. Biophys. Acta Biomembr.* **2021**, *1863*, 183629. [CrossRef]
20. da Mata, É.C.G.; Mourão, C.B.F.; Rangel, M.; Schwartz, E.F. Antiviral activity of animal venom peptides and related compounds. *J. Venom. Anim. Toxins Incl. Trop. Dis.* **2017**, *23*, 3. [CrossRef]
21. Xie, H.; Wei, J.; Qin, Q. Antiviral function of Tachyplesin I against iridovirus and nodavirus. *Fish Shellfish Immunol.* **2016**, *58*, 96–102. [CrossRef]
22. Amir, F.; Siddiqui, Z.I.; Farooqui, S.R.; Anwer, A.; Khan, S.; Azmi, M.I.; Mehmankhah, M.; Dohare, R.; Khan, L.A.; Kazim, S.N. Impact of length of replication competent genome of hepatitis B virus over the differential antigenic secretion. *J. Cell. Biochem.* **2019**, *120*, 17858–17871. [CrossRef] [PubMed]
23. Samal, J.; Kandpal, M.; Vivekanandan, P. Hepatitis B "e" antigen-mediated inhibition of HBV replication fitness and transcription efficiency in vitro. *Virology* **2015**, *484*, 234–240. [CrossRef] [PubMed]
24. Kiruthika, S.; Bhat, R.; Dash, R.; Rathore, A.S.; Vivekanandan, P.; Jayaram, B. A novel piperazine derivative that targets hepatitis B surface antigen effectively inhibits tenofovir resistant hepatitis B virus. *Sci. Rep.* **2021**, *11*, 11723. [CrossRef] [PubMed]
25. Kandpal, M.; Samal, J.; Biswas, B.; Negi, A.; Mishra, V.C.; Tyagi, N.; Raina, V.; Vivekanandan, P. Enhanced hepatitis B virus (HBV) pre-genomic RNA levels and higher transcription efficiency of defective HBV genomes. *J. Gen. Virol.* **2015**, *96*, 3109–3117. [CrossRef]
26. Ahluwalia, S.; Choudhary, D.; Tyagi, P.; Kumar, V.; Vivekanandan, P. Vitamin D signaling inhibits HBV activity by directly targeting the HBV core promoter. *J. Biol. Chem.* **2021**, *297*, 101233. [CrossRef]
27. Biswas, B.; Kandpal, M.; Vivekanandan, P. A G-quadruplex motif in an envelope gene promoter regulates transcription and virion secretion in HBV genotype B. *Nucleic Acids Res.* **2017**, *45*, 11268–11280. [CrossRef]
28. Samal, J.; Kandpal, M.; Vivekanandan, P. A simple and rapid method for the quantitation of secreted hepatitis B virions in cell culture models. *Indian J. Med. Microbiol.* **2015**, *33*, 290–292. [CrossRef]
29. Colantonio, A.D.; Epeldegui, M.; Jesiak, M.; Jachimowski, L.; Blom, B.; Uittenbogaart, C.H. IFN-α is constitutively expressed in the human thymus, but not in peripheral lymphoid organs. *PLoS ONE* **2011**, *6*, e24252. [CrossRef]
30. Schuster, S.; Overheul, G.J.; Bauer, L.; van Kuppeveld, F.J.M.; van Rij, R.P. No evidence for viral small RNA production and antiviral function of Argonaute 2 in human cells. *Sci. Rep.* **2019**, *9*, 13752. [CrossRef]
31. Kuroki, A.; Tay, J.; Lee, G.H.; Yang, Y.Y. Broad-Spectrum antiviral peptides and polymers. *Adv. Healthc. Mater.* **2021**, *10*, e2101113. [CrossRef]
32. Donkers, J.M.; Appelman, M.D.; van de Graaf, S.F.J. Mechanistic insights into the inhibition of NTCP by myrcludex B. *JHEP Rep. Innov. Hepatol.* **2019**, *1*, 278–285. [CrossRef] [PubMed]
33. Smolders, E.J.; Burger, D.M.; Feld, J.J.; Kiser, J.J. Review article: Clinical pharmacology of current and investigational hepatitis B virus therapies. *Aliment. Pharmacol. Ther.* **2020**, *51*, 231–243. [CrossRef] [PubMed]

Disclaimer/Publisher's Note: The statements, opinions and data contained in all publications are solely those of the individual author(s) and contributor(s) and not of MDPI and/or the editor(s). MDPI and/or the editor(s) disclaim responsibility for any injury to people or property resulting from any ideas, methods, instructions or products referred to in the content.

pharmaceutics

Article

Choosing an Optimal Solvent Is Crucial for Obtaining Cell-Penetrating Peptide Nanoparticles with Desired Properties and High Activity in Nucleic Acid Delivery

Abhijit Biswas [1,†], Maria Maloverjan [1,†], Kärt Padari [2], Aare Abroi [1], Margus Rätsep [3], Sebastian K. T. S. Wärmländer [4,5], Jüri Jarvet [4,5], Astrid Gräslund [4,5], Vambola Kisand [3], Rünno Lõhmus [3] and Margus Pooga [1,*]

1 Institute of Technology, University of Tartu, 1 Nooruse Street, 50411 Tartu, Estonia
2 Institute of Molecular and Cell Biology, University of Tartu, 23b Riia Street, 51010 Tartu, Estonia
3 Institute of Physics, University of Tartu, 1 Wilhelm Ostwaldi, 51014 Tartu, Estonia
4 Chemistry Section, Stockholm University, Arrhenius Laboratories, SE-10691 Stockholm, Sweden
5 CellPept Sweden AB, Kvarngatan 10B, SE-11847 Stockholm, Sweden
* Correspondence: mpooga@ut.ee; Tel.: +372-737-4836
† These authors contributed equally to this work.

Abstract: Cell-penetrating peptides (CPPs) are highly promising transfection agents that can deliver various compounds into living cells, including nucleic acids (NAs). Positively charged CPPs can form non-covalent complexes with negatively charged NAs, enabling simple and time-efficient nanoparticle preparation. However, as CPPs have substantially different chemical and physical properties, their complexation with the cargo and characteristics of the resulting nanoparticles largely depends on the properties of the surrounding environment, i.e., solution. Here, we show that the solvent used for the initial dissolving of a CPP determines the properties of the resulting CPP particles formed in an aqueous solution, including the activity and toxicity of the CPP–NA complexes. Using different biophysical methods such as dynamic light scattering (DLS), atomic force microscopy (AFM), transmission and scanning electron microscopy (TEM and SEM), we show that PepFect14 (PF14), a cationic amphipathic CPP, forms spherical particles of uniform size when dissolved in organic solvents, such as ethanol and DMSO. Water-dissolved PF14, however, tends to form micelles and non-uniform aggregates. When dissolved in organic solvents, PF14 retains its α-helical conformation and biological activity in cell culture conditions without any increase in cytotoxicity. Altogether, our results indicate that by using a solvent that matches the chemical nature of the CPP, the properties of the peptide–cargo particles can be tuned in the desired way. This can be of critical importance for in vivo applications, where CPP particles that are too large, non-uniform, or prone to aggregation may induce severe consequences.

Keywords: cell-penetrating peptides; solvent; nanoparticle formation; nucleic acid delivery

Citation: Biswas, A.; Maloverjan, M.; Padari, K.; Abroi, A.; Rätsep, M.; Wärmländer, S.K.T.S.; Jarvet, J.; Gräslund, A.; Kisand, V.; Lõhmus, R.; et al. Choosing an Optimal Solvent Is Crucial for Obtaining Cell-Penetrating Peptide Nanoparticles with Desired Properties and High Activity in Nucleic Acid Delivery. *Pharmaceutics* **2023**, *15*, 396. https://doi.org/10.3390/pharmaceutics15020396

Academic Editor: Wei Huang

Received: 2 January 2023
Revised: 16 January 2023
Accepted: 19 January 2023
Published: 24 January 2023

Copyright: © 2023 by the authors. Licensee MDPI, Basel, Switzerland. This article is an open access article distributed under the terms and conditions of the Creative Commons Attribution (CC BY) license (https://creativecommons.org/licenses/by/4.0/).

1. Introduction

Gene therapy is an explosively developing field of nanomedicine and holds promise for treating various diseases that are undruggable today [1–4]. Although viral delivery systems have historically been highly important and widely applicable to the stable expression of various proteins, they have some inherent disadvantages, such as immunogenicity and potential carcinogenicity [5–8]. However, the development of non-viral delivery systems for nucleic acids (NAs) is challenging because of their limited ability to cross the plasma membrane. Low cellular uptake and inefficient endosomal escape are the main barriers to the widespread application of NAs since only a small proportion, usually, less than 0.1% of the cargo, reaches the cytoplasm and can be biologically active [9]. Thus, the development of safe and efficient delivery vehicles is necessary for the wide clinical application of therapeutic NAs [10].

In recent decades, multiple non-viral delivery systems have been developed, including physical approaches, such as electroporation and microinjection, and chemical delivery vectors, such as lipids, polymers, and peptides [11]. Among other delivery vehicles, cell-penetrating peptides (CPPs) have emerged as potential tools for the cellular delivery of multiple types of cargo molecules [12–16]. CPPs are peptides that are less than 30 amino acids long, which are typically cationic and/or amphipathic and can be associated with their cargo molecules both via covalent bonding and by non-covalent complex formation [17–20]. They have been shown to mediate the cellular uptake of biologically relevant cargo molecules of different sizes and chemistry, including small interfering RNA (siRNA) [21,22], antisense oligonucleotides (ASOs) [23,24], plasmid DNA (pDNA) [25,26], messenger RNA (mRNA) [27], and proteins [28], both in vitro and in vivo.

To become a candidate for clinical use, a pharmaceutical formulation must have a well-defined molecular composition, appropriate size, and suitable structural properties [29]. For the association of cationic CPPs with negatively charged molecules, such as NAs, simple mixing is often sufficient, as electrostatic interactions enable the rapid formation of stable complexes that assemble into nanoparticles. This approach is highly attractive due to its simplicity and low time consumption. However, the resulting complexes/particles tend to be heterogeneous, which is characteristic of the majority of self-assembling systems. Uneven physicochemical characteristics of the particle population, especially aggregates, can cause toxicity and affect the biodistribution of the cargo [29,30]. Thus, it is essential to prepare well-characterised, highly homogeneous nanoparticle formulations, which is especially important for in vivo applications [29].

One of the key factors that define the properties of the resulting particles is the solvent used for their preparation [31], as well as for the initial dissolution of the components. The solubility of a compound depends on its relationship with a solvent. Polar solvents such as water, methanol (MeOH), ethanol (EtOH), and isopropanol (iPrOH) dissolve polar and ionic compounds, while the solubility of non-polar compounds, such as lipids or aromatic compounds, is usually poor in polar solvents. However, the addition of semi-polar solvents, such as dimethyl sulfoxide (DMSO), dimethylformamide (DMF), etc., to polar solvents can enhance the solubility of non-polar compounds in polar solvents and stabilise solutions of amphiphiles, such as fatty-acid-modified CPPs.

The self-assembly of biologically active molecules into functionally active complexes is an important phenomenon for imparting their activity in biological systems [31,32]. Cosolvents that help to solubilise and later mask hydrophobic regions play a vital role in the self-assembly of biomacromolecules in aqueous solutions in in vitro assays [33–35]. In natural environments, e.g., in the cell cytosol, there are many cosolvents, i.e., osmolytes, such as the denaturant urea and trimethylamine N-oxide (TMAO) [36,37], which facilitate the self-assembly of protein complexes with other proteins, NAs, or lipids, and regulate their functioning by building up hydrogen bonds with water [33,37]. However, the identification of a particular cosolvent that enables the generation of functional biologically active soft materials with desired properties is not a straightforward or easy task.

In this work, we dissolved CPPs in different solvents and examined their effect on the CPP-mediated cellular delivery of RNA. We mainly analysed the ability of differently dissolved PepFect14 (PF14) peptides to deliver splice-correcting oligonucleotide (SCO) and siRNA into mammalian cell lines. In addition, we characterised PF14 complexes with SCO that assemble into colloidal nanoparticles using a battery of biophysical in vitro methods. The used prototypic fatty acid-modified CPP, i.e., PF14, is an amphipathic stearylated peptide from the PepFect family that was first synthesised by Ezzat et al. in 2011 [38]. It carries a positive charge of +5 that enables the formation of non-covalent complexes with NAs, which are internalised by cells via scavenger receptor-mediated endocytosis [38–40]. Peptides from the PepFect family have shown high efficiency in the delivery of SCO [39,41], siRNA [42], miRNA [43–46], and other types of NA molecules into mammalian cells both in vitro and in vivo and have been extensively studied by our group for several years [39,40,47,48].

2. Materials and Methods

2.1. Materials

SCO-705 (5′-CCUCUUACCUCAGUUACA-3′), SCO-654 (5′-GCUAUUACCUAAACCCAG-3′), and Cy5-SCO-654, all with a phosphorothioate backbone and 2′OMe modification, were purchased from Metabion (Planegg, Germany). The siRNA targeting luciferase (5′-GGACGAGGACGAGCACUUCTT-3′) and the negative control siRNA (5′-AGGUAGUGUAAUCGCCUUGTT-3′) were obtained from Microsynth (Balgach, Switzerland). The following CPPs were used: PF14, hPep3, and CADY (see Table S1). PF14 was obtained from PepMic (Suzhou, China), CADY was kindly provided by G. Divita and S. Deshayes, Montpellier CNRS, and hPep3 by T. Lehto, ITUT. For all peptides, 1 mM stock solutions were prepared and stored at −20 °C. Lipofectamine RNAiMAX Reagent was obtained from Invitrogen (Waltham, MA, USA). The used organic solvents were of p.a. grade with low water content (>99.5%): EtOH (Chem-Lab, Zedelgem, Belgium), 2-propanol and 2-butanol (Riedel-de Häen, Honeywell, Wabash, IN, USA), DMSO (AppliChem, Darmstadt, Germany), and THF (Lachner, Eurasburg, Germany). Trimethylene carbonate, i.e., 1,3-dioxan-2-one, was sourced from Apollo (Stockport, UK), and ethylene carbonate and glycerol carbonate were sourced from Acros Organics (Geel, Belgium). Other reagents, such as $CaCl_2$, $MgCl_2$, MeOH, DMF, etc., were obtained from Merck-Sigma-Aldrich (Darmstadt, Germany) if not stated otherwise. The cell culture lysis solution contained 1% (*v:v*) Triton X-100, 50 mM of Tris (pH 7.5), 150 mM of NaCl, and 1 mM of EDTA. Ready-to-use Cell Proliferation Colorimetric Reagent, WST-8, was purchased from BioVision (Milpitas, CA, USA). The luciferase activity measurement protocol was based on Rocha et al., 2016 [49] and Helmfors et al., 2015 [50]. The silicon wafers used for scanning electron microscopy and atomic force microscopy were obtained from Micro to Nano (Haarlem, The Netherlands).

2.2. PF14-SCO Nanoparticle Formation

For the splice-correction assay, we prepared CPP-SCO nanoparticles by mixing them at a 5:1 molar ratio (MR) in Milli-Q (MQ) water in 1/10th of the final volume [51]. After mixing by pipetting, the solutions were incubated at room temperature (RT) for 15 min. Then, if needed, divalent metal ions in the form of $CaCl_2$ or $MgCl_2$ (at a final concentration of 3 mM) were added, and the solutions were incubated for 15 min more. Finally, the particle-containing solutions were diluted 10× with pre-warmed cell culture media and applied to the cells. For the gene silencing assay, the CPPs were complexed with siRNA at a 34:1 MR, following the same protocol as described for CPP-SCO complexes. In all experiments, a cell culture medium containing 10% (*v:v*) of MQ water was used as a negative control solution. In the gene silencing experiments, the CPP-siRNA particles that contained non-targeting siRNA (siNEG) were used as a negative control, and the particles prepared using siRNA and Lipofectamine RNAiMAX (Invitrogen) were used as a positive control.

2.3. Dynamic Light Scattering Analysis

The hydrodynamic diameters and zeta potentials of the nanoparticles were analysed using a Zetasizer Nano ZSP instrument (Malvern Panalytical, Malvern, UK). To measure the size of the PF14 particles, the 1 mM stock was diluted 100-fold in MQ water at RT and allowed to equilibrate for at least 30 min. The CPP-SCO nanoparticle solutions were prepared as described previously at 5:1 MR with or without the addition of $CaCl_2$ or $MgCl_2$. After 15 min of incubation at RT, the solutions were diluted 10× with MQ water to reach the final volume of 1 mL and again incubated at RT for 15 min before measurement. Three to five consecutive measurements were performed depending on the homogeneity of the particle sizes.

2.4. Cell Culture

For the splice-correction assay, two human beta-thalassemia reporter cell lines, i.e., HeLa pLuc 705 [52] and HeLa EGFP 654 [53,54], were used. For the gene-silencing assay, a luciferase-expressing cell line, i.e., U87 MG-Luc2 [55], was used. The cells were cultured at 37 °C in Dulbecco's Modified Eagle Medium (DMEM) (Corning, Corning, NY, USA) containing 4.5 g/L of glucose and supplemented with 10% (*v:v*) of fetal bovine serum (FBS) and 1% (*v:v*) of a penicillin–streptomycin solution (100 U/mL penicillin and 100 g/mL streptomycin). The cells were grown on 6 cm cell culture dishes (Corning) in a humid atmosphere that contained 5% CO_2. The cells were split every second day. To detatch the cells, trypsin-EDTA solution (Corning) was used. The cells were washed with Dulbecco's Phosphate-Buffered Saline (DPBS) without calcium and magnesium (Corning). To conduct the experiments, the cells were plated on 96-well (Sarstedt, Nümbrecht, Germany) or 24-well (CytoOne, Hamburg, Germany) plates. All of the operations using cells were performed in a sterile cell culture hood.

2.5. Luminescence Measurement

For the luminescence measurements, HeLa pLuc 705 and U87 MG-Luc2 cells were plated on a 96-well plate at 10,000 cells/well and 8000 cells/well, respectively, 24 h before transfection and incubated with nanoparticles, as described above. After 24 h (for HeLa pLuc 705 cells) or 48 h (for U87 MG-Luc2 cells) of incubation with nanoparticle-containing solutions, the cells were washed with PBS and lysed by adding 20 μL of cell culture lysis buffer [56] containing 1% Triton-X100 and freezing the solutions at −25 °C. After thawing the solutions at room temperature, 70 μL of the luciferase substrate solution [56] per well was added, and the solutions were mixed and transferred to a white 96-well plate (Greiner Bio-One, Kremsmünster, Austria). Finally, the luminescence intensity was measured with an Infinite M200 PRO microplate reader (Tecan, Männedorf, Switzerland), using Magellan 7 software.

2.6. Cytotoxicity Assay

The effects of the peptides and peptide-containing nanoparticles on the viability of the cells were measured using the WST-8 assay (APExBIO, Boston, MA, USA). The cells were seeded on a 96-well plate the day before transfection. On the next day, the cells were transfected as described above. After 24 h (for the SCO assay) or 48 h (for the siRNA assay) of incubation, the media was replaced with fresh media that contained 10% of WST-8 solution, and the plate was incubated at 37 °C for 3 h before taking the measurement of the absorbance. The absorbance of the solutions was recorded at a 450 nm wavelength with a reference wavelength of 600 nm using an Infinite M200 PRO microplate reader (Tecan). The absorbance of the wells without cells but with medium and WST-8 was subtracted from all other values. The absorbance of the untreated cells was taken for 100% viability. As a positive control, the cells were incubated with 10% DMSO solution in the medium.

2.7. Confocal Microscopy

For the confocal microscopy analysis, HeLa EGFP 654 cells were plated in a 24-well plate at 50,000 cells/well on cover glasses with a diameter of 12 mm (Menzel-Gläser, Braunschweig, Germany) 24 h before transfection. On the next day, the medium was replaced with a nanoparticle-containing medium. The nanoparticles were prepared as described above. Here, we used SCO-654:Cy5-SCO-654 at a 9:1 ratio (180 nM SCO-654 and 20 nM Cy5-SCO-654), 1 μM PF14 dissolved in EtOH90/DMSO10/TMC0.4, and 3 mM $CaCl_2$. The cells were incubated with the solutions for 24 h at 37 °C. On the next day, the cells were washed with PBS and fixed by incubation with 3% paraformaldehyde (PFA) for 30 min at RT. The cell nuclei were stained with 4,6-diamidino-2-phenylindole (DAPI) 0.5 μg/mL solution in PBS. The cells were washed with PBS and mounted to glass slides in 30% glycerol. The specimens were analysed using an Olympus FluoView FV1000 (Olympus, Tokyo, Japan) confocal microscope. For each solution, two specimens were prepared, and at least three images per specimen were obtained (with circa 10 layers per image with a step size of 1.2 μm). Cy5 ($\lambda_{ex/em}$ = 633/666 nm), EGFP ($\lambda_{ex/em}$ = 488/530 nm), and DAPI ($\lambda_{ex/em}$ = 405/470 nm) signals were analysed, and a 60× objective with water immersion was used.

2.8. Flow Cytometry

For the flow cytometry, HeLa EGFP 654 cells were plated in a 96-well plate 24 h prior to transfection. On the next day, the cells were transfected with nanoparticles prepared of 100 nM SCO-654, 10 nM Cy5-SCO-654, 500 nM PF14, and 3 mM $CaCl_2$ or $MgCl_2$, as described above. The cells were incubated with the nanoparticles for 24 h at 37 °C and then washed with PBS, detached with a trypsin-EDTA solution, and resuspended in PBS. The cell nuclei were stained with 0.5 μg/mL DAPI for the gating out of dead cells. Next, flow cytometry of the cell suspensions was performed with an Attune NxT flow cytometer (Thermo Fisher Scientific, Waltham, MA, USA), using DAPI ($\lambda_{ex/em}$ = 405/470 nm), Cy5 ($\lambda_{ex/em}$ = 633/666 nm), and EGFP ($\lambda_{ex/em}$ = 488/530 nm) channels. For each sample, 10,000 events were analysed.

2.9. Transmission Electron Microscopy (TEM)

For TEM imaging, stock solutions of PF14 were diluted 10-fold in MQ water and drop-casted on carbon-coated copper grids. The specimens were air-dried for 5 min and stained with 2% aqueous uranyl acetate solution. PF14-SCO particles were prepared as described above using a 10:1 MR (PF14:SCO). Next, the samples were air dried and analysed with a Tecnai G2 Spirit transmission electron microscope (FEI, Hillsboro, OR, USA) operating at a 120 kV accelerating voltage, and images were captured using an Orius SC1000 camera and analysed with ImageJ software (NIH, Bethesda, MD, USA).

2.10. Atomic Force Microscopy (AFM)

For the AFM analysis of the nanoparticles, PF14-SCO complexes were prepared at a 10:1 MR (PF14:SCO; 1 μM SCO) in MQ water with or without the addition of 3 mM $CaCl_2$ or $MgCl_2$, which was conducted in the same way as the nanoparticles prepared for the cell-based assay. Then, 5 μL of the solution was spotted on a silicon wafer and air-dried at ambient temperature. The samples were examined with a Veeco Dimension Edge AFM unit (Veeco Instruments Inc., Plainview, NY, USA), and the data were analysed using Gwyddion software (Czech Metrology Institute, Jihlava, Czech Republic).

2.11. Scanning Electron Microscopy (SEM)

To analyse the morphology of both the bare peptide and the peptide–SCO complexes with or without the addition of $CaCl_2$ or $MgCl_2$, we prepared the complexes in the same way as for AFM. The samples were analysed using a NOVA scanning electron microscope (FEI, Hillsboro, OR, USA), and the images were analysed using ImageJ software.

2.12. Circular Dichroism Spectroscory (CD)

The circular dichroism spectra of the PF14 peptide dissolved in different solvents were recorded in a Chirascan-plus CD-spectrometer (Applied Photophysics, Leatherhead, UK). The peptides were diluted in MQ water to reach a concentration in the range of 60 to 80 μM for the measurement. The spectra were recorded between 195 nm and 250 nm using a quartz cuvette with a path length of 1 mm. The spectra were smoothed and plotted using GraphPad Prism 8.0 (GraphPad Software, Inc., San Diego, CA, USA).

2.13. Energy-Dispersive X-ray Spectrometry (EDX)

The qualitative and quantitative elemental analyses of the peptide–oligonucleotide–metal ion-containing nanoparticles were examined using EDX analysis. The nanoparticle specimens were prepared as described above for SEM and AFM. The EDX spectra were recorded in an Oxford Instruments X-Max system coupled to a Carl Zeiss SUPRA 55VP FE-

SEM instrument (Zeiss AG, Oberkochen, Germany). INCA software was used to evaluate the data with respect to the carbon content of the samples, i.e., all atomic percentages were calculated as the % of carbon atoms.

2.14. Statistical Analysis

All of the statistical analyses were carried out using GraphPad Prism 5.0 software. Each dataset represents mean ± SD. The statistical significance of the differences between the datasets was analysed using one-way ANOVA with post-hoc Tukey's or Dunnett's test at a significance level of 0.05.

3. Results and Discussion

3.1. PF14 Dissolved in Polar Organic Solvents Assembles into Nanoparticles in Water

The dissolution of PF14 or its analogues, where the stearyl moiety is replaced with different saturated fatty acids in water, leads to the formation of long micelles and/or aggregates of different sizes [57]. On the other hand, the condensation of NA molecules of different sizes by complex formation with PepFect (PF) and NickFect (NF) peptides in water leads to the formation of nanoparticles with a diameter in the range of 50 to 100 nm [58]. However, in addition to the nanoparticles, the PF14-NA solution also contains very large particles/aggregates, especially when complexes are formed at high concentrations of the constituents [29]. Large aggregates and conglomerates of nanoparticles are considered to strongly contribute to the toxicity and immunogenicity of nanoparticle formulations [59]. We presume that aggregates of PFs that form when the peptide is dissolved in water do not completely dissociate during complex formation with NAs, and some of these produce very large particles, leading to unwanted effects that mostly manifest in in vivo experiments. We hypothesised that harnessing a suitable organic solvent for dissolving PF14 could reduce the formation of large aggregates in aqueous solutions since the peptide then retains its microenvironment [36]. This, in turn, could also suppress the formation of aggregates in PF14-NA nanoparticle formulations. In principle, if we can reduce the toxicity and other side effects of CPP-NA nanoparticles in vivo, we could even accept a small loss in activity if needed.

However, choosing a proper solvent to dissolve a CPP that is later used for the preparation of nanoparticles is a challenging task. The Food and Drug Administration of the USA (FDA) divides solvents into four classes [60] according to their safety for patients and environmental impacts. We excluded the application of class I solvents, which are defined as highly toxic and include substances such as benzene and dichloroethene. Class II solvents display lower but substantial toxicity and can be included in pharmaceutical products at very limited concentrations, and their examples include MeOH, toluene, tetrahydrofuran (THF), etc. For this study, class III solvents were primarily used, which are considered safe for humans when applied in amounts not exceeding 50 mg per day (for some of them, this amount can be higher). Examples of such solvents include EtOH, acetic acid, and DMSO.

First, we dissolved PF14 in the series of organic solvents commonly used at moderate concentrations in experiments with cultured cells and in vivo, such as EtOH, DMSO, iPrOH, and MeOH. All of these solvents dissolved PF14 quickly and completely up to a 10 mM concentration, which was the highest concentration tried. Water, in contrast, dissolved PF14 more slowly, even at a 1 mM concentration, which is the most commonly used concentration of stock solutions of this type of peptide.

Next, we diluted 1 mM PF14 stock solutions into MQ water to the 10 µM concentration that is commonly used for the preparation of its complexes with NAs to be delivered into cells. Analysis of the size and homogeneity of the nanoparticles formed by PF14 dissolved in different solvents by dynamic light scattering (DLS) showed a very high variation between the nanoparticles prepared with different solvents (Figure 1). The PF14 stocks in EtOH and iPrOH produced particles with diameters ranging from 100 to 200 nm and a rather fair homogeneity, whereas the stocks in water, DMSO, and MeOH produced larger particles/aggregates that were around 300 nm large and had high polydispersity indexes

(Figure 1). The numerical values of all particle diameters and polydispersity indexes can be found in the Supplementary materials (Table S2).

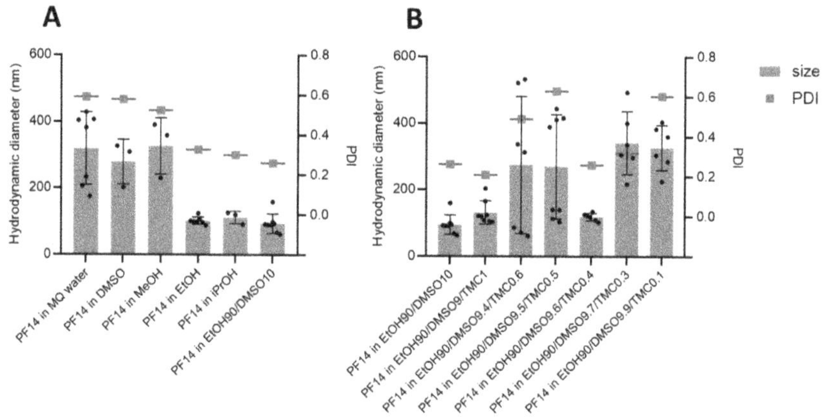

Figure 1. Hydrodynamic diameter and polydispersity indexes (PDI) of particles formed by differently dissolved PF14. (**A**) Particles formed by PF14 dissolved in MQ water, alcohols, and the mixture of EtOH and DMSO. (**B**) Particles formed by PF14 dissolved in a mixture of EtOH and DMSO with addition of TMC at different percentages. The 1 mM PF14 stock solutions in different solvent mixtures were diluted 10× with MQ water and incubated at room temperature for 30 min before measurement with Malvern Zetasizer Nano. Each dataset mean ± SD of at least three measurements with individual measurements indicated as black dots. The polydispersity indexes are presented as the average of at least three independent measurements. All proportions are given as *v:v*.

Since EtOH, which is the most commonly used alcohol in drug formulations because of its low toxicity, produced the smallest PF14 particles among the used solvents, we selected the respective stock solution for further refinement, i.e., for supplementation with DMSO. DMSO is a polar aprotic organic solvent that successfully dissolves various compounds, including hydrophobic substances that are insoluble in aqueous solutions. DMSO can legally be used in therapeutic formulations up to a 10% (*v:v*) concentration, and its side effects are considered not significant. According to the FDA, both EtOH and DMSO are class III solvents, i.e., considered the safest with minimal toxicity. Moreover, DMSO is also known to facilitate the cellular internalisation of many compounds, including macromolecules, such as CPPs [61] and their constructs with cargo [62]. The dissolution of PF14 in the EtOH/DMSO (9/1, *v/v*) mixture and further 10-fold dilution in water led to the formation of even slightly smaller peptide nanoparticles than with the pure ethanol, i.e., with about a 100 nm diameter (Figure 1A). This suggests that DMSO may contribute to the dissolution of PF14 in EtOH by weakening the hydrophobic peptide–peptide interactions, e.g., the interactions between the stearyl groups. In addition, after the dilution of the PF14 stock solution in water, DMSO could also contribute to the microenvironment around/inside the peptide particles [36], whose formation is triggered by condensation in an aqueous milieu that leads to phase separation [63].

The phase separation induced by the transfer of PF14 from an organic solvent phase into an aqueous environment is less efficient than in the case of peptides with fluoroalkane modification [64]. In order to induce a more pronounced phase separation, we analysed the effect of the inclusion of excipients into the PF14 particles. Organic carbonates that show excellent protein solvation and stabilisation properties [65] and have limited solubility in aqueous solutions were, therefore, analysed for the ability to facilitate the assembly of PF14 nanoparticles and increase their homogeneity. Among the analysed organic carbonates, the cyclic compounds contributed to the homogeneity of PF14 particles. Trimethylene carbonate had the strongest stabilising effect on the peptide nanoparticles, followed by ethylene carbonate, whereas glycerol-1,3-carbonate did not contribute to the formation of particles (data not shown).

Trimethylene carbonate (TMC) is a cyclic ester with a six-member ring structure, which is most often used for the preparation of biodegradable polymers used for biomedical applications [66]. The polymerisation of TMC results in the formation of poly(trimethylene carbonate) (PTMC), a highly promising component for soft-tissue engineering and drug delivery [67]. However, here, to assist the formation of PF14 nanoparticles in water, we added a very low amount of TMC (below 1%) to the mixture of EtOH and DMSO, which was used to dissolve the PF14 peptide. We varied the concentration of TMC in the peptide stock solution from 0.1% to 1% and analysed the hydrodynamic diameter of the resulting PF14 particles (Figure 1B). The addition of 1% TMC to the PF14 solution produced reasonably low hydrodynamic diameters and high homogeneity in the water particles (PDI around 0.2). Lowering the TMC concentration, surprisingly, increased the diameter of the PF14 particles and strongly increased their heterogeneity, especially at the 0.5–0.6% concentration. The homogeneity and suitable size of the particles were regained at a 0.4% concentration of TMC, which we considered optimal for PF14 because the NA cargo should not be too tightly condensed into nanoparticles with the peptide, which could happen at a 1% TMC concentration. Thus, after translocation into the cells, oligonucleotides would more probably dissociate from not-overly-stable PF14 nanoparticles in order to be active. Therefore, PF14 dissolved in EtOH90/DMSO10/TMC0.4 was used to analyse the properties of the particles formed by the peptide upon complexation with SCO and siRNA (Table S3). The advantageous properties of PF14 nanoparticles were not compromised by complexation with SCO or siRNA or by the further supplementation of particles with divalent metal ions, i.e., Ca^{2+} and Mg^{2+} (Table S3). The metal ions have earlier been shown to increase the efficiency of PF14-NA complexes multiple fold and were, therefore, included in this study [51]. Particles of PF14 in complex with SCO were of similar size to the particles that consisted of the peptide only (112 and 119 nm, respectively). Surprisingly, the particles prepared using PF14 and siRNA had lower hydrodynamic diameters (87 nm) than the peptide particles, suggesting the reorganisation of the latter upon complex formation with oligonucleotides. The addition of $CaCl_2$ and $MgCl_2$ to the CPP–NA complexes did not significantly alter their size, except for the PF14-siRNA-Mg^{2+} particles, which were substantially larger than the corresponding particles without $MgCl_2$. All of the nanoparticles had a positive zeta potential (Table S3), with SCO-containing complexes possessing higher charge than the siRNA-containing ones.

3.2. Dissolution of PF14 in the Mixture of Polar Organic Solvents Does Not Compromise the Ability of Peptide to Transduce Oligonucleotides into Cells

DLS analyses indicated that PF14 dissolved in various organic solvents led to the formation of smaller and more homogeneous particles in an aqueous milieu as compared to the peptide dissolved directly in water. In order to test whether the nanoparticles formed by NAs and PF14 dissolved in organics can mediate the efficient delivery of the cargo molecules and induce biological effects in cells, we first harnessed a model system based on splicing switching. The delivery of splicing switching oligonucleotide SCO-705 into HeLa pLuc 705 reporter cells abolishes the aberrant splicing of pre-mRNA and thereby rescues the expression of luciferase, whose activity correlates very well with the concentration of SCO in cell cytosol [52]. First, we prepared PF14 solutions in alcohol–DMSO (9/1) mixtures and formed the respective nanoparticles with SCO. Surprisingly, all of the PF14 solutions prepared in the alcohol–DMSO mixture yielded a slightly higher efficiency for splicing correction in HeLa pLuc 705 cells than the particles prepared with water-dissolved peptides when the particles were complemented with Ca^{2+} or Mg^{2+} ions (Figure S1). On the contrary, the particles containing only PF14 and SCO showed rather similar effects on splicing switching in the case of all of the different peptide stock solutions used. As all of the alcohol/DMSO mixtures resulted in similar activity of SCO-PF14 nanoparticles, we kept using EtOH as the most biocompatible one.

Next, in the same way, we analysed the PF14 stock solutions prepared in MQ, EtOH, EtOH90/DMSO10, and EtOH90/DMSO/TMC, with varying percentages of TMC (Figure 2). As can be seen in Figure 2A, all of the SCO-PF14 nanoparticles prepared with PF14 dissolved in organic solvents showed activity similar to the ones prepared using water stock. The inclusion of biocompatible divalent cations, Ca^{2+} and Mg^{2+}, into the nanoparticles strongly increased the SCO delivery efficiency of PF14, as we observed earlier [51]. Although the increase was observed in the case of all the PF14 stocks used, it was more prominent and consistent in the case of the solutions that contained TMC, and respective SCO-PF14-Ca^{2+}/Mg^{2+} particles led to a significantly higher splicing correction than the ones prepared with water-dissolved PF14. Although both DLS (Figure 1) and splicing correction assay (Figure 2A) showed that PF14 stocks with 0.4% and 1% of TMC provided equally small and homogeneous particles, the high activity of the cargo and good reproducibility of the results, we selected PF14 dissolved in an EtOH90/DMSO9.6/TMC0.4 mixture to be used in further experiments.

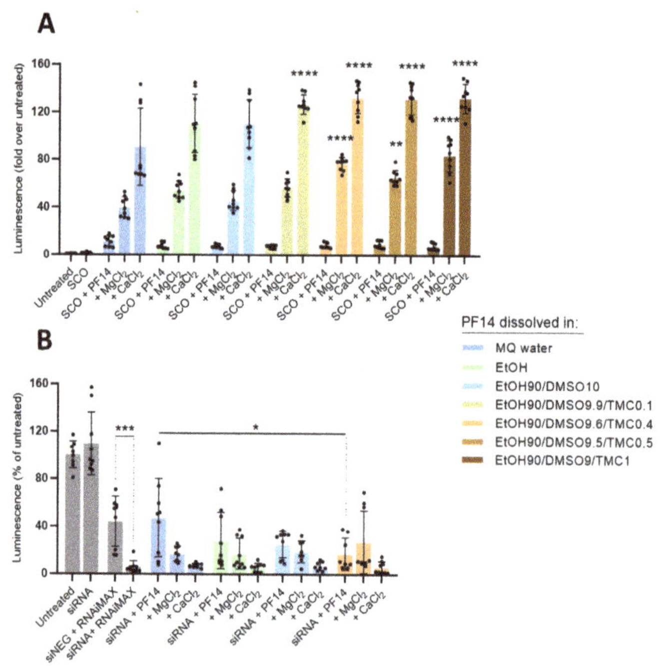

Figure 2. Effect of nanoparticles prepared from SCO or siRNA and differently dissolved PF14 on splicing correction and luciferase silencing, respectively. The 1 mM stock solutions of PF14 were prepared by dissolving the peptide in different solvents and their mixtures, as indicated in legends (proportions are given as *v:v*). (**A**) HeLa pLuc 705 cells were incubated for 24 h with solutions containing SCO alone (100 nM), nanoparticles of SCO and PF14 taken at MR 5 (PF14:SCO) with or without the addition of 3 mM $CaCl_2$ or $MgCl_2$. (**B**) U87 MG-Luc2 cells were incubated for 48 h with solutions containing siRNA alone (15 nM), nanoparticles of siRNA and PF14 taken at MR 34 (PF14:siRNA) with or without the addition of 3 mM $CaCl_2$ or $MgCl_2$. As a positive control, siRNA was transfected with Lipofectamine RNAiMAX. In all cases, as a negative control, the cells were incubated with a medium containing 10% (*v:v*) of MQ water ("Untreated"). siNEG—negative, i.e., non-targeting siRNA. Each dataset represents mean ± SD of technical replicates (shown as black dots) from three independent experiments. Data were analysed using one-way ANOVA with post-hoc Tukey's test. Asterisks indicate statistically significant difference compared to the same solution from "MQ water" group (**A**) or between indicated datasets (**B**), * p-value < 0.05, ** p-value < 0.005, *** p-value < 0.0005, **** p-value < 0.0001.

Next, we assessed whether PF14 dissolved in the polar organic solvent mixtures could perform at the same level as the water-dissolved peptides in the delivery of another NA oligomer with different biological effects, siRNA. Silencing RNA introduced into luciferase- overexpressing U87 MG-Luc2 cells with PF14 strongly suppressed luciferase expression, even at a 15 nM siRNA concentration (Figure 2B). The nanoparticles formed by siRNA and PF14 dissolved in the EtOH–DMSO mixture were slightly more efficient than those assembled from peptide stocks made in water or EtOH, reducing target protein expression by about 70%. Importantly, Ca^{2+} and mostly also Mg^{2+} ions increased the luciferase-silencing effect of PF14-siRNA nanoparticles for all of the differently dissolved PF14 samples. Importantly, none of the tested solutions significantly affected the viability of HeLa pLuc 705 or U87 MG-Luc2 cells (Figure S2).

Based on these results, we can conclude that the dissolution of PF14 in the mixture of the selected polar organic solvents does not compromise the cellular delivery of oligonucleotides, sometimes providing even slightly higher biological effects, and the efficiency of the formed nanoparticles can be substantially increased by the inclusion of divalent metal ions into the nanoparticles.

3.3. Electron Microscopy Confirms the Beneficial Effect of Dissolving PF14 in Organic Solvents, Compared to Dissolving in Water

Now, when both the DLS (Figure 1) and activity data (Figure 2 and Figure S1) showed that PF14 dissolved in organic solvents yields particles with advantageous physical properties and delivers NAs into cells with an efficiency comparable with or even superior to water-dissolved PF14 we analysed the TEM particles formed in the water milieu using the following PF14 stocks: MQ, EtOH, EtOH90/DMSO10, and EtOH90/DMSO9.6/TMC0.4 (Figure 3). In good concordance with the DLS analysis data, PF14 dissolved in water was mostly detected in electron microphotos as large aggregates or conglomerates of very large elongated micelles up to 500 nm or even larger sizes (Figure 3A). PF14 dissolved in EtOH assembled into more regular particles upon dilution in water, i.e., under the conditions used to assemble PF14 nanoparticles with oligonucleotide cargo (Figure 3B). The majority of the peptide was organised in spherical nanoparticles of about 30 to 100 nm diameters, and some smaller particles assembled into grape or necklace-like clusters, whereas very large aggregates were not detected in the TEM images.

The solvent mixture of EtOH and DMSO (9/1) seemed to be almost optimal for the dissolution of PF14 since rather homogeneous nanoparticles with <100 nm diameter formed upon the dilution in water, and no aggregates or clusters of nanoparticles were detected in microphotos (Figure 3C). The particles formed by PF14 dissolved in EtOH90/DMSO9.6/TMC0.4 were of similar size and morphology compared to the EtOH90/DMSO10 stock (Figure 3D). The numerical values of all particles' diameters calculated from electron micrographs of Figure 3 can be found in the Supplementary Materials (see Table S4).

3.4. Confocal Microscopy Analysis Corroborates Efficient Oligonucleotide Delivery and Splicing Correction with PF14 Dissolved in the Mixture of Organic Solvents

In order to visualise the transfection and intracellular distribution of SCO, as well as the splicing correction triggered by it, we performed confocal microscopy of HeLa EGFP 654 cells. In this cell line, analogous to HeLa pLuc 705, the mutated intron of beta-globin is included in the genome, interfering with the expression of EGFP protein. The pre-mRNA of EGFP contains an aberrant splicing that induces mutation in position 654, and added SCO-654 blocks the mutated site, restoring normal splicing and rescuing EGFP expression.

In addition to the particles of 1 μM PF14 and 200 nM SCO (Figure 4C–E), particles with the addition of 3 mM $CaCl_2$ were included (Figure 4F–H). As expected, all of the PF14 stocks tested induced equally efficient transduction of SCO-Cy5 (red) into cells, independent of the addition of $CaCl_2$. However, a substantial increase in EGFP expression induced by SCO-654 could be observed when $CaCl_2$ was added to the particles. In the microscopy images, no significant changes in the transfection or expression efficiency or in the intracellular distribution of SCO could be detected when different PF14 stock solutions were used. As expected, the morphology of the cells was not affected by any of the solutions used (Figure S3).

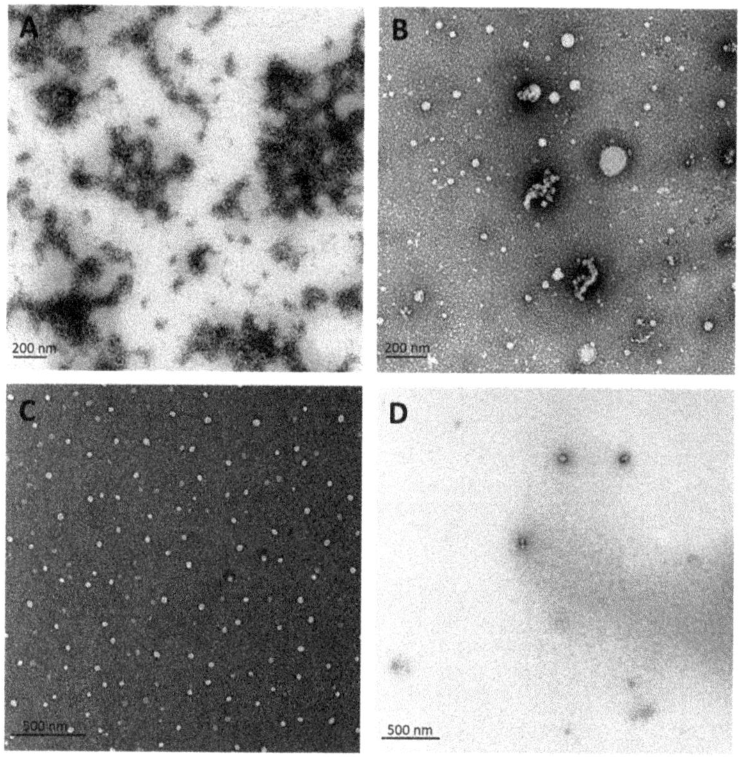

Figure 3. Morphology of particles formed by differently dissolved PF14 as analysed by TEM. The 1 mM stock solutions of PF14 were prepared by dissolving the peptide in different solvents and their mixtures: MQ water (**A**), EtOH (**B**), EtOH90/DMSO10 (**C**) or in EtOH90/DMSO9.6/TMC0.4 (**D**), and resulting particles were analysed by TEM. All proportions are given as *v:v*.

For better quantification of the induced splicing correction in HeLa EGFP 654 cells, we next applied flow cytometry analysis. The incubation of HeLa EGFP 654 cells with nanoparticles analogous to the ones used in the confocal microscopy experiment (Figure S4) revealed significant advantages of $CaCl_2$/$MgCl_2$-supplemented SCO-PF14 nanoparticles prepared with PF14 dissolved in organic solvent mixtures compared to water-dissolved PF14 (Figure S4A). In the case of PF14 dissolved in MQ water, nanoparticles supplemented with the salts provided an average 4.34× increase in EGFP expression compared to the untreated cells. However, in the case of PF14 dissolved in EtOH90/DMSO and EtOH90/DMSO9.6/TMC0.4, salt-supplemented nanoparticles provided 6× and 6.93× increase in fluorescence, respectively, being significantly more efficient than their counterparts prepared with water-dissolved PF14. It is worth mentioning that, unlike PF14 complexes with siRNA and SCO-705, PF14-SCO-654 nanoparticles benefited equally from the addition of $CaCl_2$ and $MgCl_2$ (Figure S4A). As expected, there were no remarkable differences in Cy5 fluorescence intensity between the differently dissolved PF14 solutions, suggesting that the internalisation of SCO-delivering nanoparticles was not substantially influenced. The only exceptions were $MgCl_2$-supplemented PF14-SCO particles prepared with PF14 dissolved in EtOH, which yielded slightly lower transfection than other peptide specimens (Figure S4B).

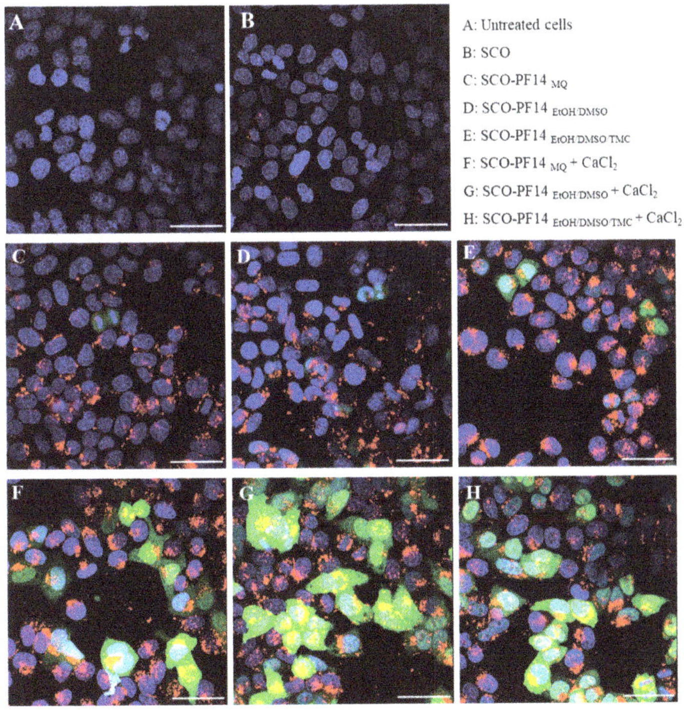

Figure 4. Internalisation and biological effect of nanoparticles containing SCO, differently dissolved PF14 and $CaCl_2$. HeLa EGFP 654 reporter cells were incubated with nanoparticles of 180 nM SCO-654, 20 nM Cy5-SCO-654, 1 μM PF14 and 3 mM $CaCl_2$ for 24 h. Cells were fixed, and specimens were analysed with Olympus FluoView FV1000 confocal microscope. Cells were either left untreated (**A**), incubated with the two SCOs (**B**), the SCO-PF14 nanoparticles (**C–E**) or the SCO-PF14-Ca^{2+} nanoparticles (**F–H**). The following PF14 stocks were used: PF14 dissolved in MQ water (**C**,**F**); PF14 dissolved in EtOH90/DMSO10 mixture (**D**,**G**); and PF14 dissolved in EtOH90/DMSO9.6/TMC0.4 mixture (**E**,**H**). Rescued EGFP expression is shown in green. Cell nuclei were visualised with DAPI (blue). For tracking cellular association and internalisation of oligonucleotide, Cy5-labelled SCO was used (red). The merged images of all confocal layers are presented. Scale bar: 50 μm.

Altogether, these results indicate that PF14 dissolved in organic solvents, especially in their mixtures, can be superior to water-dissolved PF14 in a fluorescence rescue assay based on splicing correction/switching.

3.5. *Other Organic Solvents Can Also Be Beneficial for Dissolving PF14, and Hydrophobic CPPs*

In addition to the typically used alcohols, such as MeOH, EtOH, iPrOH, butanol, and DMSO, we also tested whether other commonly used organic polar aprotic solvents, such as DMF and THF alone or in combination with different alcohols could be used to dissolve PF14 and prepare biologically highly active nanoparticles with SCO (Figure S5). In general, SCO-PF14 nanoparticles supplemented with Ca^{2+} showed higher splice-correcting effects if DMF and THF were used for the preparation of peptide stock solutions compared to the particles formed using peptides dissolved in water. DMF- and THF-containing PF14 solutions had transfection efficiency comparable to PF14 dissolved in EtOH90/DMSO9.6/TMC0.4, enabling up to a 130× increase in reporter protein activity/luminescence when supplemented with $CaCl_2$. Surprisingly, a stock prepared in MeOH90/DMF10 almost completely lacked activity, suggesting that this particular solvent system is not suitable for dissolving PF14 for biological assays. Other DMF- and THF-containing PF14 stock solutions had similar activity, with pure DMF and THF10/EtOH90 being superior to other combinations. However, it should be mentioned that the FDA classifies both DMF and THF as class II solvents and, therefore, the harnessing of colloidal nanoparticles that contain these solvents is limited, especially when considering the high efficiency of nanoparticles formed with EtOH solutions of PF14. Correspondingly, the presence of DMF and THF in the respective solutions of PF14-SCO nanoparticles interfered to some extent with the viability of both HeLa pLuc 705 and U87 MG-Luc2 cells (Figure S6), with 100% THF used as a solvent being the most cytotoxic for both cell lines and 100% DMF also showing higher-than-average toxicity in HeLa pLuc 705.

After achieving remarkable changes in the activity of PF14 in different co-solvent systems in SCO transfection, we were interested in assessing whether this strategy of dissolving is also applicable to other peptides. For this, we examined two other well-known hydrophobic and fatty-acid-modified CPPs, which have been shown to self-assemble into nanoparticles with oligonucleotide cargo analogously to PFs, and could, in principle, aggregate in aqueous solutions. We included in our experiments hPep3 peptide, which showed promising results during SCO transfection due to its increased hydrophobicity [68], and CADY, which is a known efficient siRNA transporter [69–71]. We prepared nanoparticles of hPep3 and SCO in an analogous way to PF14-SCO nanoparticles, also using MR 5 (hPep3:SCO, 100 nM SCO) and adding 3 mM $CaCl_2$ or $MgCl_2$. All three hPep3 stocks dissolved in organic solvents showed SCO transfection efficiency and reporter-protein expression in cells at a similar level with water-dissolved peptides (Figure S7A). However, the reporter-protein expression achieved with the hPep3–SCO complexes remained lower compared to PF14-mediated delivery. Nanoparticles of CADY with siRNA were prepared at MR34 and complemented with 3 mM Ca^{2+}/Mg^{2+}, as mentioned previously, for PF14 complexes. All of the solutions of CADY, both prepared in water and organics, showed similar siRNA transfection and target protein knock-down (Figure S7B). No substantial cytotoxicity was detected in the case of hPep3 and CADY nanoparticles (Figure S8). We did not include in our experiments cationic CPPs that dissolve very well in water and are not prone to self-aggregate since we considered that these do not benefit from dissolution in an organic solvent.

3.6. Nanoparticles of PF14 and Its Complexes with SCO Are Spherical

After optimising the dissolution of PF14 regarding the activity of nanoparticles prepared with SCO and analysing their hydrodynamic diameter, homogeneity, and zeta potential using DLS, we continued with more detailed analyses of the properties of such particles. First, we assessed whether the changes in the secondary structure of PF14 were introduced upon association with SCO. Using circular dichroism spectroscopy (CD), we confirmed that in water, PF14 folds into an α-helical structure notwithstanding the solvent used for the initial dissolution of the peptide (Figure S9). There are, however, clear differences between the CD spectra for the different particles. The spectra for purely α-helical peptides cross the zero line around 203 nm, such as the spectrum of PF14 in complex with SCOs and 3 mM $CaCl_2$ (Figure S10C). The spectra that cross the zero line below 203 nm, as most of the spectra seen in Figures S9 and S10 do, reflect peptides with a combination of α-helical and random-coil secondary structures. This is particularly clear for PF14 dissolved in MQ water only (Figure S9A), where the spectrum crosses the zero line below the 200 nm mark. Furthermore, typical α-helical CD spectra display a strong minimum at 208 nm and a weaker minimum at 222 nm. The [θ222]/[θ208] ratio is around 0.7 for isolated α-helical structures, while higher values for the [θ222]/[θ208] ratio—up to around 1:1 ratio—indicate that coil–coil interactions (also known as helix supercoiling) have formed [72]. The unusual spectrum shown in Figure S10D likely reflects a combination of α-helical and β-sheet structures. It should be noted that the overall intensities are different for the different samples, even though the peptide concentrations are the same. The lower signal intensities observed in, e.g., Figure S9A,C are most likely caused by the formation of large aggregates that either precipitate out of the solution or are too large to be penetrated by the CD light beam, in both cases, effectively reducing the observable peptide concentration. Surprisingly, upon interaction with SCO, the structure of PF14 was not reorganised, and it retained an α-helical structure, as characterised by CD spectroscopy (Figure S10). We believe that upon association with SCO (as well as other types of bioactive NA molecules), PF14 nanoparticles may partly dissociate to form nanocomplexes using electrostatic and hydrophobic interactions, whereas the secondary structure of the peptide is not reorganised.

Earlier, we showed by DLS and TEM analyses that PF14 dissolved in a solvent mixture of optimal composition (EtOH90/DMSO9.6/TMC0.4) formed nanoparticles upon dilution into an aqueous solution (Figures 1 and 3). However, negative staining of specimens with aqueous uranyl acetate of low pH could influence the morphology of the detected particles [73], and the particles detected by DLS in solution may not be stable [74]. Therefore, to analyse whether the morphology of the self-assembled nanostructures formed upon the complexation of SCO with PF14 is favourable for the uptake by cells, we harnessed a battery of various microscopic techniques. We applied atomic force microscopy, scanning electron microscopy, and transmission electron microscopy (Figure 5) for a detailed analysis of nanoparticles' morphology since these are commonly used to characterise nanoparticles in parallel with DLS. All, AFM, SEM, and TEM analyses confirmed that the PF14-SCO nanoparticles were spherical and with a diameter in the range of 40–60 nm (Figure 5, Table S5). The high homogeneity and small size of PF14-SCO nanoparticles detected by these biophysical methods are in very good concordance with their ability to trigger biological responses in living cells (Figure 2).

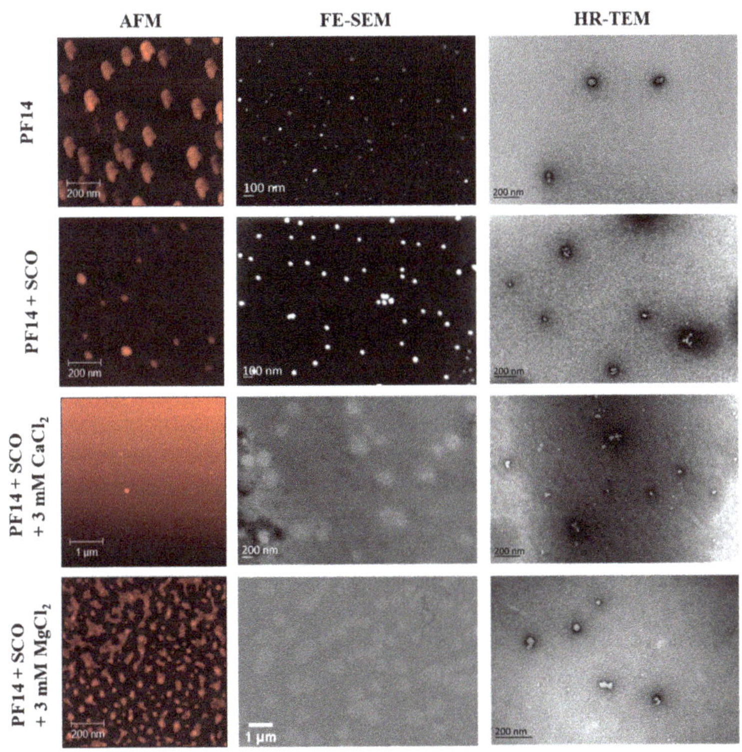

Figure 5. Morphology of particles formed by PF14 dissolved in EtOH90/DMSO9.6/TMC0.4 and its complexes with SCO and divalent metal ions added in the form of $CaCl_2$ and $MgCl_2$. Nanoparticles were prepared by mixing SCO and PF14 at MR 10 and adding $CaCl_2$ and $MgCl_2$ after 15 min. After a total of 30 min of incubation, solutions were analysed by AFM, field emission SEM (FE-SEM), or high-resolution TEM (HR-TEM).

The inclusion of biologically relevant divalent cations Ca^{2+} and Mg^{2+} into PF14 nanoparticles with oligonucleotides strongly increased the biological effect of the latter inside the cells (Figure 2), consistent with earlier reports [48,51,75–81]. Therefore, we next analysed the morphology of metal ion-complemented PF14-SCO nanoparticles by AFM, SEM, and TEM (Figure 5, Table S5). After the inclusion of Ca^{2+} ions, the size of PF14-SCO nanoparticles increased to about 100–200 nm, and they retained a rather spherical shape, as seen in the AFM and SEM images (Figure 5). In the specimens stained with uranyl acetate (UA), TEM visualised the clusters of particles of similar size and also small nanoparticles (Figure 5). We consider it plausible that in the acidic milieu of uranyl acetate (pH around 4.5), used to stain the specimen, the PF14-SCO-Ca^{2+} nanoparticles could partly dissociate. The small nanoparticles detected in the UA-stained PF14-SCO-Ca^{2+} specimens have a similar size to the PF14-SCO particles formed without Ca^{2+}/Mg^{2+}, as seen in the TEM microphotos. This suggests that calcium ions are less strongly associated with PF14-SCO-Ca^{2+} nanoparticles compared to the interaction of PF14 with SCO. Therefore, in acidic conditions, calcium ions can leave the PF14-SCO-Ca^{2+} particle, leading to the (re)formation of PF14-SCO particles. Our suggestion is supported by a similar behaviour of PF14-SCO-Mg^{2+} nanoparticles. Their size is about 100 nm, as measured by DLS [51] (Table S3), but increases to several hundreds of nanometres when analysed by AFM and SEM at a neutral pH (Figure 5). However, in an acidic milieu, the size of PF14-SCO-Mg^{2+} nanoparticles decreases to about 50 nm, which is analogous to Ca^{2+} ions-containing particles (Figure 5). We assume that a partial protonation of SCO in an acidic milieu reduces its negative charge and could explain the facilitated dissociation of Ca^{2+}/Mg^{2+} ions from PF14-SCO-Ca^{2+}/Mg^{2+} particles and their further disintegration to smaller PF14-SCO nanoparticles. The biological effect of Mg ions-containing PF14-SCO nanoparticles was lower than that of the Ca-supplemented form. Expectedly, PF14-SCO-Mg nanoparticles were less well defined and differed in morphology, as revealed by AFM and SEM analyses (Figure 5, Table S5), corroborating results published earlier [51].

In order to analyse whether supplemented Ca^{2+} and Mg^{2+} ions are indeed associated into PF14-SCO nanoparticles, we used energy-dispersive X-ray spectroscopy (EDX) (Figure S11 and Table S6). EDX analysis confirmed the presence of Ca^{2+} and Mg^{2+} ions in nanoparticles assembled of PF14, SCO and $CaCl_2$ or $MgCl_2$, respectively.

We may conclude that when PF14 was dissolved in an optimal solvent mixture, it assembled into spherical nanoparticles after dilution in an aqueous media, and the aggregation of the peptide was not detected by the used battery of biophysical analysis methods. The spherical morphology of PF14 particles was retained after complex formation with the oligonucleotide cargo, and even after supplementation of PF14-SCO nanoparticles with Ca^{2+} or Mg^{2+} ions that strongly increase the bioavailability of cargo inside living cells.

4. Conclusions

In the current study, we have elaborated a new approach for formulating nonaggregating and uniformly sized CPP–oligonucleotide nanoparticles. Our results showed that the cosolvent approach used to dissolve hydrophobic CPP in organic solvents of high biocompatibility enabled better control of the size and aggregation of the nanoparticles compared with dissolving the peptide in water. Alcohols, such as EtOH or iPrOH, form hydrogen bonds with water and help to weaken the interactions between the peptide and water, modulating the morphology of peptide-formed particles [82,83], yielding smaller and more uniform particles than the ones formed by water-dissolved peptides. In addition to alcohols, we added DMSO to the peptide stock solutions, which is a highly potent solvent for dissolving hydrophobic compounds. As shown earlier, DMSO, in addition to increasing solubility, can also facilitate cell permeability of nanoparticles by forming hydrophilic pores in the lipid bilayer [84,85], thus isolating the nanoparticles from the hydrophobic environment of the plasma membrane [86]. Finally, we added a small amount of biodegradable organic carbonate, TMC [66], to the aforementioned mixtures in order to induce more efficient phase separation of CPP nanoparticles or their complexes with oligonucleotide from an aqueous milieu. We found that TMC, in combination with alcohol and DMSO, yielded peptide nanoparticles of highly suitable size and increased the homogeneity of the nanoparticle population, and enhanced the biological effect of CPP-NA particles in living cells.

The solvent mixture that we optimised here to dissolve PF14 enabled us to assemble highly uniform nanoparticles of peptides and oligonucleotides, which provided comparable or, in some cases, increased expression of the reporter protein under cell culture conditions, compared to water, a solvent commonly used for dissolving peptides. Furthermore, an important advantage of this solvation method is that it allows the long-term storage of the peptide stock solutions without changes/loss of its efficiency/activity. We observed more consistent results while using non-freezing stocks prepared in organic solvent mixtures compared to water stocks that have to be either stored at +4 °C or thawed upon each use, with both of these approaches facilitating aggregation and changes in the properties of the peptide.

The biological effect of oligonucleotide cargo was further increased by the inclusion of Ca^{2+} and Mg^{2+} ions into the aforementioned CPP–oligonucleotide nanoparticles. Importantly, the complementation of nanoparticles with $CaCl_2$ only marginally increased the size of the particles, making these highly biocompatible for in vitro and possibly for in vivo applications. Added calcium ions seem to dissociate from the complemented nanoparticles in an acidic milieu, which could even be relevant to the escape of CPP–ON nanoparticles from entrapment in acidic endosomes. The applicability of the strategy proposed here for the cellular delivery of NA molecules with higher molecular weight, such as mRNA and pDNA, is not clear yet and is the focus of our future studies.

Supplementary Materials: The following supporting information can be downloaded at: https://www.mdpi.com/article/10.3390/pharmaceutics15020396/s1, Table S1: CPPs used in the study, their sequences, charge and number of amino acid residues; Table S2: Characterisation of particles forming from PF14 dissolved in different solvents by DLS; Table S3: Hydrodynamic diameter, polydispersity index and zeta potential of nanoparticles forming upon complexation of PF14 with SCO (at MR 5) or siRNA (at MR 34) with addition of Ca^{2+} or Mg^{2+}; Table S4: Average diameter of nanoparticles formed by differently dissolved PF14, which was calculated from HR-TEM images using the ImageJ software; Table S5: Average diameter of nanoparticles prepared of 100 nM SCO, 1 µM PF14 and $CaCl_2$ or $MgCl_2$, which was calculated from AFM, FE-SEM, and HR-TEM images using ImageJ software; Table S6: Numerical results of EDX analysis of nanoparticles assembled from PF14 in EtOH90/DMSO9.6/TMC0.4, SCO and $CaCl_2$ or $MgCl_2$; Figure S1: Effect of nanoparticles prepared from SCO-705 and differently dissolved PF14 on splicing correction; Figure S2: Viability of HeLa pLuc 705 (A) or U87 MG-Luc2 (B) cells after incubation with complexes of differently dissolved PF14 and SCO-705 (A) or siRNA (B); Figure S3: Bright field images of cells after incubation with nanoparticles containing SCO, differently dissolved PF14 and $CaCl_2$; Figure S4: Transfection of and splicing correction with SCO-654 delivered with differently dissolved PF14; Figure S5: The effect of nanoparticles prepared of SCO-705, differently dissolved PF14 and Ca^{2+}/Mg^{2+} ions on splicing correction; Figure S6: The effect of nanoparticles prepared of SCO (A) or siRNA (B), differently dissolved PF14 and Ca^{2+}/Mg^{2+} ions on cell viability; Figure S7: Effect of cosolvent on tranfection of SCO and siRNA by hPep3 and CADY; Figure S8: Effect of nanoparticles prepared of SCO and hPep3 (A) or siRNA and CADY (B), on viability of HeLa pLuc 705 or U87 MG-Luc2 cells, respectively; Figure S9: Circular dichroism (CD) spectra reflecting the secondary structure of PF14 dissolved in MQ water (A), EtOH (B), EtOH90/DMSO10 (C), and in EtOH90/DMSO9.6/TMC0.4 (D); Figure S10: Circular dichroism (CD) spectra reflecting the secondary structure of PF14 (A), complex of PF14 with SCO (B); complex of PF14 with SCO and 3 mM $CaCl_2$ (C); complex of PF14 with SCO and 3 mM $MgCl_2$ (D) dissolved in EtOH; Figure S11: EDX spectra confirming the presence of Ca^{2+} (A) and Mg^{2+} (B) ions in nanoparticles assembled from PF14, SCO and $CaCl_2$ or $MgCl_2$, respectively.

Author Contributions: Conceptualisation, M.P.; methodology, M.P., A.G., V.K. and R.L.; validation, M.P.; formal analysis, M.M.; investigation, A.B., M.M., A.A., M.R., S.K.T.S.W., J.J. and K.P.; resources, M.P.; data curation, M.M. and M.P.; writing—original draft preparation, A.B., M.M. and M.P.; writing—review and editing, M.M. and M.P.; visualisation, M.M. and K.P.; supervision, M.P.;

project administration, M.P.; funding acquisition, M.P. All authors have read and agreed to the published version of the manuscript.

Funding: This research was funded by the Estonian Research Council (Eesti Teadusagentuur, Estonia), grant numbers PRG1169, PRG1198 and PRG1506, and the Institute of Technology (Tartu Ülikooli Tehnoloogiainstituut, University of Tartu) basic financing grant PLTTI20912 to Margus Pooga.

Institutional Review Board Statement: Not applicable.

Informed Consent Statement: Not applicable.

Data Availability Statement: All data supporting the findings of this study are available within the article and its supplemental information or from the corresponding author upon reasonable request.

Acknowledgments: We thank U. Mäeorg for fruitful discussions, S. Deshayes for CADY, T. Lehto for hPep3 peptides, K. Koppel for preparation of the TEM specimens, and R. Raid for recording TEM microphotos. We highly appreciate help of D. Lubenets at performing CLSM and FACS experiments. We are grateful to R. Sihna Roy for the help in EDX experiment.

Conflicts of Interest: The funders had no role in the design of the study; in the collection, analyses, or interpretation of data; in the writing of the manuscript; or in the decision to publish the results. Authors A.G., J.J., M.P., M.M. and S.K.T.S.W. have applied for patent PCT/SE2022/050657 based on some of the results presented in this manuscript. This patent application is publicly available on the wipo website. Authors A.G., J.J. and S.K.T.S.W. are shareholders in the company CellPept Sweden AB. This company had no role in the design of the study; in the collection, analyses, or interpretation of data; in the writing of the manuscript; or in the decision to publish the results.

References

1. Damase, T.R.; Sukhovershin, R.; Boada, C.; Taraballi, F.; Pettigrew, R.I.; Cooke, J.P. The Limitless Future of RNA Therapeutics. *Front. Bioeng. Biotechnol.* **2021**, *9*, 628137. [CrossRef] [PubMed]
2. Gonçalves, G.A.R.; de Melo Alves Paiva, R. Gene Therapy: Advances, Challenges and Perspectives. *Einstein* **2017**, *15*, 369–375. [CrossRef] [PubMed]
3. Dunbar, C.E.; High, K.A.; Joung, J.K.; Kohn, D.B.; Ozawa, K.; Sadelain, M. Gene Therapy Comes of Age. *Science* **2018**, *359*, eaan4672. [CrossRef] [PubMed]
4. U.S. Food and Drug Administration. What Is Gene Therapy? Available online: https://www.fda.gov/vaccines-blood-biologics/cellular-gene-therapy-products/what-gene-therapy (accessed on 1 January 2023).
5. Thomas, C.E.; Ehrhardt, A.; Kay, M.A. Progress and Problems with the Use of Viral Vectors for Gene Therapy. *Nat. Rev. Genet.* **2003**, *4*, 346–358. [CrossRef]
6. Nayak, S.; Herzog, R.W. Progress and Prospects: Immune Responses to Viral Vectors. *Gene Ther.* **2010**, *17*, 295–304. [CrossRef] [PubMed]
7. Kofron, M.D.; Laurencin, C.T. Bone Tissue Engineering by Gene Delivery. *Adv. Drug Deliv. Rev.* **2006**, *58*, 555–576. [CrossRef] [PubMed]
8. Davé, U.P.; Jenkins, N.A.; Copeland, N.G. Gene Therapy Insertional Mutagenesis Insights. *Science* **2004**, *303*, 333. [CrossRef]
9. Al-Dosari, M.S.; Gao, X. Nonviral Gene Delivery: Principle, Limitations, and Recent Progress. *AAPS J.* **2009**, *11*, 671. [CrossRef]
10. Piotrowski-Daspit, A.S.; Kauffman, A.C.; Bracaglia, L.G.; Saltzman, W.M. Polymeric Vehicles for Nucleic Acid Delivery. *Adv. Drug Deliv. Rev.* **2020**, *156*, 119–132. [CrossRef]
11. Ramamoorth, M.; Narvekar, A. Non Viral Vectors in Gene Therapy—An Overview. *J. Clin. Diagn. Res.* **2015**, *9*, GE01–GE06. [CrossRef]
12. Filion, M.C.; Phillips, N.C. Toxicity and Immunomodulatory Activity of Liposomal Vectors Formulated with Cationic Lipids toward Immune Effector Cells. *Biochim. Biophys. Acta (BBA)-Biomembr.* **1997**, *1329*, 345–356. [CrossRef]
13. Kurrikoff, K.; Veiman, K.-L.; Künnapuu, K.; Peets, E.M.; Lehto, T.; Pärnaste, L.; Arukuusk, P.; Langel, Ü. Effective In Vivo Gene Delivery with Reduced Toxicity, Achieved by Charge and Fatty Acid-Modified Cell Penetrating Peptide. *Sci. Rep.* **2017**, *7*, 17056. [CrossRef] [PubMed]
14. Xie, J.; Bi, Y.; Zhang, H.; Dong, S.; Teng, L.; Lee, R.J.; Yang, Z. Cell-Penetrating Peptides in Diagnosis and Treatment of Human Diseases: From Preclinical Research to Clinical Application. *Front. Pharmacol.* **2020**, *11*, 697. [CrossRef]
15. Saar, K.; Lindgren, M.; Hansen, M.; Eiríksdóttir, E.; Jiang, Y.; Rosenthal-Aizman, K.; Sassian, M.; Langel, Ü. Cell-Penetrating Peptides: A Comparative Membrane Toxicity Study. *Anal. Biochem.* **2005**, *345*, 55–65. [CrossRef] [PubMed]
16. El Andaloussi, S.; Lehto, T.; Mäger, I.; Rosenthal-Aizman, K.; Oprea, I.I.; Simonson, O.E.; Sork, H.; Ezzat, K.; Copolovici, D.M.; Kurrikoff, K. Design of a Peptide-Based Vector, PepFect6, for Efficient Delivery of SiRNA in Cell Culture and Systemically In Vivo. *Nucleic Acids Res.* **2011**, *39*, 3972–3987. [CrossRef]

17. Pooga, M.; Langel, Ü. Classes of Cell-Penetrating Peptides. In *Cell-Penetrating Peptides*; Springer: New York, NY, USA, 2015; pp. 3–28. [CrossRef]
18. Kristensen, M.; Birch, D.; Mørck Nielsen, H. Applications and Challenges for Use of Cell-Penetrating Peptides as Delivery Vectors for Peptide and Protein Cargos. *Int. J. Mol. Sci.* **2016**, *17*, 185. [CrossRef]
19. Kalafatovic, D.; Giralt, E. Cell-Penetrating Peptides: Design Strategies beyond Primary Structure and Amphipathicity. *Molecules* **2017**, *22*, 1929. [CrossRef]
20. Biswas, A.; Chakraborty, K.; Dutta, C.; Mukherjee, S.; Gayen, P.; Jan, S.; Mallick, A.M.; Bhattacharyya, D.; Sinha Roy, R. Engineered Histidine-Enriched Facial Lipopeptides for Enhanced Intracellular Delivery of Functional SiRNA to Triple Negative Breast Cancer Cells. *ACS Appl. Mater. Interfaces* **2019**, *11*, 4719–4736. [CrossRef]
21. Porosk, L.; Arukuusk, P.; Põhako, K.; Kurrikoff, K.; Kiisholts, K.; Padari, K.; Pooga, M.; Langel, Ü. Enhancement of SiRNA Transfection by the Optimization of Fatty Acid Length and Histidine Content in the CPP. *Biomater. Sci.* **2019**, *7*, 4363–4374. [CrossRef]
22. Beloor, J.; Zeller, S.; Choi, C.S.; Lee, S.-K.; Kumar, P. Cationic Cell-Penetrating Peptides as Vehicles for SiRNA Delivery. *Ther. Deliv.* **2015**, *6*, 491–507. [CrossRef]
23. Boisguérin, P.; Deshayes, S.; Gait, M.J.; O'Donovan, L.; Godfrey, C.; Betts, C.A.; Wood, M.J.; Lebleu, B. Delivery of Therapeutic Oligonucleotides with Cell Penetrating Peptides. *Adv. Drug Deliv. Rev.* **2015**, *87*, 52–67. [CrossRef] [PubMed]
24. Lehto, T.; Ezzat, K.; Wood, M.J.A.; EL Andaloussi, S. Peptides for Nucleic Acid Delivery. *Adv. Drug Deliv. Rev.* **2016**, *106*, 172–182. [CrossRef] [PubMed]
25. Oba, M.; Ito, Y.; Umeno, T.; Kato, T.; Tanaka, M. Plasmid DNA Delivery Using Cell-Penetrating Peptide Foldamers Composed of Arg-Arg-Aib Repeating Sequences. *ACS Biomater. Sci. Eng.* **2019**, *5*, 5660–5668. [CrossRef] [PubMed]
26. Kato, T.; Yamashita, H.; Misawa, T.; Nishida, K.; Kurihara, M.; Tanaka, M.; Demizu, Y.; Oba, M. Plasmid DNA Delivery by Arginine-Rich Cell-Penetrating Peptides Containing Unnatural Amino Acids. *Bioorg. Med. Chem.* **2016**, *24*, 2681–2687. [CrossRef]
27. Yokoo, H.; Oba, M.; Uchida, S. Cell-Penetrating Peptides: Emerging Tools for MRNA Delivery. *Pharmaceutics* **2021**, *14*, 78. [CrossRef] [PubMed]
28. Dinca, A.; Chien, W.-M.; Chin, M.T. Intracellular Delivery of Proteins with Cell-Penetrating Peptides for Therapeutic Uses in Human Disease. *Int. J. Mol. Sci.* **2016**, *17*, 263. [CrossRef] [PubMed]
29. Freimann, K.; Arukuusk, P.; Kurrikoff, K.; Pärnaste, L.; Raid, R.; Piirsoo, A.; Pooga, M.; Langel, Ü. Formulation of Stable and Homogeneous Cell-Penetrating Peptide NF55 Nanoparticles for Efficient Gene Delivery in Vivo. *Mol. Ther.-Nucleic Acids* **2018**, *10*, 28–35. [CrossRef]
30. Sharma, A.; Madhunapantula, S.V.; Robertson, G.P. Toxicological Considerations When Creating Nanoparticle Based Drugs and Drug Delivery Systems? *Expert Opin. Drug Metab. Toxicol.* **2012**, *8*, 47–69. [CrossRef]
31. Ravichandran, R. Nanoparticles in Drug Delivery: Potential Green Nanobiomedicine Applications. *Int. J. Green Nanotechnol. Biomed.* **2009**, *1*, B108–B130. [CrossRef]
32. Whitesides, G.M.; Mathias, J.P.; Seto, C.T. Molecular Self-Assembly and Nanochemistry: A Chemical Strategy for the Synthesis of Nanostructures. *Science* **1991**, *254*, 1312–1319. [CrossRef]
33. Edwards-Gayle, C.J.C.; Hamley, I.W. Self-Assembly of Bioactive Peptides, Peptide Conjugates, and Peptide Mimetic Materials. *Org. Biomol. Chem.* **2017**, *15*, 5867–5876. [CrossRef] [PubMed]
34. Chen, Y.; Tao, K.; Ji, W.; Kumar, V.B.; Rencus-Lazar, S.; Gazit, E. Histidine as a Key Modulator of Molecular Self-Assembly: Peptide-Based Supramolecular Materials Inspired by Biological Systems. *Mater. Today* **2022**, *60*, 106–127. [CrossRef]
35. Chen, Y.; Guerin, S.; Yuan, H.; O'Donnell, J.; Xue, B.; Cazade, P.-A.; Haq, E.U.; Shimon, L.J.W.; Rencus-Lazar, S.; Tofail, S.A.M.; et al. Guest Molecule-Mediated Energy Harvesting in a Conformationally Sensitive Peptide–Metal Organic Framework. *J. Am. Chem. Soc.* **2022**, *144*, 3468–3476. [CrossRef]
36. van der Vegt, N.F.; Nayar, D. The Hydrophobic Effect and the Role of Cosolvents. *J. Phys. Chem. B* **2017**, *121*, 9986–9998. [CrossRef]
37. Nandakumar, A.; Ito, Y.; Ueda, M. Solvent Effects on the Self-Assembly of an Amphiphilic Polypeptide Incorporating α-Helical Hydrophobic Blocks. *J. Am. Chem. Soc.* **2020**, *142*, 20994–21003. [CrossRef]
38. Ezzat, K.; El Andaloussi, S.; Zaghloul, E.M.; Lehto, T.; Lindberg, S.; Moreno, P.M.; Viola, J.R.; Magdy, T.; Abdo, R.; Guterstam, P. PepFect 14, a Novel Cell-Penetrating Peptide for Oligonucleotide Delivery in Solution and as Solid Formulation. *Nucleic Acids Res.* **2011**, *39*, 5284–5298. [CrossRef]
39. Juks, C.; Padari, K.; Margus, H.; Kriiska, A.; Etverk, I.; Arukuusk, P.; Koppel, K.; Ezzat, K.; Langel, Ü.; Pooga, M. The Role of Endocytosis in the Uptake and Intracellular Trafficking of PepFect14–Nucleic Acid Nanocomplexes via Class A Scavenger Receptors. *Biochim. Biophys. Acta (BBA)-Biomembr.* **2015**, *1848*, 3205–3216. [CrossRef] [PubMed]
40. Ezzat, K.; Helmfors, H.; Tudoran, O.; Juks, C.; Lindberg, S.; Padari, K.; El-Andaloussi, S.; Pooga, M.; Langel, Ü. Scavenger Receptor-Mediated Uptake of Cell-Penetrating Peptide Nanocomplexes with Oligonucleotides. *FASEB J.* **2012**, *26*, 1172–1180. [CrossRef]
41. Andaloussi, S.E.L.; Lehto, T.; Lundin, P.; Langel, U. Application of PepFect Peptides for the Delivery of Splice-Correcting Oligonucleotides. *Methods Mol. Biol.* **2011**, *683*, 361–373. [CrossRef]
42. Srimanee, A.; Arvanitidou, M.; Kim, K.; Hällbrink, M.; Langel, Ü. Cell-Penetrating Peptides for SiRNA Delivery to Glioblastomas. *Peptides* **2018**, *104*, 62–69. [CrossRef]

43. Urgard, E.; Brjalin, A.; Langel, Ü.; Pooga, M.; Rebane, A.; Annilo, T. Comparison of Peptide-and Lipid-Based Delivery of MiR-34a-5p Mimic into Ppc-1 Cells. *Nucleic Acid Ther.* **2017**, *27*, 295–302. [CrossRef] [PubMed]
44. Carreras-Badosa, G.; Maslovskaja, J.; Periyasamy, K.; Urgard, E.; Padari, K.; Vaher, H.; Tserel, L.; Gestin, M.; Kisand, K.; Arukuusk, P. NickFect Type of Cell-Penetrating Peptides Present Enhanced Efficiency for MicroRNA-146a Delivery into Dendritic Cells and during Skin Inflammation. *Biomaterials* **2020**, *262*, 120316. [CrossRef] [PubMed]
45. Lorents, A.; Urgard, E.; Runnel, T.; Wawrzyniak, P.; Langel, U.; Pooga, M.; Rebane, A. Functional Delivery of MiR-146a to Human Primary Keratinocytes and Skin Equivalents with Cell-Penetrating Peptides. In *Allergy*; Wiley-Blackwell: Hoboken, NJ, USA, 2014; Volume 69, p. 618.
46. Urgard, E.; Lorents, A.; Klaas, M.; Padari, K.; Viil, J.; Runnel, T.; Langel, K.; Kingo, K.; Tkaczyk, E.; Langel, Ü. Pre-Administration of PepFect6-MicroRNA-146a Nanocomplexes Inhibits Inflammatory Responses in Keratinocytes and in a Mouse Model of Irritant Contact Dermatitis. *J. Control. Release* **2016**, *235*, 195–204. [CrossRef]
47. Juks, C.; Lorents, A.; Arukuusk, P.; Langel, Ü.; Pooga, M. Cell-Penetrating Peptides Recruit Type A Scavenger Receptors to the Plasma Membrane for Cellular Delivery of Nucleic Acids. *FASEB J.* **2017**, *31*, 975–988. [CrossRef]
48. Lorents, A.; Kodavali, P.K.; Oskolkov, N.; Langel, Ü.; Hällbrink, M.; Pooga, M. Cell-Penetrating Peptides Split into Two Groups Based on Modulation of Intracellular Calcium Concentration. *J. Biol. Chem.* **2012**, *287*, 16880–16889. [CrossRef] [PubMed]
49. Rocha, M.C.; Fabri, J.H.T.M.; de Godoy, K.F.; de Castro, P.A.; Hori, J.I.; da Cunha, A.F.; Arentshorst, M.; Ram, A.F.J.; van den Hondel, C.A.M.J.J.; Goldman, G.H.; et al. Aspergillus Fumigatus MADS-Box Transcription Factor RlmA Is Required for Regulation of the Cell Wall Integrity and Virulence. *G3* **2016**, *6*, 2983–3002. [CrossRef]
50. Helmfors, H.; Eriksson, J.; Langel, Ü. Optimized Luciferase Assay for Cell-Penetrating Peptide-Mediated Delivery of Short Oligonucleotides. *Anal. Biochem.* **2015**, *484*, 136–142. [CrossRef]
51. Maloverjan, M.; Padari, K.; Abroi, A.; Rebane, A.; Pooga, M. Divalent Metal Ions Boost Effect of Nucleic Acids Delivered by Cell-Penetrating Peptides. *Cells* **2022**, *11*, 756. [CrossRef]
52. Kang, S.-H.; Cho, M.-J.; Kole, R. Up-Regulation of Luciferase Gene Expression with Antisense Oligonucleotides: Implications and Applications in Functional Assay Development. *Biochemistry* **1998**, *37*, 6235–6239. [CrossRef]
53. Sazani, P.; Kang, S.-H.; Maier, M.A.; Wei, C.; Dillman, J.; Summerton, J.; Manoharan, M.; Kole, R. Nuclear Antisense Effects of Neutral, Anionic and Cationic Oligonucleotide Analogs. *Nucleic Acids Res.* **2001**, *29*, 3965–3974. [CrossRef]
54. Cellosaurus Cell Line HeLa EGFP-654 (CVCL_VS42). Available online: https://www.cellosaurus.org/CVCL_VS42 (accessed on 4 January 2023).
55. Cellosaurus Cell Line U-87 MG-Luc2 (CVCL_5J15). Available online: https://www.cellosaurus.org/CVCL_5J15 (accessed on 4 January 2023).
56. Siebring-van Olst, E.; Vermeulen, C.; de Menezes, R.X.; Howell, M.; Smit, E.F.; van Beusechem, V.W. Affordable Luciferase Reporter Assay for Cell-Based High-Throughput Screening. *J. Biomol. Screen.* **2013**, *18*, 453–461. [CrossRef] [PubMed]
57. Lehto, T.; Vasconcelos, L.; Margus, H.; Figueroa, R.; Pooga, M.; Hällbrink, M.; Langel, Ü. Saturated Fatty Acid Analogues of Cell-Penetrating Peptide PepFect14: Role of Fatty Acid Modification in Complexation and Delivery of Splice-Correcting Oligonucleotides. *Bioconjugate Chem.* **2017**, *28*, 782–792. [CrossRef]
58. Margus, H.; Arukuusk, P.; Langel, Ü.; Pooga, M. Characteristics of Cell-Penetrating Peptide/Nucleic Acid Nanoparticles. *Mol. Pharm.* **2016**, *13*, 172–179. [CrossRef]
59. Blanco, E.; Shen, H.; Ferrari, M. Principles of Nanoparticle Design for Overcoming Biological Barriers to Drug Delivery. *Nat. Biotechnol.* **2015**, *33*, 941–951. [CrossRef]
60. Center for Drug Evaluation and Research. Q3C Tables and List Rev. 4. Available online: https://www.fda.gov/regulatory-information/search-fda-guidance-documents/q3c-tables-and-list-rev-4 (accessed on 27 November 2022).
61. Wang, H.; Zhong, C.-Y.; Wu, J.-F.; Huang, Y.-B.; Liu, C.-B. Enhancement of TAT Cell Membrane Penetration Efficiency by Dimethyl Sulphoxide. *J. Control. Release* **2010**, *143*, 64–70. [CrossRef]
62. Gros, E.; Deshayes, S.; Morris, M.C.; Aldrian-Herrada, G.; Depollier, J.; Heitz, F.; Divita, G. A Non-Covalent Peptide-Based Strategy for Protein and Peptide Nucleic Acid Transduction. *Biochim. Biophys. Acta (BBA)-Biomembr.* **2006**, *1758*, 384–393. [CrossRef] [PubMed]
63. Sun, Y.; Lau, S.Y.; Lim, Z.W.; Chang, S.C.; Ghadessy, F.; Partridge, A.; Miserez, A. Phase-Separating Peptides for Direct Cytosolic Delivery and Redox-Activated Release of Macromolecular Therapeutics. *Nat. Chem.* **2022**, *14*, 274–283. [CrossRef]
64. Rong, G.; Wang, C.; Chen, L.; Yan, Y.; Cheng, Y. Fluoroalkylation Promotes Cytosolic Peptide Delivery. *Sci. Adv.* **2020**, *6*, eaaz1774. [CrossRef] [PubMed]
65. Mohajeri, S.; Chen, F.; de Prinse, M.; Phung, T.; Burke-Kleinman, J.; Maurice, D.H.; Amsden, B.G. Liquid Degradable Poly(Trimethylene-Carbonate-Co-5-Hydroxy-Trimethylene Carbonate): An Injectable Drug Delivery Vehicle for Acid-Sensitive Drugs. *Mol. Pharm.* **2020**, *17*, 1363–1376. [CrossRef] [PubMed]
66. Fukushima, K. Poly(Trimethylene Carbonate)-Based Polymers Engineered for Biodegradable Functional Biomaterials. *Biomater. Sci.* **2016**, *4*, 9–24. [CrossRef]
67. Tian, H.; Tang, Z.; Zhuang, X.; Chen, X.; Jing, X. Biodegradable Synthetic Polymers: Preparation, Functionalization and Biomedical Application. *Prog. Polym. Sci.* **2012**, *2*, 237–280. [CrossRef]

68. Bazaz, S.; Lehto, T.; Tops, R.; Gissberg, O.; Gupta, D.; Bestas, B.; Bost, J.; Wiklander, O.P.B.; Sork, H.; Zaghloul, E.M.; et al. Novel Orthogonally Hydrocarbon-Modified Cell-Penetrating Peptide Nanoparticles Mediate Efficient Delivery of Splice-Switching Antisense Oligonucleotides In Vitro and In Vivo. *Biomedicines* **2021**, *9*, 1046. [CrossRef]
69. Crombez, L.; Aldrian-Herrada, G.; Konate, K.; Nguyen, Q.N.; McMaster, G.K.; Brasseur, R.; Heitz, F.; Divita, G. A New Potent Secondary Amphipathic Cell–Penetrating Peptide for SiRNA Delivery Into Mammalian Cells. *Mol. Ther.* **2009**, *17*, 95–103. [CrossRef] [PubMed]
70. Tai, W.; Gao, X. Functional Peptides for SiRNA Delivery. *Adv. Drug Deliv. Rev.* **2017**, *110–111*, 157–168. [CrossRef]
71. Rydström, A.; Deshayes, S.; Konate, K.; Crombez, L.; Padari, K.; Boukhaddaoui, H.; Aldrian, G.; Pooga, M.; Divita, G. Direct Translocation as Major Cellular Uptake for CADY Self-Assembling Peptide-Based Nanoparticles. *PLoS ONE* **2011**, *6*, e25924. [CrossRef] [PubMed]
72. Barbar, E.; Nyarko, A. NMR Characterization of Self-Association Domains Promoted by Interactions with LC8 Hub Protein. *Comput. Struct. Biotechnol. J.* **2014**, *9*, e201402003. [CrossRef] [PubMed]
73. Asadi, J.; Ferguson, S.; Raja, H.; Hacker, C.; Marius, P.; Ward, R.; Pliotas, C.; Naismith, J.; Lucocq, J. Enhanced Imaging of Lipid Rich Nanoparticles Embedded in Methylcellulose Films for Transmission Electron Microscopy Using Mixtures of Heavy Metals. *Micron* **2017**, *99*, 40–48. [CrossRef]
74. Kaasalainen, M.; Aseyev, V.; von Haartman, E.; Karaman, D.Ş.; Mäkilä, E.; Tenhu, H.; Rosenholm, J.; Salonen, J. Size, Stability, and Porosity of Mesoporous Nanoparticles Characterized with Light Scattering. *Nanoscale Res. Lett.* **2017**, *12*, 74. [CrossRef]
75. Baoum, A.; Xie, S.-X.; Fakhari, A.; Berkland, C. "Soft" Calcium Crosslinks Enable Highly Efficient Gene Transfection Using TAT Peptide. *Pharm. Res.* **2009**, *26*, 2619–2629. [CrossRef]
76. Graham, F.L.; van der Eb, A.J. A New Technique for the Assay of Infectivity of Human Adenovirus 5 DNA. *Virology* **1973**, *52*, 456–467. [CrossRef] [PubMed]
77. Baoum, A.; Ovcharenko, D.; Berkland, C. Calcium Condensed Cell Penetrating Peptide Complexes Offer Highly Efficient, Low Toxicity Gene Silencing. *Int. J. Pharm.* **2012**, *427*, 134–142. [CrossRef] [PubMed]
78. Melikov, K.; Hara, A.; Yamoah, K.; Zaitseva, E.; Zaitsev, E.; Chernomordik, L.V. Efficient Entry of Cell-Penetrating Peptide Nona-Arginine into Adherent Cells Involves a Transient Increase in Intracellular Calcium. *Biochem. J.* **2015**, *471*, 221–230. [CrossRef] [PubMed]
79. Goldshtein, M.; Forti, E.; Ruvinov, E.; Cohen, S. Mechanisms of Cellular Uptake and Endosomal Escape of Calcium-SiRNA Nanocomplexes. *Int. J. Pharm.* **2016**, *515*, 46–56. [CrossRef]
80. Hori, S.; Yamamoto, T.; Waki, R.; Wada, S.; Wada, F.; Noda, M.; Obika, S. Ca^{2+} Enrichment in Culture Medium Potentiates Effect of Oligonucleotides. *Nucleic Acids Res.* **2015**, *43*, e128. [CrossRef] [PubMed]
81. Delos Santos, R.C.; Bautista, S.; Lucarelli, S.; Bone, L.N.; Dayam, R.M.; Abousawan, J.; Botelho, R.J.; Antonescu, C.N. Selective Regulation of Clathrin-Mediated Epidermal Growth Factor Receptor Signaling and Endocytosis by Phospholipase C and Calcium. *Mol. Biol. Cell* **2017**, *28*, 2802–2818. [CrossRef] [PubMed]
82. Maltseva, D.; Gudbrandsdottir, R.; Kizilsavas, G.; Horinek, D.; Gonella, G. Location and Conformation of the LKα14 Peptide in Water/Ethanol Mixtures. *Langmuir* **2021**, *37*, 469–477. [CrossRef]
83. Kinoshita, M.; Okamoto, Y.; Hirata, F. Peptide Conformations in Alcohol and Water: Analyses by the Reference Interaction Site Model Theory. *J. Am. Chem. Soc.* **2000**, *122*, 2773–2779. [CrossRef]
84. Notman, R.; Noro, M.; O'Malley, B.; Anwar, J. Molecular Basis for Dimethylsulfoxide (DMSO) Action on Lipid Membranes. *J. Am. Chem. Soc.* **2006**, *128*, 13982–13983. [CrossRef]
85. de Menorval, M.-A.; Mir, L.M.; Fernández, M.L.; Reigada, R. Effects of Dimethyl Sulfoxide in Cholesterol-Containing Lipid Membranes: A Comparative Study of Experiments in Silico and with Cells. *PLoS ONE* **2012**, *7*, e41733. [CrossRef]
86. Gironi, B.; Kahveci, Z.; McGill, B.; Lechner, B.-D.; Pagliara, S.; Metz, J.; Morresi, A.; Palombo, F.; Sassi, P.; Petrov, P.G. Effect of DMSO on the Mechanical and Structural Properties of Model and Biological Membranes. *Biophys. J.* **2020**, *119*, 274–286. [CrossRef]

Disclaimer/Publisher's Note: The statements, opinions and data contained in all publications are solely those of the individual author(s) and contributor(s) and not of MDPI and/or the editor(s). MDPI and/or the editor(s) disclaim responsibility for any injury to people or property resulting from any ideas, methods, instructions or products referred to in the content.

pharmaceutics MDPI

Article

Fluid-Phase Endocytosis and Lysosomal Degradation of Bovine Lactoferrin in Lung Cells

Edward John Sayers [1], Iwan Palmer [1,2], Lucy Hope [3], Paul Hope [3], Peter Watson [2] and Arwyn Tomos Jones [1,*]

1 School of Pharmacy and Pharmaceutical Sciences, Cardiff University, King Edward VII Avenue, Redwood Building, Cardiff CF10 3NB, Wales, UK; sayersej@cardiff.ac.uk (E.J.S.); palmerg2@cardiff.ac.uk (I.P.)
2 Cardiff School of Biosciences, Cardiff University, Museum Avenue, The Sir Martin Evans Building, Cardiff CF10 3AX, Wales, UK; watsonpd@cardiff.ac.uk
3 Virustatic, M-SParc, Gaerwen, Isle of Anglesey LL60 6AG, Wales, UK; lucyhope@virustatic.com (L.H.); paulhope@virustatic.com (P.H.)
* Correspondence: jonesat@cardiff.ac.uk

Abstract: The iron-binding protein lactoferrin and the cell-penetrating peptides derived from its sequence utilise endocytosis to enter different cell types. The full-length protein has been extensively investigated as a potential therapeutic against a range of pathogenic bacteria, fungi, and viruses, including SARS-CoV-2. As a respiratory antiviral agent, several activity mechanisms have been demonstrated for lactoferrin, at the extracellular and plasma membrane levels, but as a protein that enters cells it may also have intracellular antiviral activity. Characterisation of lactoferrin's binding, endocytic traffic to lysosomes, or recycling endosomes for exocytosis is lacking, especially in lung cell models. Here, we use confocal microscopy, flow cytometry, and degradation assays to evaluate binding, internalisation, endocytic trafficking, and the intracellular fate of bovine lactoferrin in human lung A549 cells. In comparative studies with endocytic probes transferrin and dextran, we show that lactoferrin binds to negative charges on the cell surface and actively enters cells via fluid-phase endocytosis, in a receptor-independent manner. Once inside the cell, we show that it is trafficked to lysosomes where it undergoes degradation within two hours. These findings provide opportunities for investigating both lactoferrin and derived cell-penetrating peptides activities of targeting intracellular pathogens.

Keywords: lactoferrin; endocytosis; intracellular trafficking; lysosomal degradation

Citation: Sayers, E.J.; Palmer, I.; Hope, L.; Hope, P.; Watson, P.; Jones, A.T. Fluid-Phase Endocytosis and Lysosomal Degradation of Bovine Lactoferrin in Lung Cells. *Pharmaceutics* **2022**, *14*, 855. https://doi.org/10.3390/pharmaceutics14040855

Academic Editors: Prisca Boisguérin and Sébastien Deshayes

Received: 18 February 2022
Accepted: 6 April 2022
Published: 13 April 2022

Copyright: © 2022 by the authors. Licensee MDPI, Basel, Switzerland. This article is an open access article distributed under the terms and conditions of the Creative Commons Attribution (CC BY) license (https://creativecommons.org/licenses/by/4.0/).

1. Introduction

Lactoferrin (LF) is a highly conserved, cationic glycoprotein of the transferrin family, which is synthesised by exocrine glands and secreted by neutrophils in infection and inflammation sites [1]. Several studies have shown that LF can internalise into cells, and a number of suggested receptors have been implicated in this process [2], taking the protein via, for example, clathrin-coated vesicles to different organelles [3]. The N-terminus of human LF (hLF) has a 22 amino acid sequence that was identified as a cell-penetrating peptide (CPP) that, when examined as a standalone peptide, utilises endocytosis to gain cell entry [4]. This sequence (38-KCFQWQRNMRKVRGPPVSCIKR-59) sits as a member of a huge list of naturally derived and artificial CPPs that have the ability to overcome the plasma membrane of cells to gain entry, either via direct translocation, or by hijacking one or more endocytic pathways [5]. Classical examples include the Tat peptide and penetratin, which are derived from the HIV-Tat protein and *Drosophila* antennapedia, respectively; synthetic variants such as octaarginine have also been described as CPPs.

Despite being highly conserved, there are significant differences in the amino acid sequences of LF from different species [6] which could impact cell binding and uptake; including in the aforementioned N-terminus. hLF contains the N-terminal sequence GRRRR, which, as a pentapeptide, was found to be crucial for heparin binding [7] and can be internalised rapidly and localised to the nucleus in a number of cell lines [8]. A pentapeptide derived from this region of bovine LF (bLF) (40-RRWQW-44) has been also shown to act as a CPP and is able to internalise plasmid DNA into A549 (human alveolar basal epithelial cells) [9]. A 25-amino-acid peptide from bLF (37-FKCRRWQWRMKKLGAPSITCVRRAF-61), equivalent to the hLF peptide described above (residues 38–59), has also been shown to rapidly internalise in a range of human breast cancer cells, remaining in the cytoplasm and not entering the nucleus [10].

The properties of LF and its peptide derivatives have been utilised as a delivery mechanism, either using the protein/peptide directly [9,11] or using the protein as a targeting/cell binding entity (reviewed here [12]). For instance, hLF along with chitosan have been conjugated to polyethyleneimine polyplexes to deliver miRNA into the brain [13], while prostate cancer cells have been targeted using LF as the targeting entity of gold nanocages carrying plasmid DNA [14].

Lactoferrin is currently of significant interest as it plays an important role in host defence against viral attack [15,16]. It can directly bind to viral proteins [17–19], and bind to heparan sulphate proteoglycans (HSPGs) on the host cell surface [20,21], with both mechanisms inhibiting viral attachment to, and entry into, target cells. In the context of CPPs, many utilise HSPGs to gain cell entry to endosomes, and bLF has been shown to disrupt the intracellular trafficking of viruses [22], which may well occur at the level of the endosome.

In vitro studies have shown that bLF inhibits early-phase viral infection by several human coronaviruses [20,23], including SARS-CoV-2, in a range of cell types [24–26]. Given the ongoing COVID-19 health crisis caused by the SARS-CoV-2 virus, there has been much interest in the role LF plays as a broad-spectrum antiviral in the management of COVID-19 patients, and also as a prophylactic treatment [27]. However, more research into the intracellular trafficking of bLF is required for this potential to be realised.

Very few recent studies have investigated in detail the endocytosis of intact bLF, despite being more easily available as a natural glycosylated protein than hLF, and thus having important therapeutic potential. In this study, we performed detailed endocytosis, trafficking, and intracellular stability analyses of a fluorescent conjugate of bLF. Our studies show that the protein binds to the surface of lung A549 cells but is displaced by washing with high concentrations of negatively charged heparin. These cells do not express the lactoferrin receptor, and bLF is subsequently endocytosed through the fluid-phase where it is trafficked to the lysosome and degraded within two hours. These studies should pave the route to further analysis of the corresponding CPP sequence derived from this protein, allowing better understanding of its mechanism of internalisation and potential as a cell-penetrating moiety that may also have other therapeutic applications, beyond the infectious disease space.

2. Materials and Methods

2.1. Materials

Bovine lactoferrin (bLF, Ingredia, Arras, France) was dissolved in distilled water to either 10 mg/mL or 20 mg/mL, and was passed through a 0.22 μm syringe filter for sterilisation. Reconstituted bLF was stored at −20 °C and used within 2 weeks.

Triton X-100, Tween-20, heparin, glycine, sodium dodecyl sulfate (SDS), and sodium bicarbonate were purchased from Sigma Aldrich, Gillingham, UK. Staurosporine, Alexa647 succinimidyl ester, transferrin-alexa488 (Tf488), Dextran-alexa488/-alexa647 (Dex488 or Dex647), and BSA (fraction V) were obtained from Fisher Scientific, Loughborough, UK.

2.2. Cell Culture

Human epithelial lung cells, A549 (CLL-185), were routinely cultured in DMEM (Fisher Scientific, Loughborough, UK) containing 10% foetal bovine serum (Thermo Fisher, Paisley, UK) under tissue culture conditions (37 °C, 5% CO_2 in a humidified incubator). Cells were passaged twice weekly using 0.25% Trypsin/EDTA (Fisher Scientific, Loughborough, UK) for a maximum of 20 passages from defrosting and kept below 80% confluency during routine cell culture. For all experiments, cells were seeded at 15.8×10^3 cells.cm^{-2} in complete medium (DMEM + 10% FBS) under tissue culture conditions, with the final timepoint of 48 h after seeding.

2.3. Labelling bLF with Alexa647

bLF was labelled using the method described previously [28]. Briefly, bLF was reacted with Alexa647 succinimidyl ester at a molar ratio of 1 to 4 for 4 h in 0.1 M sodium bicarbonate buffer pH 8.0. Unreacted dye was removed, and the protein transferred into PBS by successively passing through two Zeba spin desalting columns (Sigma Aldrich, Gillingham, UK) preloaded with PBS according to the manufacturer's instructions. The protein concentration and degree of labelling was calculated by UV-Vis absorbance (Agilent, Stockport, UK), measuring Alexa647 at 650 nm and the Alexa647-compensated protein peak at 280 nm, giving a final concentration of 23.07 µM bLF647, with an average of 1.93 moles fluorophores per mole protein. The bLF647 was filter-sterilised before use.

2.4. Viability Assays

A549 cells were seeded in 75 µL complete medium per well in a black, flat bottomed, 96-well tissue-culture-treated plate (Fisher Scientific, Loughborough, UK) and incubated under tissue culture conditions for 24 h. Cells were then treated by mixing in an additional 25 µL of drug/diluent control in complete medium. bLF was added to the cells at a maximal final concentration of 4 mg/mL and as a positive toxic control, staurosporine was alternatively added at a maximum final concentration of 50 µg/mL, and both were serially diluted to 1:1 in complete medium; 0.2% Triton X-100 was used as an additional positive control for cell death. Cells were incubated for 20 h under tissue culture conditions before 20 µL CellTitre Blue (Promega, Southampton, UK) was added and returned to tissue culture conditions for 4 h. Fluorescence intensity was measured using a plate reader (Tecan, Theale, UK). A one-way ANOVA was performed to determine significant changes in viability.

2.5. Uptake Experiments

Following a 30 min starvation in serum-free medium (SFM), cells were incubated in 24-well plates with either 250 nM bLF647, 20 nM Tf488, or 50 µg/mL Dex647 in SFM for between 10 and 360 min and analysed by flow cytometry. Alternatively, cells in 24-well plates were incubated for 15 min with between 5 and 250 nM bLF647 or Tf488, or between 5 and 250 µg/mL Dex647 in SFM following a 30 min starvation, and were analysed by flow cytometry.

2.6. Cell Binding Experiments

Cells grown in 24-well plates or imaging dishes were chilled on ice for 10 min before washing and incubating with ice-cold 250 nM bLF647 in the presence of either 25 µM unlabelled bLF, 25 µM unlabelled Tf, 1 or 5 mg/mL Heparin for 1 h in SFM before being washed in SFM and analysed by flow cytometry. In addition, cells incubated with 250 nM bLF647 for 1 h were subsequently washed with 5 min washes of either 1 or 5 mg/mL heparin before washing in SFM and being analysed by flow cytometry or microscopy.

2.7. Endocytic Trafficking and Recycling Experiments

Cells were seeded into either 6-well plates (Western blotting), 24-well plates (flow cytometry), or imaging dishes (MatTek, Ashland, MA, USA, for confocal microscopy). The morning before the experiment, cells in imaging dishes were incubated with 100 µg/mL Dex488 or Dex647 for 3 h in complete medium under tissue culture conditions before being washed in complete medium and returned to tissue culture conditions to label late endosomes/lysosomes [28]. To perform the uptake/recycling assays, they were incubated with 500 nM bLF647 (flow cytometry and microscopy) or 1 µM unlabelled bLF (Western blotting) in SFM for 1 h before being washed in complete medium and incubated for a chase period of 0–24 h. Samples were analysed as described below.

2.8. Flow Cytometry

After treatment, cells were washed three times in PBS and lifted using Accutase (Thermo Scientific, Paisley, UK) for cell binding experiments, or 0.5% trypsin for other experiments. Cell detachment media was removed by centrifugation at 400× *g* for 2 min with the cells resuspended in ice-cold PBS (cell binding experiments), or ice-cold PBS-containing 2% FBS. Cells were subsequently analysed with a FACSVerse (BD Bioscience, Wokingham, UK) using the 488 or 633 nm laser lines for excitation. A single-cell population was obtained through double gating (FSC v SSC, then FSC-W v FSC-H), with the median readout obtained for population fluorescence intensity. A single experiment represents the mean of the median intensity value from two replicates (samples were run in duplicate). The final values represent the mean of at least three independent experiments.

2.9. Confocal Microscopy

After treatment, cells were washed three times in prewarmed/prechilled phenol-red-free DMEM (imaging medium) and imaged immediately on a Leica SP5 laser scanning confocal microscope. Images were obtained using a 100× 1.4 NA objective, raster size of 1024 × 1024, producing a pixel size of 137 nm. Where both Alexa488 (excited at 488 nm using Argon laser) and Alexa647 (excited at 633 nm using HeNe laser) were imaged, imaging was set to sequentially capture images between lines.

2.10. Quantification of Microscopy Images

Images were quantified using a custom macro in FIJI [29]. Data was also obtained for the total integrated density of the whole the bLF647, Tf488, or Dex647 channel. A minimum of 10 single sections were obtained per independent experiment with the images obtained 2 μm above the glass surface.

2.11. Western Blotting

Following the uptake experiments, cells were lysed using NP40 lysis buffer (Invitrogen, Paisley, UK)-containing protease inhibitors (cOmplete Mini, Roche, St Albans, UK), centrifuged at 11,000× *g*, and supernatant proteins were quantified in a BCA assay. Samples (30 μg) were mixed with loading buffer and loaded onto an anyKD SDS-PAGE gel (BD Biosciences, Wokingham, UK), and transferred onto PDVF using TransBlot Turbo (BD Biosciences, Wokingham, UK). Bands were detected using an anti-lactoferrin polyclonal, pre-conjugated to HRP (1:30,000, A10-126P, Cambridge Bioscience, Cambridge, UK) in 5% BSA/0.075% Tween 20/PBS for 1 h after blocking in 5% BSA/0.075% Tween 20/PBS for 1 h. The membrane was washed 3 × 20 min in 0.075% Tween 20/PBS and detected using chemiluminescence on a ChemiDoc (BD Biosciences, Wokingham, UK). The membrane was stripped using 2 × 10 min stripping buffer (15 g/L glycine, 1 g/L SDS, 0.01% (*v*/*v*) Tween 20, pH 2.2) followed by 2 × 10 min washes in PBS and 2 × 10 min washes in PBST (0.025% Tween 20/PBS). After reimaging on the ChemiDoc to check for removal, membranes were re-blocked in 5% BSA/PBST and incubated with Tubulin-HRP (1:50,000, ab21058, Abcam, Cambridge, UK) in 5% BSA/PBST. The membrane was washed 3 × 5 min in PBST, and the bands were detected using the ChemiDoc.

3. Results

In A549 cells, a model cell line for the lower respiratory system, bLF is nontoxic up to 20 μM (Figure 1). Above this, there is a nonsignificant reduction in viability up to 50 μM (p = 0.23). This is in contrast to the control compound staurosporine, that showed a significant drop in viability when incubated with these cells ($p < 0.0001$).

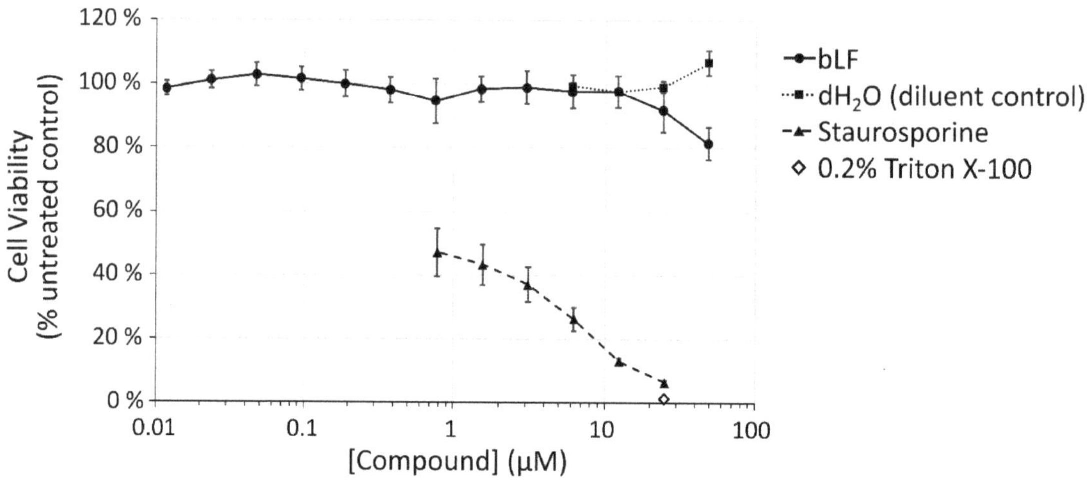

Figure 1. Viability analysis of bLF incubated with A549 over 24 h. A549 cells were incubated with bLF, diluent control (dH_2O), staurosporine, or 0.2% Triton X-100 for 24 h in serum-containing medium before analysis using CellTitre Blue metabolic assay. Error bars represent SEM.

Lactoferrin has been shown to interact with many different receptors [2], of which, the lactoferrin receptor (LfR) intelectin-1 is expressed predominately in the intestine, and is not shown to be expressed in A549 cells via RNAseq analysis [30] (Table S1 in Supplementary Materials). Other receptors implicated in endocytosis of this protein are, however, expressed in this cell line; therefore, we looked at the surface binding of bLF647 in the presence of various competitors. bLF647 was incubated with cells at 4 and 37 °C for 1 h and visualised by confocal microscopy or quantified by flow cytometry (Figure 2). At 4 °C, bLF647 showed moderate binding to the cell surface with no internalisation, whereas at 37 °C, punctate structures could be seen inside the cell, indicative of endocytosis. Coincubation with 100× excess of unlabelled bLF647 showed a significant decrease in the binding to the cell, while a 100× molar excess of unlabelled Tf showed an unexpected increase in plasma membrane bLf647 binding when analysed and quantified by flow cytometry. In comparison, and as expected, a 100× excess of unlabelled Tf completely inhibited Tf488 binding, while a 100× excess unlabelled bLF had no effect on Tf488 binding (Supplementary Figure S1).

Heparin sulphate proteoglycans have been implicated in the binding of LF and CPPs [2,31]. Coincubation with an excess of heparin or washing the cells after a 1 h incubation of bLF647 showed a significant reduction in binding, indicating roles for HSPG or surface charge in bLF binding.

The internalisation properties of bLF647 were compared with transferrin, a model for receptor-mediated endocytosis, and dextran, representing a model for fluid-phase endocytosis. Cells were incubated with bLF647 for different time periods and compared to Tf488 and Dex647 (Figure 3A). Transferrin was rapidly endocytosed into the cells before reaching a steady state after 160 min where uptake and recycling balance each other. Since fluid-phase endocytosis is an ongoing process, and not receptor-dependent, Dex647 endocytosis is a continuous process. Like dextran, bLF647 cell-associated fluorescence increased over the entire time period; however, the kinetic profile was very different. Cell uptake can also be compared as a function of concentration and assays were also conducted with increasing concentrations of the three analysed probes. Here, as in Figure 3B, fluid-phase endocytosis (Dex647) was linear with respect to concentration, while receptor-mediated uptake (Tf488) can be equated to receptor–ligand binding kinetics.

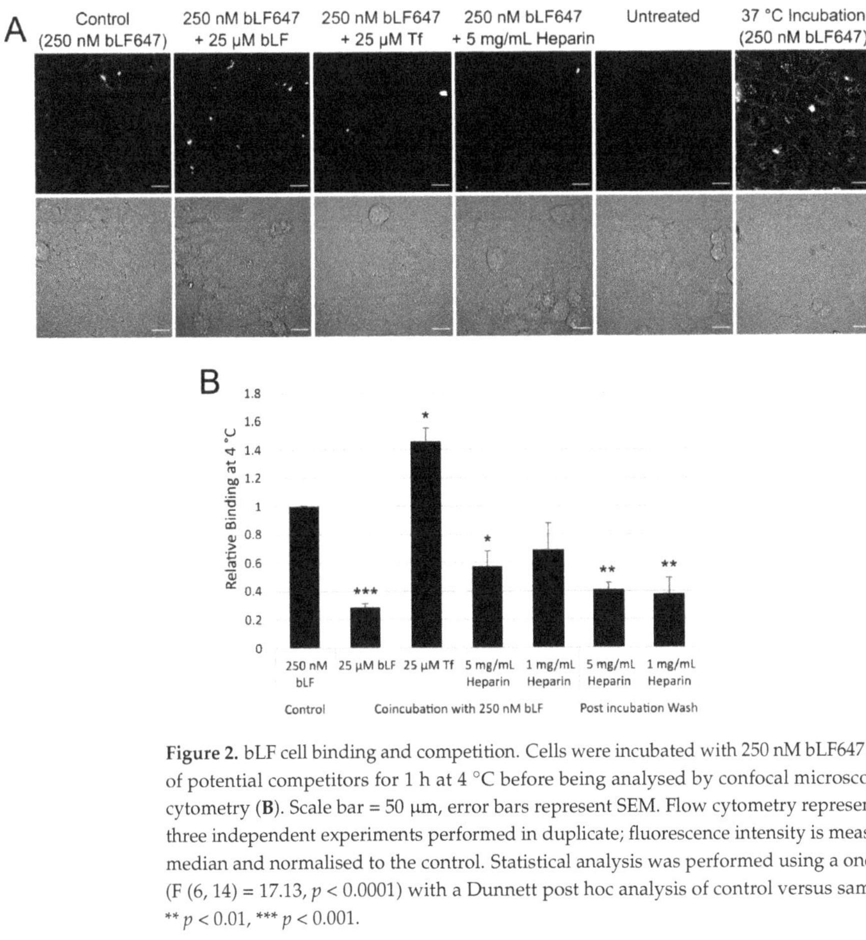

Figure 2. bLF cell binding and competition. Cells were incubated with 250 nM bLF647 in the presence of potential competitors for 1 h at 4 °C before being analysed by confocal microscopy (**A**) or flow cytometry (**B**). Scale bar = 50 μm, error bars represent SEM. Flow cytometry represents the mean of three independent experiments performed in duplicate; fluorescence intensity is measured using the median and normalised to the control. Statistical analysis was performed using a one-way ANOVA ($F (6, 14) = 17.13$, $p < 0.0001$) with a Dunnett post hoc analysis of control versus samples * $p < 0.05$, ** $p < 0.01$, *** $p < 0.001$.

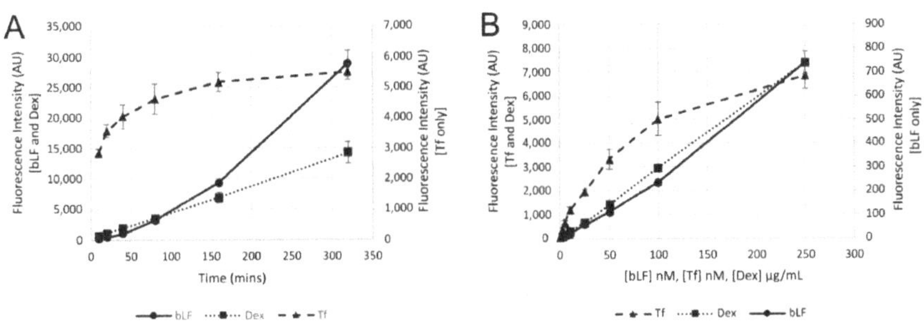

Figure 3. Kinetic analysis of the uptake of bLF647 in A549 cells. Uptake in A549 cells of bLF647, Dex647, and Tf488 analysed against time (**A**) or concentration (**B**). Cells were incubated with 250 nM bLF647, 50 μg/mL Dex647, or 20 nM Tf488 for between 10 min and 6 h; or cells were incubated with differing concentrations of bLF647, Dex647, and Tf488 for 15 min, and analysed by flow cytometry. Data represents the mean of three independent experiments with fluorescence intensity calculated as the mean from a duplicate of median cell intensity values. Error bars represent SEM.

To determine the trafficking characteristics once bLF is endocytosed within the cell, the protein was pulsed for 1 h, chased, and imaged at intervals up to 24 h in cells with labelled lysosomes (Figure 4A). After 1 h, there was little evidence that bLF had reached the lysosomes; however, its prominence in this organelle increased over time with the majority of the bLF647 signal located in lysosomes. This is in contrast to Tf488 where there was little colocalisation after 1 h with the probe itself recycling out of the cell by the next timepoint (Figure 4B). Some dextran had reached the lysosome after 1 h and there was then an increased and, by 24 h, substantial Dex488-647 colocalisation (Figure 4C).

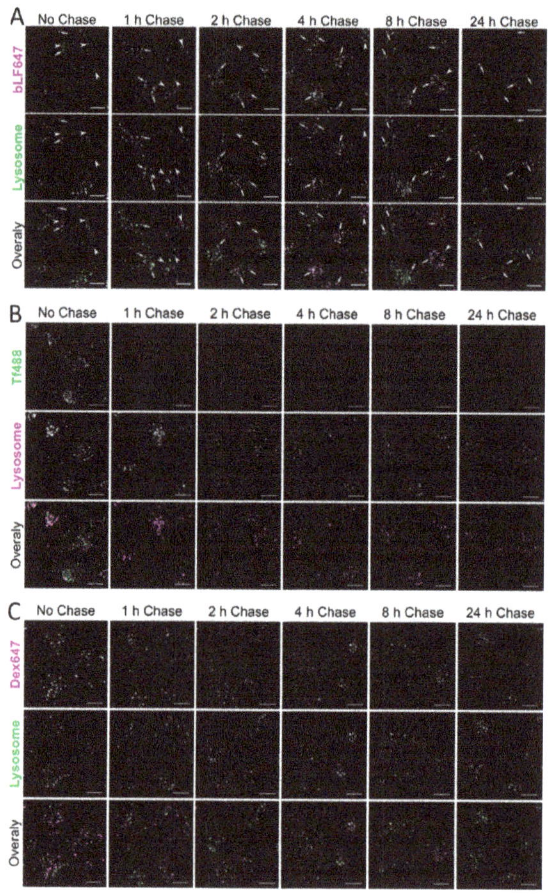

Figure 4. Trafficking of bLF647, Tf488, and Dex647 to lysosomes by confocal microscopy. Cells were incubated with 500 nM bLF647 (**A**), 20 nM Tf488 (**B**), or 100 µg/mL Dex647 (**C**) for 1 h in SFM, washed in serum-containing medium, and incubated for the chase period before being washed and imaged by confocal microscopy. Lysosomes (green, middle row) were prelabelled using a pulse–chase protocol. Arrows represent endocytic structures with colocalised bLF and endolysosomal structures (white), arrow heads indicate bLF647 endocytic structures not colocalised with the lysosome. Scale bar = 10 µm, representative images from three independent experiments quantified in Figure 5. See Supplementary Figures S2–S4 for uncropped images.

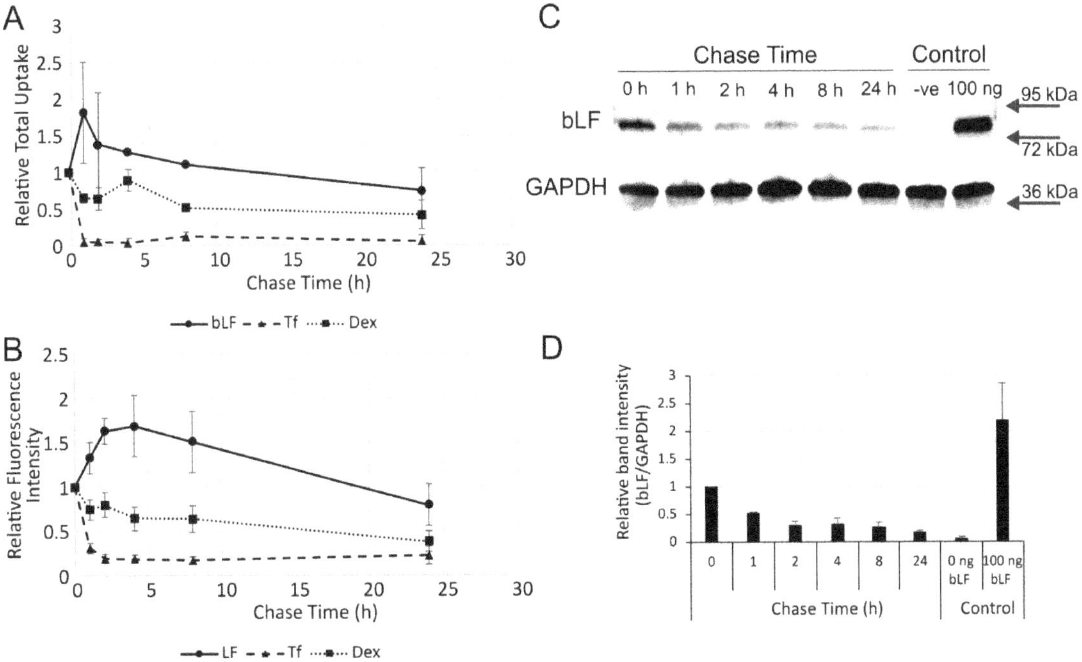

Figure 5. Quantification of trafficking. Cells were incubated with 500 nM bLF647, 100 μg/mL Dex647, or 20 nM Tf488 for 1 h in serum-free medium before being washed and incubated for between 0 and 24 h in serum-containing medium and imaged by confocal microscopy (**A**), flow cytometry (**B**), or Western blotting ((**C**,**D**), 1 μM unlabelled bLF). To obtain relative total uptake (**A**), 10 images were quantified from each independent experiment and normalised to the 0 h chase period. Flow cytometry represents the mean of the median fluorescence from each independent experiment performed in duplicate (N = 3).

To determine the potential rates of exocytosis/recycling of the internalised bLF647, the same experiments used to generate Figure 4 were performed, but with the fluorescence analysed by flow cytometry (Figure 5B). While Tf488 rapidly recycled out of the cell, there was a slower exocytosis of internalised dextran. Similarly, after a 4 h chase time, there was a slow reduction in fluorescence intensity of internalised bLF647; however, there was an unexpected increase between 0 h and this timepoint. A very similar effect can be seen when quantifying cellular fluorescence in the microscopy images (Figure 5A). Further experiments were performed as above with the collection of cell lysates at different chase timepoints. The lysates were separated by SDS-PAGE, and Western blot analysis was performed to detect bLF. The data showed a rapid reduction in detectable bLF signal over the first 2 h of the experiment, with little subsequent reduction until the 24 h timepoint (Figure 5C,D).

4. Discussion

In this study, various endocytic aspects of bLF were investigated and, at times, compared with well-characterised markers of endocytosis, with particular interest in understanding bLF cellular dynamics and how these could relate to its antimicrobial activity. For bLF to be considered as a therapeutic, it needs to be nontoxic, and here we show little evidence of loss of viability in A549 cells after 24 h incubation. This supports previous research that showed low toxicity in these cells after 24 h at concentrations <10 μM; however, at high doses (~180 μM) and after 48 h incubation, there was evidence of toxicity [32]. Toxicity has also been studied in breast cancer cell lines where the growth rate was reduced most prominently after more than 24 h incubation at >15 µM bLF; interestingly, little toxicity was seen in the near-normal cell line MCF10-2A [10]. In vivo experiments have, however, shown bLF to be nontoxic in rats up to 2000 mg/kg/day [33].

Both protein receptors and cell surface proteoglycans have been implicated in the binding and uptake of LF. Both an excess of unlabelled bLF and the strong negative charge of heparin were able to significantly reduce, but not eliminate, the cell surface binding of bLF. This is in contrast to transferrin, where excess unlabelled transferrin almost totally reduced Tf488 binding, while heparin had no effect. This suggests that the binding and endocytosis of this protein are more complex, reflecting the fact that the protein can interact with a variety of receptors in different cell lines, whether they be proteins or carbohydrates. While the canonical LfR (also known as intelectin-1) is not expressed in A549 cells, other implicated receptors, such as LDLR, are. Additionally, in line with several studies on CPPs, the hLF N-terminus showed a strong affinity to heparan sulphate, with this being implicated in its endocytosis [34]. This region has also been shown to be important for the internalisation of hLF via the LfR in Caco-2 colon epithelial cells [35], but it remains to be determined whether this alone mediated the uptake of bLF in A549 cells. Transferrin, like LF, is an iron-binding protein which shares ~60% protein sequence similarity with LF [36]; these differences are evidenced in this N-terminal region but also contribute to LF being more cationic the Tf.

Comparative analysis of bLF endocytosis and endocytic trafficking, compared with transferrin and dextran as receptor-mediated and fluid-phase probes, respectively, identified some unique characteristics for the protein we focused on here. Dextran uptake was linear against concentration and time, while transferrin uptake eventually saturates the receptor where cell-associated fluorescence signal plateaus. bLF displayed linear uptake versus concentration, indicating a fluid phase uptake pathway; however, against time, there was an increasing rate of uptake. This could be due to a stimulatory process occurring during uptake or, more likely, it could be a dequenching effect of the fluorophore attached to the protein. In summary, uptake of bLF proved to be more similar to the fluid-phase endocytosis rather than being a receptor-mediated uptake. Interestingly, bLF was shown to internalise through a receptor-mediated process in hepatocytes [3], potentially via LfR, which was not expressed in the A549 cells studied here. Clathrin-mediated endocytosis was shown to be the main internalising factor of both holo- and apo-lactoferrin in Caco-2 cells where the LfR is also expressed [37]. Clathrin was also hypothesised to be the main endocytic pathway in THP-1 macrophage cells [38]; however, there was a depletion in uptake using inhibitors of other endocytic pathways, confirming our studies showing that it can internalise via different receptors and pathways.

Lactoferrin has been researched as an antiviral and antimicrobial agent; however, little research has been performed on where it localises in the cell once it has been internalised, or whether it persists to potentially act against the pathogen. To determine this, we looked at three complimentary techniques to assess the trafficking properties of bLF: confocal microscopy, flow cytometry, and Western blotting. Both flow cytometry and Western blotting showed a persistence of the bLF647 fluorescence signal up to 24 h, indicating they had not been recycled. This does not however refute the possibility that the fluorophore has become decoupled and continues to provide a signal. This is in contrast to Tf488, where the signal was lost following a 1 h chase period, indicating that rapid recycling had occurred. Dex647 showed a gradual loss of signal to ~50% intensity during the 24 h period, indicating that there was a slow expulsion of the fluorophore. An unexpected increase in fluorescence signal was observed in cells incubated with bLF647 after just a 1 h chase by both confocal microscopy and flow cytometry. The reason for this is unclear, but we hypothesise that this may be due to a dequenching effect of the fluorophore as has been seen previously in our studies using Cy5-mRNA-loaded lipid nanoparticles in one cell line [39] and Alexa488-labelled exosomes [40]. However, Western blot analysis of unlabelled bLF did show that there was a decrease in cell-associated bLF over the first 2 h of chasing. This indicated that the majority of the bLF was degraded in lysosomes in 2 h, rather than being exocytosed, and that the remaining fluorescent signal shown by flow cytometry and confocal microscopy emanated from the fluorophore alone or attached to bLF peptides that could not be detected by Western blotting.

In conclusion, our studies show that the protein is effectively internalised, trafficked to lysosomes, and degraded in this lung cell, and provide no evidence that it can reach the nucleus, at least not at the concentrations studied here. It remains to be determined whether the protein, as an intact entity residing at any particular point on its endocytic traffic, can act as an intracellular antimicrobial agent, as we have previously shown for CPPs [41].

Supplementary Materials: The following are available online at https://www.mdpi.com/article/10.3390/pharmaceutics14040855/s1. Table S1: Expression in A549 cells of receptors implicated in LF uptake, Figure S1: Binding of transferrin to A549 cells, Figure S2: Trafficking of bLF647 to lysosomes by confocal microscopy, Figure S3: Trafficking of Tf488 to lysosomes by confocal microscopy, Figure S4: Trafficking of Dex647 to lysosomes by confocal microscopy. References [30,42–47] are cited in the supplementary materials.

Author Contributions: Conceptualisation, all authors; methodology, E.J.S., I.P., P.W., and A.T.J.; formal analysis, E.J.S.; investigation, E.J.S.; resources, L.H., P.H., P.W., and A.T.J.; data curation, E.J.S.; writing—original draft preparation, E.J.S., I.P., and A.T.J.; writing—review and editing, all authors; supervision, P.W., A.T.J., L.H., and P.H.; project administration, I.P., L.H., P.H., P.W., and A.T.J.; funding acquisition, L.H., P.H., P.W., I.P., and A.T.J. All authors have read and agreed to the published version of the manuscript.

Funding: The work described herein was a collaboration between Cardiff University and Virustatic Limited through the Celtic Advanced Life Science Innovation Network (CALIN), an Ireland Wales Co-Operation (80885) 2014–2020 programme part funded by the European Regional Development Fund through the Welsh Government.

Institutional Review Board Statement: Not applicable.

Informed Consent Statement: Not applicable.

Data Availability Statement: The data presented within this study are available within the article; Cancer Cell Line Encyclopaedia (CCLE) data sets can be found at https://sites.broadinstitute.org/ccle/ (accessed on 27 January 2022).

Conflicts of Interest: E.J.S., I.P., P.W. and A.T.J. declare no conflict of interest. P.H. is founder of Virustatic Limited, and L.H. is an employee of Virustatic Limited; both declare no conflict of interest. The funders and company had no role in the collection, analyses, and interpretation of data.

References

1. Rosa, L.; Cutone, A.; Lepanto, M.S.; Paesano, R.; Valenti, P. Lactoferrin: A Natural Glycoprotein Involved in Iron and Inflammatory Homeostasis. *Int. J. Mol. Sci.* **2017**, *18*, 1985. [CrossRef] [PubMed]
2. Kell, D.B.; Heyden, E.L.; Pretorius, E. The Biology of Lactoferrin, an Iron-Binding Protein That Can Help Defend Against Viruses and Bacteria. *Front. Immunol.* **2020**, *11*, 1221. [CrossRef] [PubMed]
3. McAbee, D.D.; Esbensen, K. Binding and endocytosis of apo- and holo-lactoferrin by isolated rat hepatocytes. *J. Biol. Chem.* **1991**, *266*, 23624–23631. [CrossRef]
4. Duchardt, F.; Ruttekolk, I.R.; Verdurmen, W.P.R.; Lortat-Jacob, H.; Burck, J.; Hufnagel, H.; Fischer, R.; van den Heuvel, M.; Lowik, D.; Vuister, G.W.; et al. A cell-penetrating peptide derived from human lactoferrin with conformation-dependent uptake efficiency. *J. Biol. Chem.* **2009**, *284*, 36099–36108. [CrossRef]
5. Milletti, F. Cell-penetrating peptides: Classes, origin, and current landscape. *Drug Discov. Today* **2012**, *17*, 850–860. [CrossRef]
6. Teng, C.T. Lactoferrin gene expression and regulation: An overview. *Biochem. Cell Biol.* **2002**, *80*, 7–16. [CrossRef]
7. van Berkel, P.H.; Geerts, M.E.; van Veen, H.A.; Mericskay, M.; de Boer, H.A.; Nuijens, J.H. N-terminal stretch Arg2, Arg3, Arg4 and Arg5 of human lactoferrin is essential for binding to heparin, bacterial lipopolysaccharide, human lysozyme and DNA. *Biochem. J.* **1997**, *328 Pt 1*, 145–151. [CrossRef]
8. Penco, S.; Scarfi, S.; Giovine, M.; Damonte, G.; Millo, E.; Villaggio, B.; Passalacqua, M.; Pozzolini, M.; Garre, C.; Benatti, U. Identification of an import signal for, and the nuclear localization of, human lactoferrin. *Biotechnol. Appl. Biochem.* **2001**, *34*, 151–159. [CrossRef]

9. Liu, B.R.; Huang, Y.W.; Aronstam, R.S.; Lee, H.J. Identification of a Short Cell-Penetrating Peptide from Bovine Lactoferricin for Intracellular Delivery of DNA in Human A549 Cells. *PLoS ONE* **2016**, *11*, e0150439. [CrossRef]
10. Zhang, Y.; Nicolau, A.; Lima, C.F.; Rodrigues, L.R. Bovine lactoferrin induces cell cycle arrest and inhibits mTOR signaling in breast cancer cells. *Nutr. Cancer* **2014**, *66*, 1371–1385. [CrossRef]
11. Kumari, S.; Kondapi, A.K. Receptor-mediated targeted delivery of DNA using Lactoferrin nanoparticles. *Int. J. Biol. Macromol.* **2018**, *108*, 401–407. [CrossRef] [PubMed]
12. Elzoghby, A.O.; Abdelmoneem, M.A.; Hassanin, I.A.; Abd Elwakil, M.M.; Elnaggar, M.A.; Mokhtar, S.; Fang, J.Y.; Elkhodairy, K.A. Lactoferrin, a multi-functional glycoprotein: Active therapeutic, drug nanocarrier & targeting ligand. *Biomaterials* **2020**, *263*, 120355. [CrossRef] [PubMed]
13. Pereira, P.; Barreira, M.; Cruz, C.; Tomas, J.; Luis, A.; Pedro, A.Q.; Queiroz, J.A.; Sousa, F. Brain-Targeted Delivery of Pre-miR-29b Using Lactoferrin-Stearic Acid-Modified-Chitosan/Polyethyleneimine Polyplexes. *Pharmaceuticals* **2020**, *13*, 314. [CrossRef] [PubMed]
14. Almowalad, J.; Somani, S.; Laskar, P.; Meewan, J.; Tate, R.J.; Mullin, M.; Dufes, C. Lactoferrin-Bearing Gold Nanocages for Gene Delivery in Prostate Cancer Cells in vitro. *Int. J. Nanomed.* **2021**, *16*, 4391–4407. [CrossRef] [PubMed]
15. Valenti, P.; Antonini, G. Lactoferrin: An important host defence against microbial and viral attack. *Cell. Mol. Life Sci.* **2005**, *62*, 2576–2587. [CrossRef]
16. Wakabayashi, H.; Oda, H.; Yamauchi, K.; Abe, F. Lactoferrin for prevention of common viral infections. *J. Infect. Chemother.* **2014**, *20*, 666–671. [CrossRef]
17. Pietrantoni, A.; Di Biase, A.M.; Tinari, A.; Marchetti, M.; Valenti, P.; Seganti, L.; Superti, F. Bovine lactoferrin inhibits adenovirus infection by interacting with viral structural polypeptides. *Antimicrob. Agents Chemother.* **2003**, *47*, 2688–2691. [CrossRef]
18. Swart, P.J.; Kuipers, M.E.; Smit, C.; Pauwels, R.; deBethune, M.P.; de Clercq, E.; Meijer, D.K.; Huisman, J.G. Antiviral effects of milk proteins: Acylation results in polyanionic compounds with potent activity against human immunodeficiency virus types 1 and 2 in vitro. *AIDS Res. Hum. Retroviruses* **1996**, *12*, 769–775. [CrossRef]
19. Yi, M.; Kaneko, S.; Yu, D.Y.; Murakami, S. Hepatitis C virus envelope proteins bind lactoferrin. *J. Virol.* **1997**, *71*, 5997–6002. [CrossRef]
20. Lang, J.; Yang, N.; Deng, J.; Liu, K.; Yang, P.; Zhang, G.; Jiang, C. Inhibition of SARS pseudovirus cell entry by lactoferrin binding to heparan sulfate proteoglycans. *PLoS ONE* **2011**, *6*, e23710. [CrossRef]
21. Waarts, B.L.; Aneke, O.J.; Smit, J.M.; Kimata, K.; Bittman, R.; Meijer, D.K.; Wilschut, J. Antiviral activity of human lactoferrin: Inhibition of alphavirus interaction with heparan sulfate. *Virology* **2005**, *333*, 284–292. [CrossRef] [PubMed]
22. Marr, A.K.; Jenssen, H.; Moniri, M.R.; Hancock, R.E.; Pante, N. Bovine lactoferrin and lactoferricin interfere with intracellular trafficking of Herpes simplex virus-1. *Biochimie* **2009**, *91*, 160–164. [CrossRef] [PubMed]
23. Mesel-Lemoine, M.; Millet, J.; Vidalain, P.O.; Law, H.; Vabret, A.; Lorin, V.; Escriou, N.; Albert, M.L.; Nal, B.; Tangy, F. A human coronavirus responsible for the common cold massively kills dendritic cells but not monocytes. *J. Virol.* **2012**, *86*, 7577–7587. [CrossRef]
24. de Carvalho, C.A.M.; da Rocha Matos, A.; Caetano, B.C.; de Sousa Junior, I.P.; da Costa Campos, S.P.; Geraldino, B.R.; Barros, C.A.; de Almeida, M.A.P.; Rocha, V.P.; da Silva, A.M.V.; et al. In Vitro Inhibition of SARS-CoV-2 Infection by Bovine Lactoferrin. *bioRxiv* **2020**, 2020.2005.2013.093781.
25. Hu, Y.; Meng, X.; Zhang, F.; Xiang, Y.; Wang, J. The in vitro antiviral activity of lactoferrin against common human coronaviruses and SARS-CoV-2 is mediated by targeting the heparan sulfate co-receptor. *Emerg. Microbes Infect.* **2021**, *10*, 317–330. [CrossRef] [PubMed]
26. Mirabelli, C.; Wotring, J.W.; Zhang, C.J.; McCarty, S.M.; Fursmidt, R.; Pretto, C.D.; Qiao, Y.; Zhang, Y.; Frum, T.; Kadambi, N.S.; et al. Morphological cell profiling of SARS-CoV-2 infection identifies drug repurposing candidates for COVID-19. *Proc. Natl. Acad. Sci. USA* **2021**, *118*, e2105815118. [CrossRef]
27. Chang, R.; Ng, T.B.; Sun, W.Z. Lactoferrin as potential preventative and adjunct treatment for COVID-19. *Int. J. Antimicrob. Agents* **2020**, *56*, 106118. [CrossRef]
28. Moody, P.R.; Sayers, E.J.; Magnusson, J.P.; Alexander, C.; Borri, P.; Watson, P.; Jones, A.T. Receptor Crosslinking: A General Method to Trigger Internalization and Lysosomal Targeting of Therapeutic Receptor:Ligand Complexes. *Mol. Ther.* **2015**, *23*, 1888–1898. [CrossRef]
29. Schindelin, J.; Arganda-Carreras, I.; Frise, E.; Kaynig, V.; Longair, M.; Pietzsch, T.; Preibisch, S.; Rueden, C.; Saalfeld, S.; Schmid, B.; et al. Fiji: An open-source platform for biological-image analysis. *Nat. Methods* **2012**, *9*, 676–682. [CrossRef]
30. Nusinow, D.P.; Szpyt, J.; Ghandi, M.; Rose, C.M.; McDonald, E.R., 3rd; Kalocsay, M.; Jane-Valbuena, J.; Gelfand, E.; Schweppe, D.K.; Jedrychowski, M.; et al. Quantitative Proteomics of the Cancer Cell Line Encyclopedia. *Cell* **2020**, *180*, 387–402.e16. [CrossRef]
31. Gestin, M.; Dowaidar, M.; Langel, U. Uptake Mechanism of Cell-Penetrating Peptides. *Adv. Exp. Med. Biol.* **2017**, *1030*, 255–264. [CrossRef] [PubMed]
32. Tung, Y.T.; Chen, H.L.; Yen, C.C.; Lee, P.Y.; Tsai, H.C.; Lin, M.F.; Chen, C.M. Bovine lactoferrin inhibits lung cancer growth through suppression of both inflammation and expression of vascular endothelial growth factor. *J. Dairy Sci.* **2013**, *96*, 2095–2106. [CrossRef] [PubMed]

33. Yamauchi, K.; Toida, T.; Nishimura, S.; Nagano, E.; Kusuoka, O.; Teraguchi, S.; Hayasawa, H.; Shimamura, S.; Tomita, M. 13-Week oral repeated administration toxicity study of bovine lactoferrin in rats. *Food Chem. Toxicol.* **2000**, *38*, 503–512. [CrossRef]
34. Wallbrecher, R.; Verdurmen, W.P.; Schmidt, S.; Bovee-Geurts, P.H.; Broecker, F.; Reinhardt, A.; van Kuppevelt, T.H.; Seeberger, P.H.; Brock, R. The stoichiometry of peptide-heparan sulfate binding as a determinant of uptake efficiency of cell-penetrating peptides. *Cell. Mol. Life Sci.* **2014**, *71*, 2717–2729. [CrossRef] [PubMed]
35. Suzuki, Y.A.; Wong, H.; Ashida, K.Y.; Schryvers, A.B.; Lonnerdal, B. The N1 domain of human lactoferrin is required for internalization by caco-2 cells and targeting to the nucleus. *Biochemistry* **2008**, *47*, 10915–10920. [CrossRef]
36. Baker, E.N.; Baker, H.M.; Kidd, R.D. Lactoferrin and transferrin: Functional variations on a common structural framework. *Biochem. Cell Biol.* **2002**, *80*, 27–34. [CrossRef]
37. Jiang, R.; Lopez, V.; Kelleher, S.L.; Lonnerdal, B. Apo- and holo-lactoferrin are both internalized by lactoferrin receptor via clathrin-mediated endocytosis but differentially affect ERK-signaling and cell proliferation in Caco-2 cells. *J. Cell. Physiol.* **2011**, *226*, 3022–3031. [CrossRef]
38. Florian, P.; Macovei, A.; Sima, L.; Nichita, N.; Mattsby-Baltzer, I.; Roseanu, A. Endocytosis and trafficking of human lactoferrin in macrophage-like human THP-1 cells (1). *Biochem. Cell Biol.* **2012**, *90*, 449–455. [CrossRef]
39. Sayers, E.J.; Peel, S.E.; Schantz, A.; England, R.M.; Beano, M.; Bates, S.M.; Desai, A.S.; Puri, S.; Ashford, M.B.; Jones, A.T. Endocytic Profiling of Cancer Cell Models Reveals Critical Factors Influencing LNP-Mediated mRNA Delivery and Protein Expression. *Mol. Ther.* **2019**, *27*, 1950–1962. [CrossRef]
40. Roberts-Dalton, H.D.; Cocks, A.; Falcon-Perez, J.M.; Sayers, E.J.; Webber, J.P.; Watson, P.; Clayton, A.; Jones, A.T. Fluorescence labelling of extracellular vesicles using a novel thiol-based strategy for quantitative analysis of cellular delivery and intracellular traffic. *Nanoscale* **2017**, *9*, 13693–13706. [CrossRef]
41. Bahnsen, J.S.; Franzyk, H.; Sayers, E.J.; Jones, A.T.; Nielsen, H.M. Cell-penetrating antimicrobial peptides—Prospectives for targeting intracellular infections. *Pharm. Res.* **2015**, *32*, 1546–1556. [CrossRef] [PubMed]
42. Chien, Y.J.; Chen, W.J.; Hsu, W.L.; Chiou, S.S. Bovine lactoferrin inhibits Japanese encephalitis virus by binding to heparan sulfate and receptor for low density lipoprotein. *Virology* **2008**, *379*, 143–151. [CrossRef] [PubMed]
43. Grey, A.; Banovic, T.; Zhu, Q.; Watson, M.; Callon, K.; Palmano, K.; Ross, J.; Naot, D.; Reid, I.R.; Cornish, J. The low-density lipoprotein receptor-related protein 1 is a mitogenic receptor for lactoferrin in osteoblastic cells. *Mol. Endocrinol.* **2004**, *18*, 2268–2278. [CrossRef] [PubMed]
44. Shin, K.; Wakabayashi, H.; Yamauchi, K.; Yaeshima, T.; Iwatsuki, K. Recombinant human intelectin binds bovine lactoferrin and its peptides. *Biol. Pharm. Bull.* **2008**, *31*, 1605–1608. [CrossRef]
45. Curran, C.S.; Demick, K.P.; Mansfield, J.M. Lactoferrin activates macrophages via TLR4-dependent and -independent signaling pathways. *Cell. Immunol.* **2006**, *242*, 23–30. [CrossRef] [PubMed]
46. Takayama, Y.; Aoki, R.; Uchida, R.; Tajima, A.; Aoki-Yoshida, A. Role of CXC chemokine receptor type 4 as a lactoferrin receptor. *Biochem. Cell Biol.* **2017**, *95*, 57–63. [CrossRef] [PubMed]
47. Perdijk, O.; van Neerven, R.J.J.; van den Brink, E.; Savelkoul, H.F.J.; Brugman, S. Bovine Lactoferrin Modulates Dendritic Cell Differentiation and Function. *Nutrients* **2018**, *10*, 848. [CrossRef]

pharmaceutics

MDPI

Article

Highway to Cell: Selection of the Best Cell-Penetrating Peptide to Internalize the CFTR-Stabilizing iCAL36 Peptide

Quentin Seisel [1], Israpong Lakumpa [1], Emilie Josse [2], Eric Vivès [2], Jessica Varilh [3], Magali Taulan-Cadars [3] and Prisca Boisguérin [2,*]

[1] CRBM, University of Montpellier, CNRS UMR 5237, 34000 Montpellier, France; quentin.seisel@live.fr (Q.S.); israpong.lakumpa@gmail.com (I.L.)
[2] PhyMedExp, Bâtiment Crastes de Paulet, University of Montpellier, INSERM U1046, CNRS UMR 9214, 34000 Montpellier, France; emilie.josse@inserm.fr (E.J.); eric.vives@umontpellier.fr (E.V.)
[3] PhyMedExp, Institut Universitaire de Recherche Clinique, University of Montpellier, INSERM U1046, CNRS UMR 9214, 34000 Montpellier, France; jessica.varilh@inserm.fr (J.V.); magali.taulan@inserm.fr (M.T.-C.)
* Correspondence: prisca.boisguerin@inserm.fr

Abstract: Therapeutic peptides have regained interest as they can address unmet medical needs and can be an excellent complement to pharmaceutic small molecules and other macromolecular therapeutics. Over the past decades, correctors and potentiators of the cystic fibrosis transmembrane conductance regulator (CFTR), a chloride ion channel causing cystic fibrosis (CF) when mutated, were developed to reduce the symptoms of the patients. In this context, we have previously designed a CFTR-stabilizing iCAL36 peptide able to further increase the CFTR amount in epithelial cells, thereby resulting in a higher CFTR activity. In the present study, optimization of the peptidyl inhibitor was performed by coupling five different cell-penetrating peptides (CPP), which are Tat, dTat, TatRI (*retro-inverso*), MPG, and Penetratin. Screening of the internalization properties of these CPP-iCAL36 peptides under different conditions (with or without serum or endocytosis inhibitors, etc.) was performed to select TatRI as the optimal CPP for iCAL36 delivery. More importantly, using this TatRI-iCAL36 peptide, we were able to reveal for the first time an additive increase in the CFTR amount in the presence of VX-445/VX-809 compared to VX-445/VX-809 treatment alone. This finding is a significant contribution to the development of CFTR-stabilizing peptides in addition to currently used treatments (small-molecule correctors or potentiators) for CF patients.

Keywords: CFTR-CAL interaction; CFTR stabilizer; cell-penetrating peptide; p.Phe508del mutant; internalization mechanism; cystic fibrosis

Citation: Seisel, Q.; Lakumpa, I.; Josse, E.; Vivès, E.; Varilh, J.; Taulan-Cadars, M.; Boisguérin, P. Highway to Cell: Selection of the Best Cell-Penetrating Peptide to Internalize the CFTR-Stabilizing iCAL36 Peptide. *Pharmaceutics* **2022**, *14*, 808. https://doi.org/10.3390/pharmaceutics14040808

Academic Editor: Yoshiyuki Hattori

Received: 17 February 2022
Accepted: 31 March 2022
Published: 7 April 2022

Copyright: © 2022 by the authors. Licensee MDPI, Basel, Switzerland. This article is an open access article distributed under the terms and conditions of the Creative Commons Attribution (CC BY) license (https://creativecommons.org/licenses/by/4.0/).

1. Introduction

The molecular drivers of many basic cellular functions and pathways are protein–protein interactions (PPIs); therefore, they have broadly been used as targets for drug development. Besides small molecules, which are not useful for targeting large or flat PPI binding sites, peptides are increasingly becoming attractive therapeutic candidates. The advantages of therapeutic peptides lie primarily in their high efficacy and low toxicity, but also in the unlimited possibilities of introducing modifications to improve their stability and binding affinity. Accordingly, therapeutic approaches using peptides have become an emerging market in the pharmaceutical industry over the past two decades [1], with more than 60 peptide drugs approved by the Food Drug Administration and many more studied in clinical and preclinical trials [2].

Within the context of cystic fibrosis (CF), an inherited recessive autosomal disorder caused by mutations in the gene cystic fibrosis transmembrane conductance regulator (CFTR), several peptides have been proposed as therapy, such as the S18 peptide as an ENaC antagonist [3], the alpha-1 proteinase inhibitor (PI) to reduce neutrophil elastase burden [4] or antimicrobial peptides (e.g., Esculentin [5], M33 [6] and Colistin [7]).

We and others have been working for many years on the development of therapeutic peptides acting as CFTR "stabilizers" [8–12]. Indeed, CFTR functional interactions at the apical membrane are regulated by the CAL (CFTR-associated ligand) protein, which limits cell surface levels of either wild-type CFTR or of the most common disease-associated mutant p.Phe508del-CFTR. CAL (also known as PIST, GOPC, and FIG) is a protein containing two coiled-coil domains and one PDZ domain interacting with the C-terminus of CFTR [13]. As shown by Guggino and co-workers, this CFTR–CAL interaction reduced the amount of surface CFTR through endocytic recycling [14], resulting in lysosomal degradation via the proteasome [15]. Based on this observation, selective inhibitors against the CFTR–CAL interaction should provide a novel class of CFTR "stabilizers".

In line with this, we were the first to validate this approach by developing a peptide inhibitor of the CAL protein, iCAL36, which enhanced the CFTR membrane half-life in cultured airway epithelial cells and increased CFTR activity (+11%) [8,9]. Since then, other peptide-based inhibitors have been optimized in terms of their affinity [16], their proteolytic stability [10], or their cell-permeability [11].

Indeed, within the design of therapeutic peptides, the development of their optimal vectorization inside cells is a major point. Different vectors are used for the transfection of therapeutic peptides such as viral-based vectors, liposomes, nanoparticles, or cell-penetrating peptides (for reviews, see [17,18]). Cell-penetrating peptides (CPPs) were discovered thirty years ago, with the pioneering peptides Tat [19] and Penetratin [20]. To date, more than 100 different CPPs have been designed and constitute some of the most promising non-viral strategies for the delivery of peptides, proteins, and different nucleic acids [21]. They have gained increasing popularity because of their short length and efficient cell translocation ability, together with their cargoes.

Most CPPs are small, cationic peptides with typical lengths ranging from 5–30 residues, a median charge of +5, and an isoelectric point above 10 [22]. Although several different criteria have been proposed for the classification of CPPs based on their origin, sequence, function, or mechanism of uptake, no unified taxonomy of these peptides presently exists. Therefore, they can be simply categorized into three main classes according to their physico-chemical properties: cationic, amphipathic, and hydrophobic peptides [23]. Depending on the used CPP and even on the used cargo or cellular context, CPPs can pass through cell membranes via energy-dependent mechanisms or energy-independent ones without the implication of specific receptors [24].

In this study, we aimed to find out the most suitable CPP for the iCAL36 delivery in human epithelial cells. For this purpose, we selected several different CPPs, which were conjugated to the N-terminus of iCAL36. The cellular internalization (amount and localization) of these CPP-iCAL36 peptides was analyzed in Caco-2 and Calu-3 cells. Internalization mechanisms were elucidated using leakage assays, endocytosis inhibitors, endocytosis or vesicle markers, as well as via their internalization behavior in the presence of serum (model for mucus presence). Finally, we selected TatRI-iCAL36 as a more suitable CFTR stabilizer, showing for the first time an increase in p.Phe508del-CFTR after TatRI-iCAL36 treatment in epithelial 16HBE cells.

2. Materials and Methods

2.1. Peptide Synthesis

Peptide synthesis was performed using a LibertyBlue™ Microwave Peptide Synthesizer (CEM Corporation, Matthews, NC, USA) with an additional module of Discover™ (CEM Corporation, Matthews, NC, USA) combining microwave energy at 2450 MHz to the fluorenylmethoxycarbonyl (Fmoc)/tert-butyl (tBu) strategy. All chemicals or solvents were purchased from Sigma-Aldrich (Saint-Quentin-Fallavier, France) or Carlo Erba (Val-de-Reuil, France). Syntheses were conducted on a Fmoc-Ile-Wang resin (Bachem, Bubendorf, Switzerland, 0.25 mmol scale) using Fmoc-L- or Fmoc-D-amino acids (Bachem, Bubendorf, Switzerland). Tamra labeling of the peptide was performed by overnight coupling with 1.5 eq. 5(6)-carboxytetramethylrhodamine (Merck, Fontenay Sous Bois, France),

1.6 eq. HATU and 2 eq. DIEA (Merck, Fontenay Sous Bois, France) in DMF, then washing with DMF and CH_2Cl_2.

After purification by column chromatography, the peptide identity and purity were checked by LC-MS (Waters, Guyancourt, France). Unlabeled and Tamra-labeled peptides (Table 1) were used with a purity higher than 95%.

Table 1. Peptide sequences used in this study.

Name	Code	Sequence	Residues	pI
iCAL36	36	ANSRWPTSII	10	9.79
Tat-iCAL36 [1]	T36	GRKKRRQRRRPPQ-ANSRWPTSII	23	12.78
dTat-iCAL36	dT36	grkkrrqrrrppq-ANSRWPTSII	23	12.78
TatRI-iCAL36	TRI36	qpprrrqrrkkrg-ANSRWPTSII	23	12.78
Pen-iCAL36	P36	RQILIWFQNRRMKWKK-ANSRWPTSII	26	12.48
MPG-iCAL36 [1]	M36	GALFLGWLGAAGSTMGAWSQPKKKRKV-ANSRWPTSII	37	12.03

[1] Sequences presented in Seisel et al. 2019 [25]; pI was determined using the Expasy ProtParam tool.

2.2. Cell Culture Conditions

Human epithelial colorectal adenocarcinoma (Caco-2, ATCC, HTB-37) and human epithelial pulmonary adenocarcinoma (Calu-3, ATCC, HTB55) cells were maintained in DMEM 4.5 g/L glucose supplemented with UltraGlutamine (Lonza, Levallois-Perret, France), 20% v/v fetal bovine serum (FBS from Thermo Fisher Scientific Inc., Rockford, IL, USA), 1% MEM non-essential amino acids, 1% penicillin/streptomycin, 1% sodium pyruvate and 1% sodium bicarbonate (all v/v, 100×, Thermo Fisher Scientific Inc., Rockford, IL, USA). Cells were passaged once a week using trypsin (0.05%, Life Technologies) and grown in a humidified incubator with 5% CO_2 at 37 °C. All cells in experiments were used between passages 9 and 25 and were regularly tested for mycoplasma contamination. Cells were seeded at different densities (see below) for the corresponding experiments. In all cases, the medium was exchanged after 24 h, and cells were used 48 h after seeding.

Human bronchial epithelial cells 16HBEge-p.Phe508del (16HBE gene-edited CFTR p.Phe508del cells, generated by Dieter Grünert and distributed by the Cystic Fibrosis Foundation) were maintained in MEM medium supplemented with UltraGlutamine (Lonza, Levallois-Perret, France), 10% v/v fetal bovine serum (FBS from Thermo Fisher Scientific Inc., Rockford, IL, USA) and 1% v/v penicillin/streptomycin (100×, Life Technologies). Cells were passaged twice a week using trypsin (0.05%, Life Technologies) and grown in a humidified incubator with 5% CO_2 at 37 °C. The 16HBE cells in experiments were used between passages 5 and 20 and were regularly tested for mycoplasma contamination. Cells were seeded at different densities (see below) for the corresponding experiments.

2.3. Cell Cytotoxicity Measurements

An evaluation of the cytotoxicity induced by CPP-iCAL36 conjugates was performed using a Cytotoxicity Detection KitPlus (LDH, (Merck, Fontenay Sous Bois, France) following the manufacturer's instructions. In brief, 1.6×10^4 cells (160 µL) were seeded in 96-well Nunc culture plates (Thermo Fisher Scientific Inc., Rockford, IL, USA). Then, 48 h post-seeding, cells were rinsed twice with D-PBS and then incubated with 160 µL of a 1 µM, 10 µM, or 100 µM CPP-iCAL36 solution (in triplicates) in OptiMEM (Life Technologies) for 24 h at 37 °C with 5% CO_2. At least 3 wells were kept for the LDH negative control (0% viability) and for non-treated cells as a positive control (100% viability).

After a 24 h incubation period, negative controls were performed by adding Triton X-100 (Sigma-Aldrich, Saint-Quentin-Fallavier, France) (~10 min incubation at 37 °C). Afterward, 50 µL of supernatant was collected, mixed with 50 µL of the "dye solution/catalyst" mixture, and incubated in the darkness for 30 min at room temperature. The reaction was stopped by adding 25 µL of HCl (1 N) to each well before measuring the absorption at 490 nm. Relative toxicity (%) = ((exp. value − value non-treated cells)/(value triton − value non-treated cells)) × 100.

2.4. Cell Transfection Conditions for Flow Cytometry Acquisition

Here, 2×10^5 cells (1 mL) were seeded in 24-well Nunc culture plates (Thermo Fisher Scientific Inc., Rockford, IL, USA). Next, 48 h post-seeding, cells were rinsed twice with D-PBS and then incubated with 500 µL of Tamra-CPP-iCAL36 solution at the indicated conditions.

Standard transfection and kinetic measurements. Cells were incubated for 1.5 h, 3 h, or 4.5 h with solutions of Tamra-CPP-iCAL36 (1 µM or 5 µM) in OptiMEM (500 µL) at 37 °C with 5% CO_2.

Transfection in the presence of serum. Cells were incubated for 3 h with solutions of Tamra-CPP-iCAL36 at 1 µM or 5 µM in OptiMEM (500 µL) supplemented with 0%, 10% or 20% (*v*/*v*) FBS at 37 °C with 5% CO_2.

Transfection at 4 °C. Cells were pre-incubated for 30 min at 4 °C (on ice) with OptiMEM, followed by incubation for 1.5 h at 4 °C with Tamra-CPP-iCAL36 (1 µM or 5 µM) in OptiMEM (500 µL).

Transfection in the presence of endocytosis inhibitors or chloroquine. Cells were pretreated for 30 min in OptiMEM (500 µL) using one of the following conditions: chlorpromazine (CPZ, 20 µM), nystatin (NYS, 50 µM), methyl-β-cyclodextrin (MBCD, 5 mM), sodium azide/2-deoxyglucose (NaN_3/DG, 100 µM/60 µM) or chloroquine (CQ, 100 µM) at 37 °C with 5% CO_2. Afterward, Tamra-CPP-iCAL36 (1 µM or 5 µM) in OptiMEM were added, and cells were incubated for a further 1.5 h at 37 °C with 5% CO_2.

At the end of each incubation, cells were rinsed twice with D-PBS. To remove peptides sticking to the extracellular membrane, cells were treated for 10 min with 100 µL of trypsin (0.05%) at 37 °C with 5% CO_2. After the addition of 400 µL of D-PBS supplemented with 5% (*v*/*v*) FBS, cells were transferred to 1.5 mL tubes and centrifuged (10 min, 4 °C, 1500 rpm). Cell pellets were resumed in 500 µL of D-PBS supplemented with 0.5% (*v*/*v*) FBS and 0.1% (*v*/*v*) 4′,6-diamidino-2-phenylindole (DAPI solution, Sigma-Aldrich, (Saint-Quentin-Fallavier, France), transferred to round-bottom polystyrene Falcon tubes (Thermo Fisher Scientific Inc., Rockford, IL, USA) and analyzed with a Fortessa LSR cytometer (Becton Dickinson, Pont-de-Claix, France; DAPI: λex = 405 nm/λem = 425–475 nm; TAMRA: λex = 561 nm/λem = 605–625 nm). Damaged cells were excluded with the DAPI labeling. For each condition, 10,000 events were recorded.

2.5. Cell Transfection Conditions for Confocal Microscopy Imaging

Here, 3×10^5 cells (1 mL) were seeded in 35 mm FluoroDish culture dishes (WPI). Next, 48 h post-seeding, cells were rinsed twice with D-PBS and then incubated with 1 mL of Tamra-CPP-iCAL36 solution at the indicated concentration in OptiMEM at 37 °C with 5% CO_2.

Screening of Tamra-CPP-iCAL36 localization. Cells were incubated for 3 h with solutions of Tamra-CPP-iCAL36 at 1 µM (P36, M36) or 5 µM (T36, dT36, TRI36) in OptiMEM at 37 °C with 5% CO_2. Next, 10 min before the end of incubation, 10 µL of WGA-488 (final concentration: 1 µg/mL, Thermo Fisher Scientific Inc., Rockford, IL, USA) was added to label membranes.

Transfection in the presence of endocytosis or lysosome markers. Cells were incubated for 1.5 h with solutions of Tamra-TatRI-iCAL36 (TRI36, 5 µM) or Tamra-Pen-iCAL36 (P36, 1 µM), together with one of the following markers: Transferrin-Alexa488 (Transferrin-488, 50 µg/mL), cholera toxin subunit B-Alexa488 (CtB-488, 4 µg/mL), Dextran-pHrodo Green (Dextran-Green, 25 µg/mL) or LysoTracker™ Green DND-26 (1 µM) (all provided by Thermo Fisher Scientific Inc., Rockford, IL, USA) at 37 °C with 5% CO_2.

Transfection in the presence of endosome markers. Cells (150,000 c/mL) were pretreated for 16 h before peptide incubation in a complete medium with one of the following markers: CellLight™ Early Endosomes-GFP, BacMam 2.0 (EE-GFP, 50 µg/mL) or CellLight™ Late Endosomes-GFP, BacMam 2.0 (LE-GFP, 50 µg/mL) (all supplied by Thermo Fisher Scientific Inc., Rockford, IL, USA). Afterward, cells were rinsed twice with D-PBS and then incubated with Tamra-CPP-iCAL36 solutions (TRI36 at 5 μM or P36 at 1 μM) in OptiMEM for 1.5 h at 37 °C with 5% CO_2.

Next, 10 min before the end of each incubation, 10 μL of Hoechst 33342 (1 μg/mL, Sigma-Aldrich) to label the cell nuclei and 10 μL of WGA-Alexa488 (Alexa Fluor™ 488 wheat germ agglutinin conjugate, 1 mg/mL, Thermo Fisher Scientific Inc., Rockford, IL, USA) to label cell membranes were added to the cell medium. Then, cells were rinsed twice with D-PBS and then covered with 1.5 mL of FluoroBrite (Life Technologies) for imaging using a Leica SP5-SMD confocal microscope (objective Leica HCS PL Apo CS 63x/1.4 NA oil objective; Hoechst 33342: λex = 405 nm/λem = 415–485 nm; TAMRA: λex = 561 nm/λem = 571–611 nm; Alexa488: λex = 488 nm/λem = 498–548 nm). After the acquisition, the images were processed under the same conditions and parameters using ImageJ software.

2.6. Cell Transfection Conditions for CFTR Quantification

Transfection without VX pre-incubation: Next, 3×10^5 cells/well were seeded in collagen type I-coated 6-well Corning culture plates (Thermo Fisher Scientific Inc., Rockford, IL, USA). Then, 24 h post-seeding, cells were incubated with TatRI-iCAL36 diluted in serum-free medium at the indicated concentrations (1000 μL final volume). After 3.5 h of incubation, the medium was replaced with a fresh serum-supplemented medium containing VX-445 (3 μM final concentration, CliniSciences, Nanterre, France) and VX-809 (3 μM final concentration, CliniSciences, Nanterre, France), and cells were incubated for a further 24 h at 37 °C, 5% CO_2.

Transfection with VX pre-incubation: Here, 2×10^5 cells/well were seeded in collagen type I-coated 6-well Corning culture plates (Thermo Fisher Scientific Inc., Rockford, IL, USA). Then, 24 h post-seeding, cells were pre-incubated with a serum-supplemented medium containing VX-445/VX-809 (3 μM/3 μM final concentrations, CliniSciences, Nanterre, France). Again, 24 h later, the medium was replaced by TatRI-iCAL36 diluted in serum-free medium at the indicated concentrations (1000 μL final volume). After 3.5 h of incubation, the medium was again replaced with fresh serum-supplemented medium containing VX-445/VX-809 (3 μM/3 μM final concentrations), and cells were incubated for a further 24 h at 37 °C with 5% CO_2.

Western blot: Cells transfected with TatRI-iCAL36 (with or without VX pre-incubation) were washed twice with PBS and lysed in 100 μL RIPA buffer (50 mM Tris pH 8.0, 150 mM sodium chloride, 1% Triton X-100, 0.1% SDS (sodium dodecyl sulfate, Sigma-Aldrich), including protease inhibitors (SigmaFAST, Sigma-Aldrich)). Protein concentrations were determined using the Pierce BCA Protein Assay (ThermoFisher). Cell extracts were separated by 7.5% Mini-PROTEAN® TGX™ Precast Gel (Bio-Rad, Marnes-la-Coquette, France). After electrophoresis, samples were transferred onto Trans-Blot® Turbo™ Mini PVDF Transfer membranes (Bio-Rad, Marnes-la-Coquette, France). As antibodies, we used anti-CFTR-450 or anti-CFTR-432 (recognizing both CFTR R domain, Cystic Fibrosis Foundation, Chapel Hill, NC, USA, 1:1000), anti-Lamin A/C (Sigma-Aldrich, 1:2000), and anti-mouse IgG HRP (Cell Signaling, Ozyme, Saint-Cyr-L'École, France, 1:2000). Blots were revealed with the Pierce ECL plus Western blotting substrate (Thermo Fisher Scientific Inc., Rockford, IL, USA) on an Amersham 600 imager (GE Healthcare Life Science, Vélizy-Villacoublay, France). The signal intensities of the blots were quantified using ImageJ software.

2.7. Statistical Analysis

One-way or two-way analysis of variance (ANOVA) with the Bonferroni test was used to compare data from control and multiple experimental groups. A confidence interval of 95% ($p < 0.05$) was considered statistically significant. Data analysis was performed with GraphPad Prism v.6.01 (GraphPad Software Inc., San Diego, CA, USA).

3. Results and Discussion

3.1. CPP Selection for iCAL36 Coupling and Evaluation of Their Effects on Cell Viability

In one of our past studies, we designed the iCAL36 peptide as a CFTR stabilizer [8]. This peptide, which was internalized via a lipid-based transfection reagent (BioPORTER® QuikEase™ Protein Delivery Kit), induced an increase (+11%) in CFTR activity at the apical membrane of bronchial epithelia [9]. However, these results were obtained using a high iCAL36 concentration (500 µM), a value that is not compatible with a clinical application via pulmonary nebulization (equal to an extrapolated quantity of 6 g iCAL36 per day).

For that reason, we decided to couple the iCAL36 sequence with cell-penetrating peptides (CPPs) commonly used in the literature, namely Tat [19], MPG, and Penetratin (Pen), showing suitable internalization in different cell lines [26], as well as C6M1 and WRAP5 CPPs due to their secondary amphipathicity and their internalization via a direct cell membrane translocation [27,28] (Table 1).

Unfortunately, the first evaluations of C6M1-iCAL36 and WRAP5-iCAL36 revealed that both peptides were mainly stuck at or within the cell membrane, resulting in low cytoplasmatic release compared to Tat-iCAL36 and Pen-iCAL36 [25]. Therefore, both conjugates were excluded from this study.

To evaluate whether more protease-stable CPPs would increase cellular internalization, we included Tat analogs such as dTat (sequence with amino acids in isoform-D) [29] and TatRI (*retro-inverso* sequence in isoform-D) [30].

First of all, we evaluated the potential cytotoxic effects of iCAL36 (10 residues) and the five CPP-iCAL conjugates (23 to 37 residues) in human colorectal and human pulmonary epithelial cells (Caco-2 and Calu-3, respectively) using an LDH cytotoxicity kit. A screening of the literature revealed CPP concentrations varying from 1 µM to 50 µM, a range over which CPPs are normally not toxic [26,31,32]. Therefore, we evaluated the potent cytotoxic effects of the peptides by applying two commonly used concentrations (1 µM and 10 µM), as well as a more extreme concentration (100 µM) (Figure S1A,B in the Supplemental Material). The lower concentrations between 1 and 10 µM showed no significant effects on cell viability in either cell line compared to the non-treated cells (=100% cell viability). Only at the high CPP-iCAL36 concentration of 100 µM was reduced cell viability (~70–80%) observed for MPG-iCAL36, Pen-iCAL36, and dTat-iCAL36. These cytotoxic effects were not considered as a restriction for the screening of our peptides because (1) the goal of the CPP conjugation was the reduction in the applied iCAL36 concentration and (2) only peptide concentrations below 50 µM were to be used in the following experiments.

3.2. Comparison of CPP-iCAL36 Internalization and Cellular Localization

Tamra-iCAL36 and Tamra-CPP-iCAL36 peptides listed in Table 1 were tested in Caco-2 and Calu-3 cells, and cellular uptake was quantified by fluorescence measurements using flow cytometry (Figure 1A,B). In parallel, cell mortality was assessed using DAPI labeling. Compared to the cell viability assay (LDH, Figure S1A,B), we observed that the procedure of cell preparation (trypsinization, washing, and transfer to the flow cytometer) reduced cell viability to ~80%, as shown for the non-treated cells. Therefore, we defined viable cells as cells having cell viability over 70% to evaluate the effect of the peptide incubation. As shown in Figure 1A,B, for all incubation conditions (1 µM or 5 µM) in both cell lines we did not observe cell viability below 70%.

Although all 5 CPPs were supposed from the literature to have cell-penetrating properties, they did not show the same degree of cellular internalization when conjugated to the iCAL36 peptide. However, compared to the naked iCAL36 peptide, all CPP-iCAL36 conjugates increased their cellular internalization. In both cell lines, MPG-iCAL36 and Pen-iCAL36 seemed to be superior to Tat-iCAL36 and its analogs.

To visualize the CPP-iCAL36 internalization and cellular localization, we performed confocal laser scanning microscopy (CLSM) measurements on living cells. Caco-2 or Calu-3 cells were incubated with the corresponding Tamra-CPP-iCAL36 peptides for 3 h. The incubation time was set from 1.5 h to 3 h to increase the fluorescence without changing the peptide concentrations as reported previously [28].

First, we performed control experiments without peptide (NT) or with Tamra-iCAL36 to show that no auto-fluorescence nor iCAL36 (without CPP) internalization was detected, respectively (Figure 1C,D). For the five Tamra-CPP-iCAL36 peptides, a similar cellular pattern was visible—they were all located as a punctuated pattern in the cytoplasms of both cell lines, with more signals in the Caco-2 cells compared to the Calu-3 cells, confirming the flow cytometry results (Figure 1C,D).

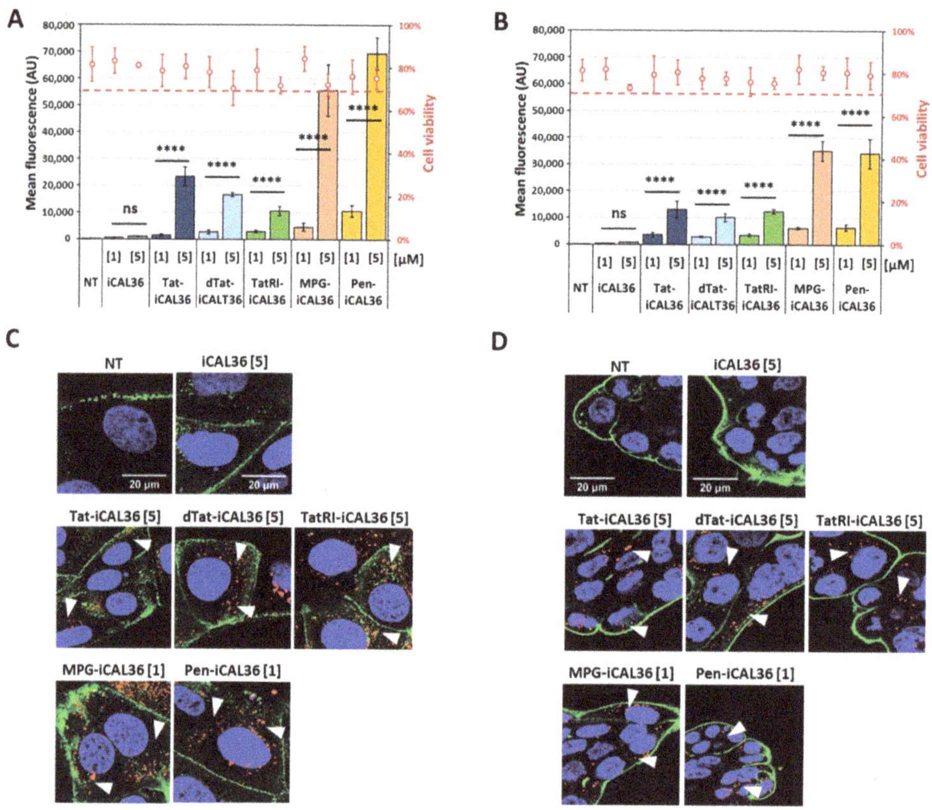

Figure 1. Comparison of CPP-iCAL36 internalization and cellular localization. (**A**) Caco-2 and (**B**) Calu-3 cells were incubated with Tamra-iCAL36 or Tamra-CPP-iCAL36 solutions (1 μM = [1] or 5 μM = [5]) in OptiMEM for 1.5 h and compared to non-treated cells (NT). After cell trypsinization (to remove membrane externally bound peptides), fluorescence and cell viability were assessed via flow cytometry. Graphical representation of data obtained from four independent measurements (mean ± SD, n = 4). A statistically significant difference was noticed between the two different used concentrations for all CPP-iCAL36 peptides (one-way ANOVA with Bonferroni test, statistical relevance given as **** $p < 0.0001$ and ns > 0.05). Living (**C**) Caco-2 and (**D**) Calu-3 cells were imaged by confocal laser scanning microscopy (CLSM) after 3 h incubation with the Tamra-iCAL36 and Tamra-CPP-iCAL36 peptides (red, see white arrows) at the indicated concentrations (1 μM = [1] or 5 μM = [5]) in OptiMEM. Next, 10 min before the end of the incubation, Hoechst dye (blue = nuclei) and WGA-Alexa488 (green = cell membrane) were added. Cells were washed and covered with DMEM FluoroBrite before imaging. White bars represent 20 μm. The results of the Tat-iCAL36 (1 μM = [1]) and MPG-iCAL36 (1 μM = [1]) peptides on Caco-2 cells were previously reported [25].

Due to the absence of net improvements compared to Tat-iCAL36, dTat-iCAL36 was excluded and we finally evaluated the time-dependent internalization of the four remaining CPP-iCAL36 conjugates in both cell lines. We determined that 30 min incubation was sufficient to measure significant cellular uptake. Longer incubation times (4.5 h) only increased the signal intensities by a factor of ~1.3 in Caco-2 cells. Only TatRI-iCAL36 revealed ~3-fold higher internalization (Figure S2A in the Supplemental Material). Calu-3 cells seemed to be more sensitive to the incubation duration with an increase in signal intensities by a factor of ~1.7–2.4 (Figure S2B in the Supplemental Material). Curiously, while it was not seen in Caco-2 cells, Tat-iCAL36 revealed a ~5.7-fold higher internalization compared to a factor of ~2.0–2.4 for the other CPP-iCAL36 peptides.

3.3. Internalization of the CPP-iCAL36 Peptides in the Presence of Serum

CF is characterized by a dysfunctional chloride CFTR channel, which leads to the production of a thick and viscous mucus layer perturbating the lung function of CF patients. The challenges in the development of CFTR stabilizers are related to the peptide retention within the mucus.

To evaluate this purpose in a simple cell model, CPP-iCAL36 transfections were performed in a serum-free medium (0%) or in a medium supplemented with 10% or 20% serum (all *v*/*v*) to mimic an environment rich in proteases and other serum proteins (Figure 2). As expected, neither the presence of the Tamra-CPP-iCAL36 peptides nor the increased serum concentrations in the transfection medium affected the cell viability of either tested cell line.

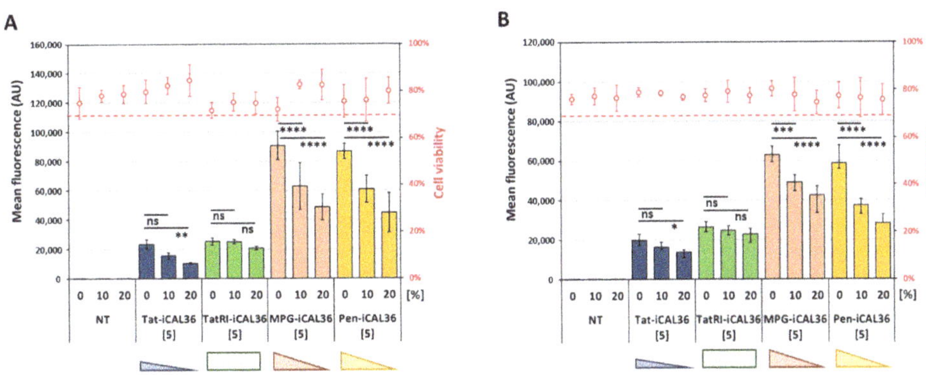

Figure 2. Internalization efficiency of CPP-iCAL36 peptides in the presence of serum. Caco-2 (**A**) and Calu-3 (**B**) cells were incubated with the Tamra-CPP-iCAL36 peptides (5 µM = [5]) for 3 h and compared to non-treated cells (NT) at the indicated serum concentration (0%, 10% or 20% FBS). After cell trypsinization (to remove membrane externally bound peptides), fluorescence and cell viability were assessed via flow cytometry. Graphical representation of data obtained from four independent measurements (mean ± SD, n = 4). A statistically significant difference was noticed between the three different used serum percentages for MPG-iCAL36 and Pen-iCAL36, whereas no statistical difference was revealed for TatRI-iCAL36 (2-way ANOVA with Bonferroni test, statistical relevance given as **** $p < 0.0001$, *** $p < 0.001$, ** $p < 0.01$, * $p < 0.05$, ns > 0.05).

In contrast, the transfection efficiency of Tamra-CPP-iCAL36 peptides decreased in a dose-dependent manner with increasing serum concentrations. In Caco-2 cells, the transfection efficiency decreased by ~30% with a 10% serum incubation and by ~50% with a 20% serum condition for Tat-iCAL36, MPG-iCAL36, and Pen-iCAL36. As expected, due to the protease stability of TatRI, the peptide TatRI-iCAL36 revealed the lowest uptake changes (no significant difference with 10% serum and 18% decrease with 20% serum) (Figure 2A).

In Calu-3 cells, the observation was similar: we observed ~40% and ~50% decreases with 10% and 20% serum, respectively, for Pen-iCAL36; and ~20% and ~30% decreases with 10% and 20% serum, respectively, for Tat-iCAL36 and MPG-iCAL36 (Figure 2B). Here again, the TatRI-iCAL36 peptide showed nearly the same internalization behavior independent of the serum concentration (6% with 10% serum and 12% with 20% serum).

In conclusion, TatRI-iCAL36 was the conjugate most able to internalize in the presence of high serum content as compared to the others. Based on these results, we selected the two peptides TatRI-iCAL36 and Pen-iCAL36 to further evaluate their cellular internalization mechanism.

3.4. *Evaluation of TatRI-iCAL36 and Pen-iCAL36 Membrane Interaction*

CPPs are considered membrane-active peptides [33]. Some of these membrane-active peptides have a structural polymorphism (helicoidal structure formation in a specific environment) that drives their cellular uptake. For example, few residues were described to drive helical or beta secondary structures essential for membrane penetration in the presence of high detergent concentrations [34], at high peptide-to-lipid ratios, or in the presence of acidic phospholipids [35]. On the other hand, penetratin has been shown to be oriented parallel to the membrane surface [36].

To evaluate the structural features of the TatRI-iCAL36 and Pen-iCAL36 peptides, circular dichroism spectra (CD) were recorded for both peptides alone and compared to those measured in the presence of large unilamellar vesicles (LUVs) reflecting the lipid composition of cell membranes (for details see Supplementary Materials and Figure S3A,B). As expected, the signal compensation due to L- and D-isoform amino acid compositions of TatRI-iCAL provided a weak CD profile that could not be associated with any specific structure. In addition, no significant change was detected in the presence of increasing LUV amounts or Trifluoroethanol (TFE), a solvent favoring helix formation (Figure S3A). For the Pen-iCAL36 peptide, we observed a random-coiled spectrum characterized by a minimum at 198 nm (Figure S3B). No obvious changes were recorded after the addition of LUVs, even if an α-helical conformation could be forced in the presence of TFE. However, the latter has no pharmacological consequences for the Pen-iCAL36 peptide, as TFE is only used as a conformational positive control.

To further evaluate the potential membrane-active properties of both peptides, we performed leakage experiments on LUVs encapsulating a quenched fluorescent dye incubated with increasing peptide concentrations (for protocol details see Supplementary Materials). In a previous study, we used this model to demonstrate the membrane activity of WRAP5-iCAL36, showing leakage of 85% [37]. However, for both peptides analyzed here, no significant LUV leakage was detected, even at high peptide concentrations (Data not shown).

Altogether, these results suggested that TatRI-iCAL36 and Pen-iCAL36 were not able to adopt a particular secondary structure (CD), resulting in the absence of membrane destabilization properties (no leakage). Both CPP-iCAL36 peptides are probably internalized via endocytosis-dependent pathways.

3.5. *Dissecting the Internalization Mechanism of TatRI-iCAL36 and Pen-iCAL36*

Since the discovery of CPPs, hundreds of studies have been performed to identify the mechanism(s) of cellular entry [38,39], which occurs mainly via endocytosis-dependent pathways (macropinocytosis, clathrin- and caveolin-dependent endocytosis) [40–42], even if in some cases direct cell membrane translocation has been reported at high CPP concentrations [40,43].

If an endocytosis-dependent pathway was activated for the peptide internalization, the fluorescent-labeled peptides would be entrapped in endosomes, where the fluorescent dye would be mainly quenched [25]. To confirm this hypothesis, we incubated the Caco-2 cells with an endosomolytic agent Chloroquine (CQ) to determine if we would obtain an increased fluorescent signal (Figure 3A). First of all, we confirmed that CQ treatment did not induce significant cytotoxicity compared to untreated Caco-2 cells (80 $\pm$ 10%).

Then, we showed that the addition of CQ doubled the intracellular fluorescence signal of Tamra-TatRI-iCAL36 and Tamra-Pen-iCAL36, demonstrating the release of peptides from endosomes. These results confirmed the involvement of an endocytosis mechanism and indicated that non-negligible fractions of the peptides were trapped in endosomes.

Because endocytosis-dependent internalization requires energy, we downregulated the energy-dependent pathways by incubating cells either at 4 °C or under ATP depletion (NaN_3/2-Deoxy-Glucose at 37 °C) to compare the Tamra-TatRI-iCAL36 and Tamra-Pen-iCAL36 internalization efficiency to the standard 37 °C incubation condition. The internalization of both peptides was reduced under ATP depletion (55% of the signal at 37 °C) and was almost lost at 4 °C (10–15% of the signal at 37 °C), confirming an endocytosis-dependent internalization (Figure 3B).

To further depict the exact internalization mechanism, we evaluated the cellular uptake of both Tamra-TatRI-iCAL36 and Tamra-Pen-iCAL36 peptides in the presence of specific chemical inhibitors of different endocytosis pathways such as chlorpromazine (CPZ, clathrin-dependent endocytosis inhibitor), nystatin (NYS, disrupting caveolar structure and function) and methyl-β-cyclodextrin (MBCD, lipid raft inhibitor) (for a review see [43]). Even if in the CPP field all of these chemical inhibitors were commonly used in several investigations to hamper endocytic processes, the maximal concentration with low cytotoxicity for each inhibitor was first determined to ensure maximal inhibition (data not shown). As expected, treatment of Caco-2 cells with CPZ, NYS or MBCD induced slight cytotoxicity (5–20% loss of viability), as observed for ATP depletion compared to non-treated cells (80 ± 10%) (Figure 3B). Curiously, endocytosis inhibitors did not lead to a strong change in the intracellular fluorescence of Tamra-TatRI-iCAL36–only slight non-significant decreases (21%) were observed in the presence of NYS and MBCD. Nearly the same results were observed for the intracellular fluorescence of Tamra-Pen-iCAL36 but with a significant reduction (28%, *p) in internalization in the presence of NYS, revealing the implication of caveolae-dependent endocytosis. These results suggested an internalization mechanism depending on multiple endocytosis-dependent pathways.

Because a number of these chemical endocytosis inhibitors lack specificity, it was possible that they could give misleading results [44,45]. Therefore, we further evaluated the internalization of both peptide conjugates by confocal microscopy using three specific endocytosis markers, namely transferrin, cholera toxin subunit B (CtB), and dextran, to visualize the clathrin-dependent, caveolin-mediated and macropinocytotic pathways, respectively [46]. As observed in Figure 3C,D, colocalization of Tamra-TatRI-iCAL36 and Tamra-Pen-iCAL36 peptides was relatively strong with all endocytosis markers (see arrows). These results were surprising because previous analyses with endocytosis inhibitors showed nearly no significant effect on internalization. On the other hand, the results were coherent with a mechanism involving multiple replaceable endocytosis-dependent pathways.

3.6. *Intracellular Fate of TatRI-iCAL36 and Pen-iCAL36*

To complete the study of the internalization mechanism of both Tamra-TatRI-iCAL36 and Tamra-Pen-iCAL36 peptides, we investigated their fate once inside the cell. A compound taken up by endocytosis will be entrapped in endosomes (called early endosomes). After a pH drop, early endosomes will be changed into late endosomes and then merge into lysosomes, where the peptides will be recycled or degraded.

To follow vesicle trafficking, we performed colocalization experiments between Tamra-TatRI-iCAL36 and Tamra-Pen-iCAL36 peptides and between specific markers of early and late endosomes, as well as lysosomes. As shown in Figure 4A,B, neither peptide fully colocalized with early endosome markers. Instead, they showed strong degrees of colocalization with late endosomes and with lysosomes (see arrows). These results suggested that after 1.5 h of incubation, the majority of the peptide internalized by endocytosis was found in late endosomes or lysosomes, confirming the internalization via endocytosis-dependent pathways.

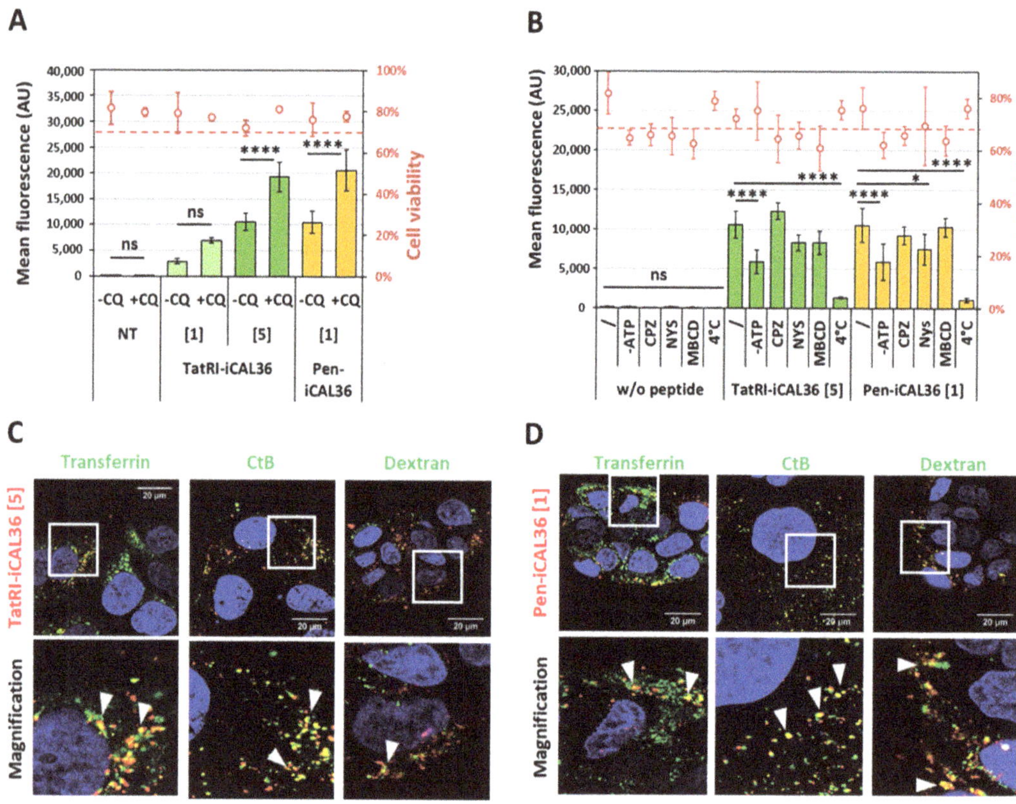

Figure 3. Dissecting the internalization mechanism of TatRI-iCAL36 and Pen-iCAL36. (**A**) Caco-2 cells were incubated with Tamra-TatRI-iCAL36 (1 μM = [1] or 5 μM) = [5] and Tamra-Pen-iCAL36 (1 μM = [1]) and compared to non-treated cells (NT) in the absence or presence of chloroquine (CQ, 100 μM). After cell trypsinization (to remove external peptides), the fluorescence and cell viability were assessed via flow cytometry. Graphical representation of data obtained from four independent measurements (mean ± SD, n = 4). A statistically significant difference was noticed between the incubation conditions without and with CQ for TatRI-iCAL36 and Pen-iCAL36 at 5 μM, whereas no statistical difference was revealed for TatRI-iCAL36 at 1 μM (one-way ANOVA with Bonferroni test, statistical relevance given as **** $p < 0.0001$ and ns > 0.05). (**B**) Caco-2 cells were incubated with Tamra-TatRI-iCAL36 (5 μM) and Tamra-Pen-iCAL36 (1 μM) at 4 °C or in the presence of either ATP-depleting reagents (NaN_3/DG) or endocytosis inhibitors (CPZ, NYS or MBCD) at 37 °C. The same conditions were used on cells without peptides (w/o peptide) to determine the effect of the endocytosis inhibition. For all conditions, fluorescence and cell viability were acquired via flow cytometry after cell trypsinization (to remove membrane externally bound peptides). Graphical representation of data obtained from four independent measurements (mean ± SD, n = 4). A statistically significant difference compared to untreated condition was noticed for ATP depletion and 4 °C incubation for TatRI-iCAL36 and Pen-iCAL36 (2-way ANOVA with Bonferroni test, statistical relevance given as * $p < 0.05$, **** $p < 0.0001$ and ns > 0.05). (**C**,**D**) Caco-2 cells were incubated with Tamra-TatRI-iCAL36 (5 μM = [5]) and Tamra-Pen-iCAL36 (1 μM = [1]) in the presence of endocytosis markers such as transferrin, cholera toxin subunit B (CtB) or Dextran. Next, 10 min before the end of the incubation, Hoechst dye (blue = nuclei) and WGA-Alexa488 (green = cell membrane) were added. Cells were washed and covered with DMEM FluoroBrite before imaging. White bars represent 20 μm.

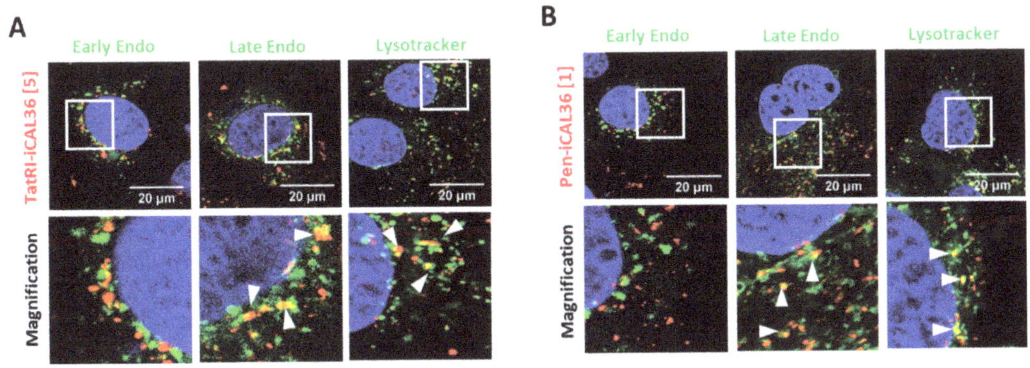

Figure 4. Dissecting vesicular entrapment of TatRI-iCAL36 and Pen-iCAL36. Caco-2 cells were incubated with Tamra-TatRI-iCAL36 (5 µM = [5] (**A**)) and Tamra-Pen-iCAL36 (1 µM = [1] (**B**)) in the presence of vesicle markers such as early and late endosomes, as well as lysosomes. Here, 10 min before the end of the incubation, Hoechst dye (blue = nuclei) and WGA-Alexa488 (green = cell membrane) were added. Cells were washed and covered with DMEM FluoroBrite before imaging. White bars represent 20 µm.

In conclusion, we could deduce that both CPP-iCAL36 peptides were internalized via an energy-dependent endocytosis mechanism, which could include multiple endocytosis-dependent pathways.

3.7. Stabilizing Effect of TatRI-iCAL36 Increases CFTR Amount

Comparing the internalization properties of Tamra-TatRI-iCAL36 and Tamra-Pen-iCAL36, we could not determine an overall better candidate between both peptides. However, because TatRI-iCAL36 internalization was less reduced in the presence of serum (Figure 2A,B) and because its sequence is shorter than Pen-iCAL36 (Table 1), we selected it for activity measurement. Indeed, the measured inhibition constant (Ki) revealed a specific interaction of Tat-RI-iCAL36 with the CAL PDZ domain, with a value of 1.5 ± 0.4 µM. More importantly, interactions of the peptide with the PDZ domains of the NHERF2 proteins (N2P1 and N2P2) were not observed (Ki > 5000 µM) (for protocol details see the Supplemental Materials), as the NHERF2-CFTR interaction was reported to stabilize the CFTR at the plasma membrane of epithelial cells [47].

Afterward, we first evaluated the internalization properties of Tamra-TatRI-iCAL36 in a human bronchial epithelial gene-edited CFTR p.Phe508del cell line (16HBEge-pPhe508del, noted 16HBEge) as a more appropriated model for the CF pathology. As shown in Figure 5A, we confirmed via confocal microscopy that the Tamra-labeled iCAL36 peptide was not internalized, and more importantly that the Tamra-TatRI-iCAL36 peptide revealed an impressive internalization profile in all cells after a 3.5 h transfection. Additionally, we confirmed that at the used concentrations (between 1 µM and 10 µM), no cytotoxic effects were induced by the peptide incubation (LDH and Cristal Violet assays, Supplementary Materials and Figure S4A,B).

The stabilizing property of TatRI-iCAL36 was evaluated indirectly. Because the peptide might inhibit the interaction of the CAL protein with the C-terminus of the mutated p.Phe508del-CFTR, CFTR degradation via the endosomal–lysosomal pathway should be reduced or abolished. As a consequence, an increase in the total CFTR amount should be detected in the 16HBEge cells. These experiments were performed in the presence of two CFTR correctors VX-445 (elexacaftor) and VX-809 (lumacaftor), which were found to be particularly effective when administrated together without significant cytotoxic effects [48]. Furthermore, a VX cocktail was always used during these experiments because p.Phe508del-CFTR can only be quantified if enough folded receptors reach the membrane.

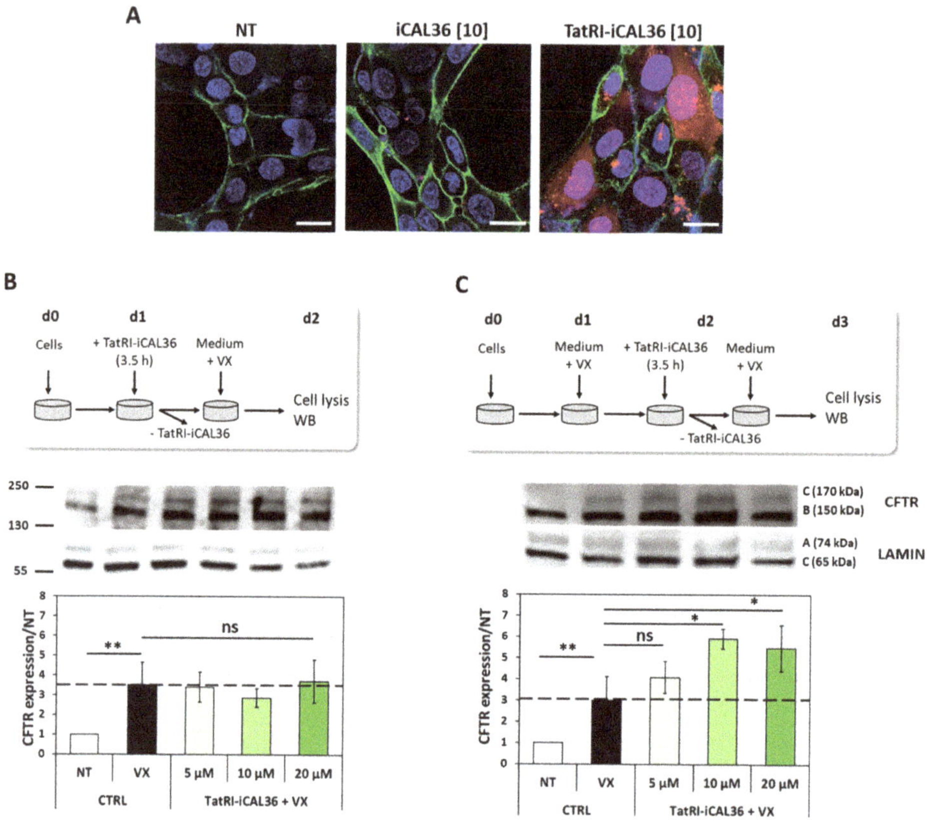

Figure 5. Dissecting the internalization and activity of TatRI-iCAL36 in 16HBEge cells. (**A**) Living 16HBEge cells were imaged by CLSM after 3 h incubation with Tamra-iCAL36 (10 µM = [1]) or Tamra-TatRI-iCAL36 (10 µM = [10])) in OptiMEM. Then, 10 min before the end of the incubation, Hoechst dye (blue) and WGA-Alexa488 (green) were added. Cells were washed and covered with DMEM FluoroBrite before imaging. White bars represent 10 µm. (**B**) The 16HBEge cells were incubated with TatRI-iCAL36 (5 µM, 10 µM, or 20 µM) in the presence of the corrector VX-445/VX-809 cocktail as indicated in the scheme. On day 2 (d2), the cells were lysed and p.Phe508del-CFTR was quantified by Western blot. Graphical representation of data obtained from three independent Western blot quantifications (mean ± SD, n = 3). A statistically significant difference was noted between non-treated (NT) and VX-445/VX-809-treated (VX) cells, whereas no statistical significance was revealed for the TatRI-iCAL incubation (one-way ANOVA with Bonferroni test, statistical relevance given as ** $p < 0.01$ and ns > 0.05). (**C**) The 16HBEge cells were pre-incubated with the corrected VX-445/VX-809 for 24 h. Afterward, the cells were incubated with TatRI-iCAL36 (5 µM, 10 µM, or 20 µM) in the presence of the VX-445/VX-809 corrector cocktail as indicated in the scheme. On day 3 (d3), cells were lysed and p.Phe508del-CFTR was quantified by Western blot. Graphical representation of data obtained from three independent Western blot quantifications (mean ± SD, n = 3). A statistically significant difference was noticed between non-treated (NT) and VX-445/VX-809-treated (VX) cells, as well as between the VX-treated and the TatRI-iCAL36-treated cells at 10 µM and 20 µM (one-way ANOVA with Bonferroni test, statistical relevance given as ** $p < 0.01$, * $p < 0.05$ and ns > 0.05).

First, we confirmed that the VX-445/VX-809 corrector cocktails increase the CFTR amount in the 16HBEge cells after 24 h incubation. Indeed, as shown in Figure 5B, we observed a 3.5-fold increase in p.Phe508del-CFTR. Curiously, using these VX-incubation conditions, we could not observe an additive effect of the TatRI-iCAL36 peptide.

We thought that the lack of peptide effect could be due to the VX incubation conditions, which were only performed after peptide incubation. The stabilizing properties of the TatRI-iCAL36 would only be observed if the correctors have induced enough p.Phe508del-CFTR folding for proper accumulation at the apical membrane of epithelial cells. Therefore, we decided to perform pre-peptide incubation with the VX-cocktail to ensure that p.Phe508del-CFTR avoided proteasomal degradation and reached the plasma membrane. Using this new incubation condition, we revealed a 3-fold increase in the CFTR amount in the 16HBEge cells after VX treatment alone. More interestingly, we observed for the first time 4-fold and 6-fold increases in the p.Phe508del-CFTR quantity after 5 μM and 10 μM peptide incubation, respectively (Figure 5C). This effect reached a plateau when we increased the TatRI-iCAL36 peptide concentration to 20 μM.

At this stage, these results are very promising in the development of TatRI-iCAL36 as a new CF treatment in combination with VX-445/VX-809 correctors.

4. Discussion

Therapeutic peptides, which address unmet medical needs in complement to low molecular weight drugs, have seen renewed interest in recent years. For the development of therapeutics against cystic fibrosis (CF), different modulators of the mutated chloride ion channel CFTR were developed and approved by the Food and Drug Agency (FDA), such as correctors (Lumicaftor, Elexacaftor) or potentiators (Ivacaftor). In parallel, CFTR–stabilizing peptides were also developed by our group [8,9,12,25] and by others [11,16,49] to further increase the CFTR amount in epithelial cells by decoupling the negatively regulating CFTR–CAL interaction.

In a previous study, we engineered iCAL36 as a CFTR stabilizer [8] and validated its CFTR-stabilizing effect using a lipid-based transfection [9]. To reduce the applied iCAL36 quantity (500 μM) for a more physiological application, we conjugated the iCAL36 stabilizer to five different cell-penetrating peptides (CPPs): Tat and its analogs dTat (D-isoform) and TatRI (*retro-inverso*), as well as MPG and Penetratin based on a previous CPP screen [25]. To determine the optimal CPP for iCAL36 transfection, we performed a precise screening of the internalization properties (depending on their concentrations, incubation time, cellular localization, and presence of serum) of these five CPP-iCAL36 peptides. All candidates were able to be internalized in the cytosol of epithelial cells compared to iCAL36 alone, even if the transfection ratios were different between the peptides. Based on their capacity to internalize in the presence of serum, TatRI-iCAL36 and Pen-iCAL36 were pre-selected to evaluate their cellular uptake mechanism (using endocytosis inhibitors, endocytosis, and vesicle markers). The internalization of both conjugates occurred mainly through energy-dependent endocytosis.

Based on the fact that TatRI-iCAL36 had a shorter sequence and was easier to synthesize, we selected this peptide as the more appropriate CPP for iCAL36 transfection. Indeed, TatRI-iCAL36 had a remarkable affinity to CAL (~1 μM) and an impressive selectivity (no interaction with the NHERF PDZ domains) compared to the iCAL36 peptide (22 μM) or other published iCAL36 analogs (~11 μM) [50]. More importantly, compared to the previously used iCAL36 peptide transfected by a lipid-based reagent, we could reduce the peptide concentration from 500 μM to 10 μM during the experiments performed to measure CFTR stabilization. Using this concentration, and in the presence of the correctors VX-445 and VX-809, we revealed a significantly increased amount of p.Phe508del-CFTR in bronchial epithelial cells. In turn, this would reduce the theoretical extrapolated dose for the patients from 6 g/day to approximately 0.3 g/day.

Taken together, the knowledge provided by this report will significantly contribute to pushing forward the development of novel CFTR stabilizers as co-therapies to the currently used Trikafta® treatment (elexacaftor VX-445/tezacaftor VX-661/ivacaftor VX-770).

Supplementary Materials: The following supporting information can be downloaded at: https://www.mdpi.com/article/10.3390/pharmaceutics14040808/s1, Reference [50] are cited in the Supplementary Materials. Materials and Methods: Circular dichroism (CD) measurements. Liposome leakage assay. Fluorescence polarization (FP) for Ki determination. Figure S1: Cellular viability of Caco-2 (A) and Calu-3 (B) cells after CPP-iCAL36 incubation. Figure S2: Time-dependent cellular internalization of CPP-iCAL36 in Caco-2 (A) and Calu-3 (B) cells. Figure S3: CD spectra of TatRI-iCAL36 (A) and Pen-iCAL36 (B). Figure S4: Evaluation of 16HBE cell toxicity (LDH1 and LDH2) and viability (Cristal Violet) after TatRI-iCAL36 incubation.

Author Contributions: Conceptualization, Q.S. and P.B.; methodology, Q.S., E.J., J.V., M.T.-C. and P.B.; software, Q.S., I.L. and E.J.; validation, M.T.-C. and P.B.; formal analysis, Q.S., I.L., E.J., E.V. and P.B.; investigation, Q.S., I.L., E.J., J.V., M.T.-C. and P.B.; resources, Q.S., I.L. and E.J.; data curation, Q.S., I.L., E.V. and E.J.; writing—original draft preparation, P.B.; writing—review and editing, Q.S., I.L., E.J., E.V., M.T.-C. and P.B.; visualization, Q.S., I.L., E.J., M.T.-C. and P.B.; supervision, M.T.-C. and P.B.; project administration, P.B.; funding acquisition, M.T.-C. and P.B. All authors have read and agreed to the published version of the manuscript.

Funding: This research was funded by the National Institutes of Health (5R01DK101541), the foundation "Vaincre la Mucoviscidose" (RF20170502030/1/1/47), the foundation "Fonds de Recherche en Santé Respiratoire/Fondation du Souffle (2017)" and the SATT AxLR (StabCFTR #814).

Institutional Review Board Statement: Not applicable.

Informed Consent Statement: Not applicable.

Acknowledgments: Peptide synthesis was performed with excellent technical support from the Synbio3 platform, which is supported by the GIS IBISA/ITMO Cancer (Plan Cancer 2014–2019) and Chimie Balard Cirimat Carnot Institute. We acknowledge the Montpellier Ressources Imagerie (MRI), a member of the national infrastructure France-BioImaging supported by the French National Research Agency (ANR-10-INBS-04, "Investments for the Future"), especially for technical support from Myriam Boyer (flow cytometry) and Baptiste Monterroso (confocal microscopy). We also thank Karidia Konate and Sébastien Deshayes for their discussion and critical reading of the manuscript.

Conflicts of Interest: The authors declare no conflict of interest.

References

1. Henninot, A.; Collins, J.C.; Nuss, J.M. The Current State of Peptide Drug Discovery: Back to the Future? *J. Med. Chem.* **2018**, *61*, 1382–1414. [CrossRef] [PubMed]
2. Erak, M.; Bellmann-Sickert, K.; Els-Heindl, S.; Beck-Sickinger, A.G. Peptide Chemistry Toolbox—Transforming Natural Peptides into Peptide Therapeutics. *Bioorg. Med. Chem.* **2018**, *26*, 2759–2765. [CrossRef] [PubMed]
3. Terryah, S.T.; Fellner, R.C.; Ahmad, S.; Moore, P.J.; Reidel, B.; Sesma, J.I.; Kim, C.S.; Garland, A.L.; Scott, D.W.; Sabater, J.R.; et al. Evaluation of a SPLUNC1-Derived Peptide for the Treatment of Cystic Fibrosis Lung Disease. *Am. J. Physiol.-Lung Cell. Mol. Physiol.* **2018**, *314*, L192–L205. [CrossRef] [PubMed]
4. Gaggar, A.; Chen, J.; Chmiel, J.F.; Dorkin, H.L.; Flume, P.A.; Griffin, R.; Nichols, D.; Donaldson, S.H. Inhaled Alpha1-Proteinase Inhibitor Therapy in Patients with Cystic Fibrosis. *J. Cyst. Fibros.* **2016**, *15*, 227–233. [CrossRef] [PubMed]
5. Cappiello, F.; Grazia, A.D.; Segev-Zarko, L.; Scali, S.; Ferrera, L.; Galietta, L.; Pini, A.; Shai, Y.; Di, Y.P.; Mangoni, M.L. Esculentin-1a-Derived Peptides Promote Clearance of Pseudomonas Aeruginosa Internalized in Bronchial Cells of Cystic Fibrosis Patients and Lung Cell Migration: Biochemical Properties and a Plausible Mode of Action. *Antimicrob. Agents Chemother.* **2016**, *60*, 7252–7262. [CrossRef]
6. Brunetti, J.; Roscia, G.; Lampronti, I.; Gambari, R.; Quercini, L.; Falciani, C.; Bracci, L.; Pini, A. Immunomodulatory and Anti-Inflammatory Activity in Vitro and in Vivo of a Novel Antimicrobial Candidate. *J. Biol. Chem.* **2016**, *291*, 25742–25748. [CrossRef]
7. Grégoire, N.; Aranzana-Climent, V.; Magréault, S.; Marchand, S.; Couet, W. Clinical Pharmacokinetics and Pharmacodynamics of Colistin. *Clin. Pharmacokinet.* **2017**, *56*, 1441–1460. [CrossRef]
8. Vouilleme, L.; Cushing, P.R.; Volkmer, R.; Madden, D.R.; Boisguerin, P. Engineering Peptide Inhibitors to Overcome PDZ Binding Promiscuity. *Angew. Chem.* **2010**, *49*, 9912–9916. [CrossRef]
9. Cushing, P.R.; Vouilleme, L.; Pellegrini, M.; Boisguerin, P.; Madden, D.R. A Stabilizing Influence: CAL PDZ Inhibition Extends the Half-Life of ΔF508-CFTR. *Angew. Chem.* **2010**, *49*, 9907–9911. [CrossRef]
10. Qian, Z.; Xu, X.; Amacher, J.F.; Madden, D.R.; Cormet-Boyaka, E.; Pei, D. Intracellular Delivery of Peptidyl Ligands by Reversible Cyclization: Discovery of a PDZ Domain Inhibitor That Rescues CFTR Activity. *Angew. Chem. Int. Ed.* **2015**, *54*, 5874–5878. [CrossRef]

11. Dougherty, P.G.; Wellmerling, J.H.; Koley, A.; Lukowski, J.K.; Hummon, A.B.; Cormet-Boyaka, E.; Pei, D. Cyclic Peptidyl Inhibitors against CAL/CFTR Interaction for Treatment of Cystic Fibrosis. *J. Med. Chem.* **2020**, *63*, 15773–15784. [CrossRef] [PubMed]
12. Amacher, J.F.; Cushing, P.R.; Weiner, J.A.; Madden, D.R. Crystallization and Preliminary Diffraction Analysis of the CAL PDZ Domain in Complex with a Selective Peptide Inhibitor. *Acta Crystallograph. Sect. F Struct. Biol. Cryst. Commun.* **2011**, *67*, 600–603. [CrossRef] [PubMed]
13. Cheng, J.; Wang, H.; Guggino, W.B. Modulation of Mature Cystic Fibrosis Transmembrane Regulator Protein by the PDZ Domain Protein CAL. *J. Biol. Chem.* **2004**, *279*, 1892–1898. [CrossRef] [PubMed]
14. Swiatecka-Urban, A.; Duhaime, M.; Coutermarsh, B.; Karlson, K.H.; Collawn, J.; Milewski, M.; Cutting, G.R.; Guggino, W.B.; Langford, G.; Stanton, B.A. PDZ Domain Interaction Controls the Endocytic Recycling of the Cystic Fibrosis Transmembrane Conductance Regulator. *J. Biol. Chem.* **2002**, *277*, 40099–40105. [CrossRef]
15. Cheng, J.; Moyer, B.D.; Milewski, M.; Loffing, J.; Ikeda, M.; Mickle, J.E.; Cutting, G.R.; Li, M.; Stanton, B.A.; Guggino, W.B. A Golgi-Associated PDZ Domain Protein Modulates Cystic Fibrosis Transmembrane Regulator Plasma Membrane Expression. *J. Biol. Chem.* **2002**, *277*, 3520–3529. [CrossRef]
16. Zhao, Y.; Cushing, P.R.; Smithson, D.C.; Pellegrini, M.; Pletnev, A.A.; Al-Ayyoubi, S.; Grassetti, A.V.; Gerber, S.A.; Guy, R.K.; Madden, D.R. Cysteine Modifiers Suggest an Allosteric Inhibitory Site on the CAL PDZ Domain. *Biosci. Rep.* **2018**, *38*, BSR20180231. [CrossRef]
17. Chellappan, D.K.; Prasher, P.; Saravanan, V.; Vern Yee, V.S.; Wen Chi, W.C.; Wong, J.W.; Wong, J.K.; Wong, J.T.; Wan, W.; Chellian, J.; et al. Protein and Peptide Delivery to Lungs by Using Advanced Targeted Drug Delivery. *Chem. Biol. Interact.* **2022**, *351*, 109706. [CrossRef]
18. Xu, J.; Khan, A.R.; Fu, M.; Wang, R.; Ji, J.; Zhai, G. Cell-Penetrating Peptide: A Means of Breaking through the Physiological Barriers of Different Tissues and Organs. *J. Control. Release* **2019**, *309*, 106–124. [CrossRef]
19. Vivès, E.; Brodin, P.; Lebleu, B. A Truncated HIV-1 Tat Protein Basic Domain Rapidly Translocates through the Plasma Membrane and Accumulates in the Cell Nucleus. *J. Biol. Chem.* **1997**, *272*, 16010–16017. [CrossRef]
20. Derossi, D.; Calvet, S.; Trembleau, A.; Brunissen, A.; Chassaing, G.; Prochiantz, A. Cell Internalization of the Third Helix of the Antennapedia Homeodomain Is Receptor-Independent. *J. Biol. Chem.* **1996**, *271*, 18188–18193. [CrossRef]
21. Langel, Ü. (Ed.) *Cell-Penetrating Peptides*; Methods in Molecular Biology; Springer New York: New York, NY, USA, 2015; Volume 1324, ISBN 978-1-4939-2805-7.
22. Kauffman, W.B.; Fuselier, T.; He, J.; Wimley, W.C. Mechanism Matters: A Taxonomy of Cell Penetrating Peptides. *Trends Biochem. Sci.* **2015**, *40*, 749–764. [CrossRef] [PubMed]
23. Milletti, F. Cell-Penetrating Peptides: Classes, Origin, and Current Landscape. *Drug Discov. Today* **2012**, *17*, 850–860. [CrossRef] [PubMed]
24. Ruseska, I.; Zimmer, A. Internalization Mechanisms of Cell-Penetrating Peptides. *Beilstein J. Nanotechnol.* **2020**, *11*, 101–123. [CrossRef] [PubMed]
25. Seisel, Q.; Pelletier, F.; Deshayes, S.; Boisguerin, P. How to Evaluate the Cellular Uptake of CPPs with Fluorescence Techniques: Dissecting Methodological Pitfalls Associated to Tryptophan-Rich Peptides. *Biochim. Biophys. Acta Biomembr.* **2019**, *1861*, 1533–1545. [CrossRef]
26. Mueller, J.; Kretzschmar, I.; Volkmer, R.; Boisguerin, P. Comparison of Cellular Uptake Using 22 CPPs in 4 Different Cell Lines. *Bioconjug. Chem.* **2008**, *19*, 2363–2374. [CrossRef]
27. Jafari, M.; Karunaratne, D.N.; Sweeting, C.M.; Chen, P. Modification of a Designed Amphipathic Cell-Penetrating Peptide and Its Effect on Solubility, Secondary Structure, and Uptake Efficiency. *Biochemistry* **2013**, *52*, 3428–3435. [CrossRef]
28. Konate, K.; Dussot, M.; Aldrian, G.; Vaissière, A.; Viguier, V.; Neira, I.F.; Couillaud, F.; Vivès, E.; Boisguerin, P.; Deshayes, S. Peptide-Based Nanoparticles to Rapidly and Efficiently "Wrap 'n Roll" SiRNA into Cells. *Bioconjug. Chem.* **2019**, *30*, 592–603. [CrossRef]
29. Meloni, B.P.; Craig, A.J.; Milech, N.; Hopkins, R.M.; Watt, P.M.; Knuckey, N.W. The Neuroprotective Efficacy of Cell-Penetrating Peptides TAT, Penetratin, Arg-9, and Pep-1 in Glutamic Acid, Kainic Acid, and in Vitro Ischemia Injury Models Using Primary Cortical Neuronal Cultures. *Cell. Mol. Neurobiol.* **2014**, *34*, 173–181. [CrossRef]
30. Tünnemann, G.; Martin, R.M.; Haupt, S.; Patsch, C.; Edenhofer, F.; Cardoso, M.C. Cargo-Dependent Mode of Uptake and Bioavailability of TAT-Containing Proteins and Peptides in Living Cells. *FASEB J.* **2006**, *20*, 1775–1784. [CrossRef]
31. El-Andaloussi, S.; Järver, P.; Johansson, H.J.; Langel, U. Cargo-Dependent Cytotoxicity and Delivery Efficacy of Cell-Penetrating Peptides: A Comparative Study. *Biochem. J.* **2007**, *407*, 285–292. [CrossRef]
32. Duchardt, F.; Ruttekolk, I.R.; Verdurmen, W.P.R.; Lortat-Jacob, H.; Bürck, J.; Hufnagel, H.; Fischer, R.; van den Heuvel, M.; Löwik, D.W.P.M.; Vuister, G.W.; et al. A Cell-Penetrating Peptide Derived from Human Lactoferrin with Conformation-Dependent Uptake Efficiency. *J. Biol. Chem.* **2009**, *284*, 36099–36108. [CrossRef] [PubMed]
33. Bechinger, B.; Aisenbrey, C. The Polymorphic Nature of Membrane-Active Peptides from Biophysical and Structural Investigations. *Curr. Protein Pept. Sci.* **2012**, *13*, 602–610. [CrossRef] [PubMed]
34. Lindberg, M.; Gräslund, A. The Position of the Cell Penetrating Peptide Penetratin in SDS Micelles Determined by NMR. *FEBS Lett.* **2001**, *497*, 39–44. [CrossRef]

35. Magzoub, M.; Eriksson, L.E.G.; Gräslund, A. Conformational States of the Cell-Penetrating Peptide Penetratin When Interacting with Phospholipid Vesicles: Effects of Surface Charge and Peptide Concentration. *Biochim. Biophys. Acta* **2002**, *1563*, 53–63. [CrossRef]
36. Balayssac, S.; Burlina, F.; Convert, O.; Bolbach, G.; Chassaing, G.; Lequin, O. Comparison of Penetratin and Other Homeodomain-Derived Cell-Penetrating Peptides: Interaction in a Membrane-Mimicking Environment and Cellular Uptake Efficiency. *Biochemistry* **2006**, *45*, 1408–1420. [CrossRef]
37. Konate, K.; Seisel, Q.; Vivès, E.; Boisguérin, P.; Deshayes, S. Fluorescent Leakage Assay to Investigate Membrane Destabilization by Cell-Penetrating Peptide. *J. Vis. Exp.* **2020**, *166*, e62028. [CrossRef]
38. Hoyer, J.; Neundorf, I. Peptide Vectors for the Nonviral Delivery of Nucleic Acids. *Acc. Chem. Res.* **2012**, *45*, 1048–1056. [CrossRef]
39. Langel, U. (Ed.) *Cell-Penetrating Peptides*; CRC Press: Boca Raton, FL, USA, 2007.
40. Brock, R. The Uptake of Arginine-Rich Cell-Penetrating Peptides: Putting the Puzzle Together. *Bioconjug. Chem.* **2014**, *25*, 863–868. [CrossRef]
41. Jones, A.T.; Sayers, E.J. Cell Entry of Cell Penetrating Peptides: Tales of Tails Wagging Dogs. *J. Control. Release* **2012**, *161*, 582–591. [CrossRef]
42. Vercauteren, D.; Rejman, J.; Martens, T.F.; Demeester, J.; De Smedt, S.C.; Braeckmans, K. On the Cellular Processing of Non-Viral Nanomedicines for Nucleic Acid Delivery: Mechanisms and Methods. *J. Control. Release* **2012**, *161*, 566–581. [CrossRef]
43. Cleal, K.; He, L.; Watson, P.D.; Jones, A.T. Endocytosis, Intracellular Traffic and Fate of Cell Penetrating Peptide Based Conjugates and Nanoparticles. *Curr. Pharm. Des.* **2013**, *19*, 2878–2894. [CrossRef] [PubMed]
44. Ivanov, A.I. Pharmacological Inhibition of Endocytic Pathways: Is It Specific Enough to Be Useful? *Methods Mol. Biol.* **2008**, *440*, 15–33. [CrossRef] [PubMed]
45. Vercauteren, D.; Vandenbroucke, R.E.; Jones, A.T.; Rejman, J.; Demeester, J.; De Smedt, S.C.; Sanders, N.N.; Braeckmans, K. The Use of Inhibitors to Study Endocytic Pathways of Gene Carriers: Optimization and Pitfalls. *Mol. Ther.* **2010**, *18*, 561–569. [CrossRef]
46. Rosazza, C.; Deschout, H.; Buntz, A.; Braeckmans, K.; Rols, M.-P.; Zumbusch, A. Endocytosis and Endosomal Trafficking of DNA After Gene Electrotransfer In Vitro. *Mol. Ther. Nucleic Acids* **2016**, *5*, e286. [CrossRef]
47. Zhang, W.; Zhang, Z.; Zhang, Y.; Naren, A.P. CFTR-NHERF2-LPA_2 Complex in the Airway and Gut Epithelia. *Int. J. Mol. Sci.* **2017**, *18*, 1896. [CrossRef] [PubMed]
48. Capurro, V.; Tomati, V.; Sondo, E.; Renda, M.; Borrelli, A.; Pastorino, C.; Guidone, D.; Venturini, A.; Giraudo, A.; Mandrup Bertozzi, S.; et al. Partial Rescue of F508del-CFTR Stability and Trafficking Defects by Double Corrector Treatment. *Int. J. Mol. Sci.* **2021**, *22*, 5262. [CrossRef]
49. Amacher, J.F.; Zhao, R.; Spaller, M.R.; Madden, D.R. Chemically Modified Peptide Scaffolds Target the CFTR-Associated Ligand PDZ Domain. *PLoS ONE* **2014**, *9*, e103650. [CrossRef]
50. Wang, Z.X. An Exact Mathematical Expression for Describing Competitive Binding of Two Different Ligands to a Protein Molecule. *FEBS Lett.* **1995**, *360*, 111–114. [CrossRef]

pharmaceutics

Article

Peptides vs. Polymers: Searching for the Most Efficient Delivery System for Mitochondrial Gene Therapy

Rúben Faria [1], Milan Paul [2], Swati Biswas [2], Eric Vivès [3], Prisca Boisguérin [3], Ângela Sousa [1] and Diana Costa [1,*]

[1] CICS-UBI—Health Sciences Research Centre, Universidade da Beira Interior, Avenida Infante D. Henrique, 6200-506 Covilha, Portugal; ruben.faria@ubi.pt (R.F.); angela@fcsaude.ubi.pt (Â.S.)
[2] Nanomedicine Research Laboratory, Department of Pharmacy, Birla Institute of Technology and Science-Pilani, Hyderabad Campus, Jawahar Nagar, Medchal, Hyderabad 500078, India; milan.paul95@gmail.com (M.P.); swati.biswas@hyderabad.bits-pilani.ac.in (S.B.)
[3] PhyMedExp, Université de Montpellier, INSERM, CNRS, 34295 Montpellier, France; eric.vives@umontpellier.fr (E.V.); prisca.boisguerin@inserm.fr (P.B.)
* Correspondence: dcosta@fcsaude.ubi.pt

Abstract: Together with the nucleus, the mitochondrion has its own genome. Mutations in mitochondrial DNA are responsible for a variety of disorders, including neurodegenerative diseases and cancer. Current therapeutic approaches are not effective. In this sense, mitochondrial gene therapy emerges as a valuable and promising therapeutic tool. To accomplish this goal, the design/development of a mitochondrial-specific gene delivery system is imperative. In this work, we explored the ability of novel polymer- and peptide-based systems for mitochondrial targeting, gene delivery, and protein expression, performing a comparison between them to reveal the most adequate system for mitochondrial gene therapy. Therefore, we synthesized a novel mitochondria-targeting polymer (polyethylenimine–dequalinium) to load and complex a mitochondrial-gene-based plasmid. The polymeric complexes exhibited physicochemical properties and cytotoxic profiles dependent on the nitrogen-to-phosphate-group ratio (N/P). A fluorescence confocal microscopy study revealed the mitochondrial targeting specificity of polymeric complexes. Moreover, transfection mediated by polymer and peptide delivery systems led to gene expression in mitochondria. Additionally, the mitochondrial protein was produced. A comparative study between polymeric and peptide/plasmid DNA complexes showed the great capacity of peptides to complex pDNA at lower N/P ratios, forming smaller particles bearing a positive charge, with repercussions on their capacity for cellular transfection, mitochondria targeting and, ultimately, gene delivery and protein expression. This report is a significant contribution to the implementation of mitochondrial gene therapy, instigating further research on the development of peptide-based delivery systems towards clinical translation.

Keywords: cell-penetrating peptides; mitochondrial gene therapy; mitochondrial DNA diseases; mitochondria targeting; nanodelivery systems; PEI-based complexes

Citation: Faria, R.; Paul, M.; Biswas, S.; Vivès, E.; Boisguérin, P.; Sousa, Â.; Costa, D. Peptides vs. Polymers: Searching for the Most Efficient Delivery System for Mitochondrial Gene Therapy. *Pharmaceutics* **2022**, *14*, 757. https://doi.org/10.3390/pharmaceutics14040757

Academic Editor: Wenbing Dai

Received: 19 February 2022
Accepted: 29 March 2022
Published: 31 March 2022

Copyright: © 2022 by the authors. Licensee MDPI, Basel, Switzerland. This article is an open access article distributed under the terms and conditions of the Creative Commons Attribution (CC BY) license (https://creativecommons.org/licenses/by/4.0/).

1. Introduction

Mitochondria play a key role in maintaining the normal functioning of cells, especially metabolic functions. This cell organelle has its own genome, called mitochondrial DNA (mtDNA). Mitochondrial DNA is a circular, double-stranded molecule with a size of around 16 kbp, and contains 37 genes [1]. These genes encode 13 polypeptides that participate in the oxidative phosphorylation chain, 2 rRNAs, and 22 tRNAs—all exclusive to the mitochondria [2,3]. In addition to being responsible for the production of ATP through oxidative phosphorylation, in recent years many other cellular processes have been discovered revealing the involvement of mitochondria [4,5]. Their participation in cellular mechanisms ranging from inflammation to regulation of stem cell generation [6,7], cell signaling, ion homeostasis, and metabolism of amino acids, lipids, cholesterol, steroids, and nucleotides has been demonstrated [8,9]. They also contribute to cell cycle control, cell growth, and apoptosis mechanisms [10]. In this way, mutations in mtDNA cause excessive cell death and promote the appearance of several pathologies, including metabolic and neurodegenerative syndromes linked to Parkinson's, Alzheimer's, amyotrophic lateral sclerosis, Huntington's disease, and diabetes [11]. Loss of influence on cell cycle control and cell death regulation, due to mutations in mtDNA, leads to cancer and autoimmune diseases [12,13]. Currently, treatments available for diseases associated with mtDNA mutations are not effective and only serve to mitigate symptoms, without providing a cure [14,15].

In line with this, the need to develop new, effective therapies arises. Since most mitochondrial diseases are associated with mutations in mtDNA, mitochondrial gene therapy emerges as a very promising approach to treat the problem at its root [16–18]. This type of therapy is focused on the delivery of genetic material to the mitochondria to suppress, alter, or complement the effect of defective genes [16]. To deliver genes of interest directly into the mitochondria, it is necessary to develop a vector that allows the encapsulation of DNA while providing targeting specificity [19]. This vehicle must also protect, transport, and direct the genetic content to the mitochondria, promoting its efficient release and, thus, guaranteeing the expression of both mitochondrial genes and respective proteins [20]. Viral vectors, such as retroviruses, adenoviruses, and lentiviruses, have been widely used in gene delivery due to their ability to transfect cells [21]. Among these vectors, adeno-associated viruses (AAVs) are the most applied in preclinical studies [21]. Due to their high loading capacity, biocompatibility, antigenicity, and lack of immune response, non-viral delivery systems based on cell-penetrating peptides (CPPs), micelles, polymers, and lipids have been commonly used in gene release studies [22–25]. Furthermore, ternary non-viral systems, for instance, constituted by polymers and/or peptides, have proven to be a valuable strategy to enhance payload encapsulation efficiency, thereby contributing to efficient gene delivery and expression [26–28].

CPPs have gained considerable interest in the gene therapy field due to their beneficial properties. CPPs are small (between 15 and 30 amino acids), and can be divided into arginine-rich and amphipathic peptides [29–32]. Because they have hydrophilic and hydrophobic domains, amphipathic peptides make it possible to formulate delivery systems that can encapsulate DNA and enable its membrane translocation and subsequent entry into cells [22,30]. Due to these characteristics, CPP-based vectors have demonstrated their ability to internalize in cells, and to deliver therapeutic molecules, in several studies [33–36].

Polymer-based systems are also widely explored as gene carriers, due to their favorable physicochemical properties, low toxicity, and tailorability [28,37–39]. In this context, one of the most used polymers has been polyethylenimine (PEI), due to its ability to transport different types of nucleic acids, regardless of their type and size [40,41]. Moreover, PEI has characteristics that allow it to go beyond the endosome/lysosome membrane [42]. PEI is a cationic polymer displaying a high positive charge density. This charge enables strong electrostatic interactions between its amine groups and DNA phosphate groups, resulting in the formation of nanoparticles in which the genetic material is mostly condensed within. PEI-based systems demonstrate high endosomolytic activity, critical to the potential success of mitochondrial gene delivery [43,44].

Currently, gene therapy with mitochondria as a therapeutic target is still a challenging strategy. However, some studies have demonstrated the feasibility of this type of therapy using mitochondrial targeting sequences (MTSs), and focused on the mitochondrial genes ATP6 (mitochondrially encoded synthase membrane subunit 6′) and ND4 (mitochondrially encoded NADH:ubiquinone oxidoreductase core subunit 4). Following this approach, their function was restored within their respective respiratory chain complexes and, consequently, the production of ATP was reestablished [45,46]. Other investigations revealed interesting therapeutic outcomes when addressing one of the most prevalent mitochondrial diseases—Leber's hereditary optic neuropathy (LHON). The introduction of functional ND1 (mitochondrially encoded NADH dehydrogenase 1 protein) and ND4 genes, usually mutated in this type of disease, provided the replacement of the normal activity of complex I in the respiratory chain [47,48].

In this study, we aimed to find out the most suitable delivery system for mitochondrial gene therapy by comparing the efficacy of peptide- and polymer-based complexes. For this, CPP- and PEI-based compounds were designed/synthesized to complex the mitochondrial gene ND1's plasmid DNA (pDNA). To ensure specific targeting of the mitochondria, moieties that confer affinity to this organelle were covalently coupled. In the case of PEI, dequalinium chloride (DQA) [49] was conjugated, resulting in the compound PEI–SA–DQA, whereas for CPPs, the MTS sequence [50] was added. DQA—a lipophilic cation—and especially its vesicular form (DQAsomes), has been demonstrated to selectively accumulate in the mitochondrial matrix [51–53]. To overcome the lack of stability exhibited by DQAsomes under conditions of low temperature and/or high salt concentration, researchers have conceived DQA-based carriers for payload release [53–55]. In this context, polymer–DQA delivery systems have been developed and optimized for mitochondrial targeting, leading to great advances in drug delivery to mitochondria [53,56–58].

Furthermore, as PEI–SA–DQA revealed a low ability to complex pND1—even at very high N/P ratios—TAT (the transcriptional activator protein in HIV-1—an 11-amino-acid peptide with 6 arginine and 2 lysine residues) was additionally included in PEI–DQA/pND1 complexes to complex pND1. These PEI-based ternary complexes were developed at various N/P ratios and adequately characterized. The biocompatibility profile was evaluated, and in vitro studies were carried out to assess the capacity of the developed complexes to reach the mitochondria. The physicochemical properties exhibited by the novel delivery systems, together with their ability to target the mitochondria and promote transgene expression, confer them with promising applicability as carriers for mitochondrial gene therapy. In addition, the comparison between various complexes based on CPPs and PEI revealed differences in their physicochemical properties, with repercussions on the capacity for mitochondrial targeting, gene delivery, and protein expression. This work demonstrated the efficacy of both peptide- and polymer-based delivery systems for mitochondrial targeting and mitochondrial gene expression. Furthermore, our study draws a comparison between peptides and polymers to reveal the most adequate delivery system to promote mitochondrial targeting and functional protein production, contributing to improvements/advances in the design of pDNA complexes for mitochondrial gene expression.

2. Materials and Methods

2.1. Materials

The following reagents were obtained from Sigma-Aldrich (St. Louis, MO, USA): trifluoroacetic acid (TFA), piperidine, oxyma, diisopropyl carbodiimide (DIC), Fmoc-amino acids, dimethylformamide (DMF), diisopropylethylamine (DIEA), dichloromethane (DCM), acetonitrile, and diethyl ether. AmphiSpheres 40™ resin was acquired from Agilent Technologies (Les Ulis, France). The Peptide Synthesizer Liberty Blue HT12™ was obtained from CEM (Matthews, NC, USA). The HPLC Pumps (321) and the FC 204 Fraction Collector were purchased from Gilson. The LKB-REC 102 was obtained from Pharmacia (Stockholm, Sweden). The HPLC System (Waters Alliance 2695) was obtained from Waters Corporation. The dialysis against water was carried out using 3k MCWO Dialysis Tubes (Spectrum laboratories Inc. Rancho Dominguez, CA, USA). The lyophilization was performed with a FreeZone 1 L Laboratory Lyophilizer (LABCONCO). DAPI was purchased from Invitrogen (Carlsbad, CA, USA), and MitoTracker Orange CMTMRos from Molecular Probes (Leiden, The Netherlands). TAT (47–57) peptide (YGRKKRRQRRR), chemically synthesized, was supplied as a lyophilized powder from Biomatik (Cambridge, ON, Canada). Commercial branched polyethylenimine (PEI) with average Mw of 10 kDa and 25 kDa, fluorescein isothiocyanate (FITC), succinic anhydride (SA), and N-hydroxysulfosuccinimide (NHS) were obtained from Sigma-Aldrich. Agarose and GreenSafe Premium were obtained from NZYTech Lda (Lisbon, Portugal). HeLa cancer cells were purchased from Invitrogen (Carlsbad, CA, USA), and human dermal fibroblasts (NHDF, Ref. C-12302, cryopreserved cells) were obtained from PromoCell (Heidelberg, Germany).

All solutions for the preparation of the peptide-based systems were freshly prepared by using ultrapure-grade water, purified with a Milli-Q system from Millipore (Billerica, MA, USA).

In this study, two different plasmids were used: the recoded plasmid DNA encoding green fluorescent protein (pGFP) (5.9 kbp)—a plasmid developed for exclusive mitochondrial translation, a kind gift from Dr. Diana Lyrawati [51]—and the plasmid pCAG-GFP-ND1 (pND1) (5.4 kbp), developed previously by our team through the cloning of the mitochondrial NADH dehydrogenase 1 protein-encoded gene (mtND1) in the pDNA vector. All details concerning gene cloning and plasmid production can be consulted elsewhere [59].

2.2. *Methods*

2.2.1. Synthesis of Peptides and PEI–DQA

The peptides used in this work were synthesized according to the protocol described in a previous work [18]. The synthesis of the PEI–DQA polymer was carried out in two steps: The first step was the synthesis of PEI–SA. To the solution of PEI (10 and 25 kDa) in DMSO, succinic anhydride (molar ratio of PEI: SA = 1:10, 1:25, respectively) was added. The reaction mixture was stirred at room temperature for 24 h. The solvent was evaporated from the crude reaction mixture and dialyzed for 48 h in water using cellulose ester membranes. The dialysate was lyophilized to obtain a fluffy white powder at a yield of ~57%. In the second and final step of this synthesis, 200 mg of PEI–SA was dissolved in 5 mL of DMF, and EDC, NHS, and triethylamine solution were slowly added dropwise under vigorous stirring for 2 h. Then, 77.18 mg of dequalinium was dissolved in 2 mL of DMF, and this solution was added dropwise into the vigorously stirred solution described above. Thereafter, the reacting mixture was evaporated by rotovap, and 3 mL of distilled water was added to completely dissolve the product. The mixture was dialyzed (3 k MWCO membrane) against distilled water for 24 h. Finally, the product was lyophilized for 24 h. The average reaction yield was ca. 65%.

2.2.2. Nuclear Magnetic Resonance Spectroscopy (NMR)

A proton nuclear magnetic resonance spectroscope (Bruker spectrometer-300 MHz, Bruker, Billerica, MA, USA) was used to examine the conjugate's chemical structure in $CDCl_3$ solution at 25 °C. The conjugates dissolved in $CDCl_3$ (10 mg/mL) were measured with 64 scans.

2.2.3. Formulation of Peptide/pDNA and PEI–DQA/TAT/pDNA Complexes

Peptide-based complexes were formulated according to the procedure described in a recent publication by our team [18].

PEI–DQA, TAT, and plasmid DNA stock solutions were prepared in sodium acetate buffer (0.1 mM sodium acetate/0.1 M acetic acid, pH 4.5) at a concentration of 0.5 mg/mL for PEI–DQA and TAT solutions, and 100 µg/mL for pDNA solutions. In ternary complexes, the determination of the N/P ratio was established by considering the proportion of charges, individually, for PEI or TAT in relation to pDNA. This parameter was defined as the molar ratio of the amine groups in the polymer or peptide—which represent the positive charges—to negatively charged phosphate groups in the pDNA. Different concentrations of PEI–DQA and TAT (50 µL) were added, drop by drop, to 150 µL of pDNA under vortex for 1 min. The mixture was left for equilibration for 30 min at room temperature to promote the formation of complexes. Thereafter, the formed complexes were centrifuged (12,000 rpm) for 20 min, and the pellet containing the complexes was recovered. The amount of non-complexed pDNA on the supernatants was visualized by horizontal electrophoresis for 30 min under 120 V in 1% agarose gel stained with GreenSafe Premium. Samples were analyzed under ultraviolet (UV) light using a FireReader Imaging System (UVITEC, Cambridge, UK).

The ability of the systems to complex pND1 was also evaluated by quantifying the intensity of the bands of the respective agarose gels, and compared with the band intensity of the initial pND1 sample of each gel. The quantification of bands' intensity was performed by densitometry using Image Lab software version 6.1. The percentage of complexation capacity (CC) was calculated according to the following formula:

$$\text{Complexation capacity } (\%) = 100 - \left(\frac{\text{Band intensity}}{\text{Band intensity of initial pND1}} \times 100\right) \quad (1)$$

2.2.4. Determination of Size and Surface Charge

The average size and the surface charges exhibited by PEI–DQA/TAT/pDNA complexes were both inferred by dynamic light scattering (DLS) in a Zetasizer Nano ZS device (Malvern Instruments, Malvern, UK), at 25 °C. The pellet with the complexes was suspended in 5% glucose with 1 mM NaCl. For size determination, a He–Ne laser at 633 nm with non-invasive backscatter (NIBS) was applied, while zeta potential values were measured by electrophoretic light scattering optics with an M3-PALS laser (phase analysis light scattering). Data were considered from 3 independent measurements, each of which was performed with 12 runs. Malvern Zetasizer software v 6.34 was employed to analyze the set of results.

2.2.5. Cell Culture

HeLa cells were grown in Dulbecco's modified Eagle's medium with Ham's F12 Nutrient Mixture (DMEM-F12) and L-glutamine supplemented with 0.5 g/L sodium bicarbonate, 1.10 g/L HEPES, 10% heat-inactivated fetal bovine serum (FBS), and 1% (*v/v*) of a mixture of penicillin (100 μg/mL) and streptomycin (100 μg/mL). Cells were maintained in a 5% CO_2 humidified atmosphere, at 37 °C until confluence was achieved.

2.2.6. Cytotoxicity Evaluation

The cytotoxicity profile of PEI–DQA/TAT/pND1 complexes was monitored in HeLa cells by MTT (3-[4,5dimethyl-thiazol-2-yl]-2,5-diphenyltetrazolium bromide) assay. Cancer cells were seeded in a 96-well plate, at a density of 1×10^4 cells/well, and were grown at 37 °C in a 95% O_2/5% CO_2 humidified atmosphere. Complexes (100 μL) were first resuspended in serum-free DMEM medium, and then applied to the well plates for 6 h. The medium was changed to end the transfection process. After incubation for periods of 24 h and 48 h, the redox activity was evaluated by MTT reduction. Measurements of absorbance at 570 nm were performed on a Bio-Rad Microplate Reader Benchmark. The medium without cells was settled as zero absorbance and considered for spectrophotometer calibration. Non-transfected cells were used as positive controls, while ethanol-treated cells were used as negative controls. The relative cell viability (%) compared to control wells was calculated by [A] test/[A] control × 100, where [A] test is the absorbance of the test sample and [A] control is the absorbance of the control sample. All measurements were performed in triplicate.

2.2.7. Fluorescence Confocal Microscopy

FITC Plasmid Labeling

For the preparation of FITC-labelled pDNA, 2 μg of pDNA, 2 μL of FITC (in sterile anhydrous dimethyl sulfoxide, 50 mg/100 μL), and 81 μL of labeling buffer (0.1 M sodium tetraborate, pH 8.5) were mixed. The mixture was stirred for 4 h at room temperature and in the dark. Two-and-a-half volumes of 100% ethanol (212.5 μL) and one volume of 3 M NaCl (85 μL) were added. FITC pDNA samples were incubated at −20 °C overnight. Thereafter, they were centrifuged at 4 °C for 30 min, and the pellet was washed with ethanol (75%).

Live Cell Imaging

The cellular internalization and mitochondrial targeting capacity of PEI–DQA/TAT/FITC–pND1 complexes were monitored by confocal laser scanning microscopy (CLSM). HeLa cells (2×10^3) were grown in μ-slide 8-plate wells until 50–60% confluence was attained. Cell nuclei and mitochondria were stained with DAPI and MitoTracker Orange dye, respectively. The complete medium was replaced with a serum-free culture medium 12 h before transfection; 1 μg of complexed FITC–pND1 was added to each well. Images of transfected cancer cells were acquired after 6 h. The LSM 710 confocal microscope (Carl Zeiss SMT, Inc., Oberkochen, Germany) was used to observe real live transfection, under a 63× oil immersion objective. Images were analyzed with LSM software (Carl Zeiss SMT, Inc., Oberkochen, Germany). Throughout the study, HeLa cells were kept at 37 °C with 5% CO_2. All images were acquired using the same parameters, with the laser and filters corresponding to the respective DAPI (445/450 nm), FITC (525/550 nm), and MitoTracker (555/580 nm) dyes. After the acquisition, the images were processed under the same conditions and parameters using ImageJ software.

2.2.8. Reverse Transcription Polymerase Chain Reaction (RT-PCR)

Reverse transcription polymerase chain reaction (RT-PCR) was used to qualitatively analyze mRNA GFP expression. HeLa cells were seeded in 12-well plates at a density of 2×10^5 cells/well. The medium was removed 24 h after transfection, and cells were washed with PBS. Untreated cells were used as controls. The cells were lysed with TRIzol (250 μL/well), incubated for 5 min at room temperature, chloroform was added and the mixture stirred to promote RNA extraction. The samples were incubated for 10 min at room temperature and then centrifuged at 10,000 rpm, at 4 °C, for 15 min. To ensure RNA precipitation, the aqueous phase was withdrawn and 125 μL of ice-cold isopropanol was added. Another centrifugation cycle (10,000 rpm, 4 °C, 15 min) was performed; 125 μL of 75% ethanol in DEPC water was added to the obtained pellet to remove the organic compounds. After centrifugation, 20 μL of DEPC water was added to rehydrate the pellet, and samples were quantified using a NanoPhotometer™. In complement, electrophoresis analysis on agarose gel (1%) was carried out. The "Xpert cDNA Synthesis Kit" from GRiSP (GRiSP, Porto, Portugal) was utilized for the cDNA synthesis. All of the instructions provided by the manufacturer were followed. The amplification of cDNA was performed through the addition of 10 μL of RNase-free water, 1 μL of reverse primer (5′-CGTTCTTGTACGTAGCCTTC-3′), 1 μL of forward primer (5′-CTGCACCACCGGAAAACTCC-3′), 12 μL of Speedy Supreme NZYTaq 2× Green Master Mix (NZYTech, Lisbon, Portugal), and 1 μL of cDNA, in each PCR reaction. After homogenization of the samples, they were placed in a T100™ thermal cycler (Bio-Rad Laboratories, Inc., Hercules, CA, USA). The following reaction conditions were established: denaturation (94 °C for 2 s), annealing (57 °C for 5 s), and extension (72 °C for 5 s) for 35 cycles. Agarose gel electrophoresis was used to analyze the PCR products. The visualization was performed on a UV FireReader Imaging System (UVITEC, Cambridge, UK).

2.2.9. Mitochondrial Isolation

Mitochondria were isolated from other cellular organelles with the Mitochondria Isolation Kit for Cultured Cells (Thermo Fisher Scientific Inc., Rockford, IL, USA). This kit ensures the separation of mitochondria with high purity and yield. The instructions provided by the manufacturer were followed. Briefly, HeLa cells (1×10^4) were transferred to Falcon tubes, and 800 μL of Reagent A was added to the cells, followed by incubation on ice for 2 min. Then, 10 μL of Reagent B was added, and the cells were vortexed for 10 s at maximum speed, incubated on ice for 5 min, and vortexed again at maximum speed every minute for 10 min. After this, Reagent C (800 μL) was added, and samples were centrifuged (20,000× *g*) for 10 min at 4 °C. Then, 500 μL of Reagent C was added to the resultant pellet containing the mitochondria. Final centrifugation at 12,000× *g* for 5 min was performed, and the supernatant was discarded. The obtained pellets full of mitochondria were resuspended in 50 μL of ice-cold PBS and mixed with 500 μL of carbonate buffer (fresh cold 0.1 M Na_2CO_3).

2.2.10. Protein Quantification

The ND1 protein produced by transfection of HeLa cells with the developed systems was identified using an ND1 ELISA kit (Biomatik, EKL54820, Wilmington, DE, USA), following the procedure described by the manufacturer. ND1 was quantified by a sandwich enzyme immunoassay. Transfected HeLa cells with different peptide- or polymer-based complexes were lysed following standard cell lysis methods. Cells were detached with trypsin and centrifuged. They were then washed three times in cold PBS, resuspended in PBS, and ultrasonicated 4 times. After this procedure, HeLa cells were centrifuged at 15,000× *g* for 10 min at 4 °C. The manufacturer provided reagent solutions that were used according to instructions. In summary, Reagent A was added to each well and incubated for 1 h at 37 °C, followed by incubation with Reagent B. Thereafter, TMB substrate solution was added to each well, and samples were incubated for 20 min at 37 °C in the dark. A sample displaying a yellow color was obtained after the addition of Stop Solution. The content of ND1 protein was inferred by measuring the absorbance in a microplate reader at 450 nm.

2.2.11. Statistical Analysis

The normality of the distribution of sample data was evaluated by running appropriate tests, such as the D'Agostino–Pearson omnibus. The statistical analysis performed was a one-way or two-way analysis of variance (ANOVA), followed by Bonferroni's multiple comparison test. A *p*-value below 0.05 was considered statistically significant. Data analysis was conducted with GraphPad Prism v.8.01 (GraphPad Software, Inc., San Diego, CA, USA).

3. Results and Discussion

3.1. Synthesis and Characterization of PEI–SA, and PEI–DQA

The synthesis of PEI conjugates was carried out following the general scheme depicted in Figure 1. The amine-terminated PEI was succinated to further react with DQA. The terminal carboxylic acid group of succinic acid provided the attachment point with DQA via the acid–amine coupling reaction. The reaction conditions produced optimal yield at each step. The analysis of synthesized PEI derivatives was carried out using spectroscopic methods. The ^{1}H NMR spectra of PEI-SA (10 and 25 kDa) are shown in Figure 2A,B, respectively. In both spectra, broad multiplet peaks were observed at δ3.12–3.42 ppm, corresponding to the methylene groups present in the primary amine group of PEI, along with the introduced succinic substituent ((PEI)–$(CH_2)_2$–NH–CO–$(CH_2)_2$–COOH). For DQA conjugates, the characteristic signals of the aromatic ring protons were present at δ7.25–7.35 ppm (Figure 3A,B). The peak located at δ3.32 ppm was identified as benzyl–CH_2 protons. The methylene moiety had a broad multiplet at δ3.63–3.72 ppm. The successful conjugation in PEI–DQA (10 and 25 kDa) was confirmed from ^{1}H NMR. Figure 3C shows the DQA spectrum. Fourier-transform infrared (FTIR) spectroscopy was carried out to confirm the success of the modification. Figure S1, available in the Supplementary Materials (SM), shows the FTIR spectra of PEI–DQA (10 kDa and 25 kDa) (A) and the FTIR spectra of DQA (B). The signals at 3300 cm^{-1} and 2822–2530 cm^{-1} indicated the N–H stretching and aliphatic C–H stretching vibrations, respectively, in all PEI–SA conjugates. The PEI–DQA (10 and 25 kDa) displayed a broad peak at 1720 cm^{-1} as the stretching vibrations of the C=O moiety. Similarly, the C-O-C (1228 cm^{-1}) and N-H (3210 cm^{-1}) stretching vibrations were observed in the PEI–DQA conjugates. The chromatogram obtained via gel permeation chromatography (GPC) is represented in Figure S2, available in the SM. The peak shift on the left due to the change in retention time of the polymer and polymer conjugates indicated the increase in molecular weight. According to the molecular weights of the conjugates obtained through GPC analysis (Table S1 in the SM), ~11 molecules of SA were attached to PEI (10 kDa), which conjugated ~9 molecules of DQA. Similarly, PEI (25 kDa) was attached to ~15 molecules of SA, which resulted in ~13 molecules of DQA attachment. The critical micelle concentration (CMC) was observed to be 12.5 μg/mL for PEI–DQA (10 and 25 kDa) conjugates. The reduced CMC showed that PEI–DQA conjugates can readily develop a core–shell structure in an aqueous environment (Figure S3 in SM).

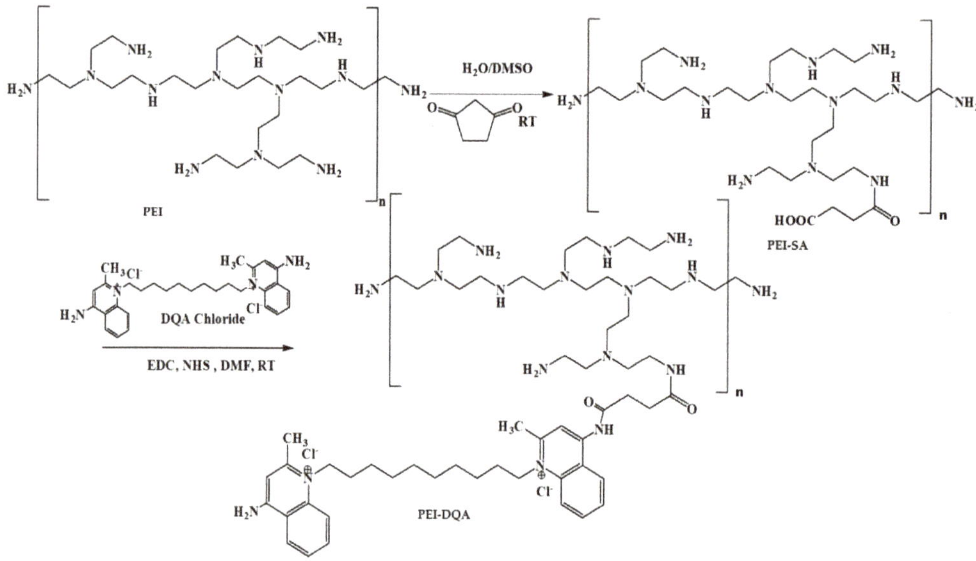

Figure 1. Scheme illustrating the synthesis of the polymer PEI–DQA. RT: room temperature.

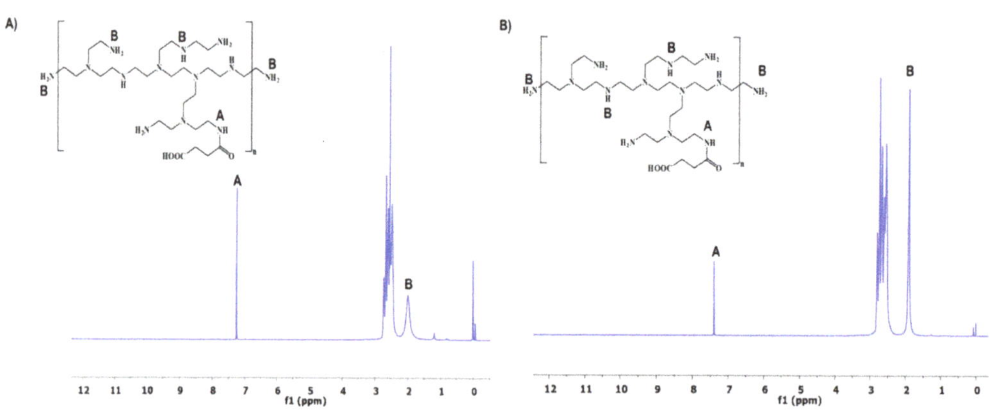

Figure 2. ^{1}H NMR spectra of the intermediate PEI–SA: PEI-10 kDa (**A**) and PEI-25 kDa (**B**).

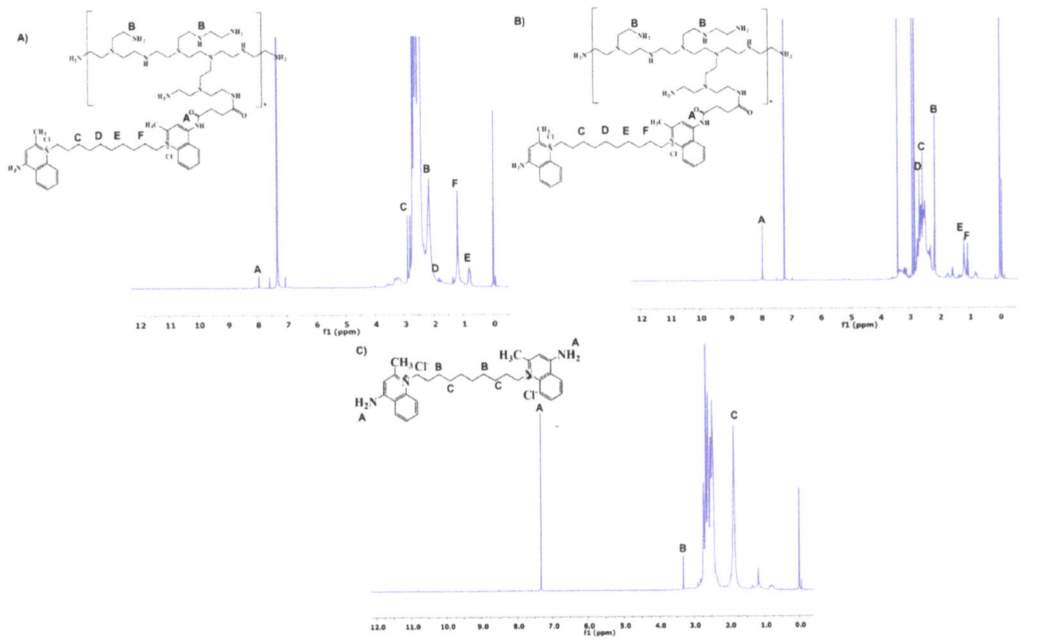

Figure 3. ^{1}H NMR spectra of PEI–DQA: PEI (10 kDa) (**A**) and PEI (25 kDa) (**B**); the spectrum of DQA (**C**).

3.2. pDNA Complexation Capacity

The formulation of PEI–DQA/TAT/pND1 systems was elaborated using a dropwise precipitation method. First, PEI–DQA solution was added to pDNA solution for 1 min, after which TAT peptide solution was added. Electrostatic interactions are the main force that allows the encapsulation of pDNA by PEI–DQA and TAT. These interactions occur due to the negative charges on pND1 that bind to the positive charges on both PEI and TAT, forming nanoscale systems. The interaction strength is dependent on the N/P ratio considered at the formulation step, and the higher the ratio, the greater the availability of positive charges to interact with pND1. Moreover, in previous research by our team, the cytotoxicity of TAT/pDNA complexes was evaluated as a function of the N/P ratio on both fibroblasts and HeLa cells [26]. The results demonstrated that TAT/pDNA complexes formulated at N/P ratios of 1, 2, 4, 8, and 10 are all biocompatible. After incubation with TAT-based complexes, a moderate increment in the cellular viability was observed as the N/P ratio increased.

Here, the complexation capacity of the systems and the influence of the N/P ratio were evaluated using agarose gel electrophoresis. The ratios of PEI–DQA to pND1 used were 10, 20, 50, 100, 200, and 500, and the ratios of TAT to pND1 were 1 and 2. Results are shown in Figures 4–6. In Figure 4 we can see that the 10 kDa and 25 kDa PEI–DQA systems had a low capacity to encapsulate pND1, even at the highest ratios (100:1 and 200:1). Figure 5 concerns 10 kDa and 25 kDa PEI–DQA systems to which TAT was added, where N/P = 1. By analyzing the images, we found that all formulated systems exhibited encapsulation capacity to some extent, compared to the initial pND1 sample. However, only systems prepared at a 500:1:1 ratio of both PEIs demonstrated high encapsulation capacity. As PEI is a compound known to have considerable cytotoxicity at high concentrations, the need arose to try to improve encapsulation by keeping PEI ratios as low as possible. Following this, we doubled the TAT ratio, which resulted in a great improvement in pND1 encapsulation for all formulated systems, maintaining PEI ratios (Figure 6). The prepared systems presented a high efficiency of pND1 encapsulation, being able to neutralize its negative charges; thus, there were no bands in the agarose gel. Moreover, pND1 CC (%) was calculated from the band intensities of the agarose gels, as described in the Materials and Methods section; the results are shown in Figure 7. PEI (10 or 25 kDa)–DQA/pND1 complexes displayed poor ability to complex pND1 at lower ratios, and it was found that CC varies with the N/P ratio (Figure 7A). It seemed that an increase in the N/P ratio led to high CC values. Additionally, in general, PEI (25 kDa)–DQA/pND1 complexes condensed pND1 to a higher extent when compared to the capacity exhibited by the corresponding PEI (10 kDa) complexes. The incorporation of TAT peptide into PEI (10 or 25 kDa)–DQA/pND1 complexes significantly increased their capacity to condense pND1 (Figure 7B,C). Considerably high CC values were obtained for all N/P ratios investigated. At an N/P ratio of TAT/pND1 of 2, practically all complexes possessed a CC of around 100%. The exception was PEI (10 kDa)–DQA/TAT/pND1 prepared at the lowest ratio (Figure 7C). From this, there was no need to consider TAT/pND1 ratios higher than 2. In this work, we also intended to compare these PEI-based systems with previously studied systems based on peptides that also have sequences to confer specificity for mitochondrial targeting [18]. Comparing the pND1 encapsulation ability of PEI–DQA/TAT/pND1 complexes with that displayed by peptide-based complexes, PEI-based complexes showed lower efficiency for pND1 encapsulation, since to encapsulate the same amount of pND1 higher N/P ratios were required (N/P ratios of 10 versus N/P ratios of 0.5 and 1). This lower capacity can be explained by the decrease in the amines available in PEI due to the addition of DQA to the polymer, reducing the charges available to interact with pND1. Nevertheless, our study proceeded in analyzing the properties of PEI–DQA/TAT/pND1 systems and their potential for mitochondrial targeting/gene delivery.

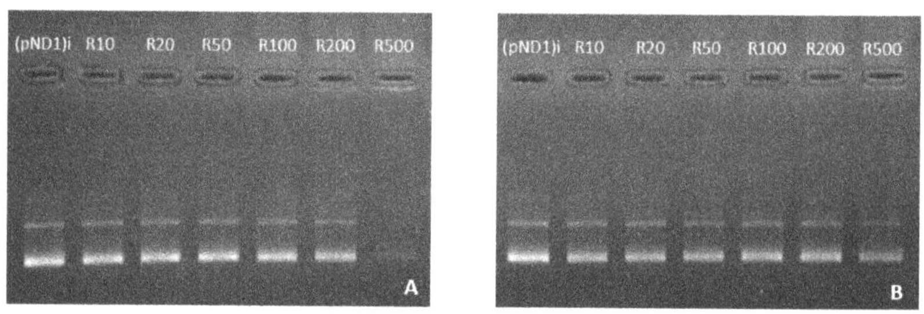

Figure 4. Agarose gel electrophoresis of the initial pND1 solution ((pND1)i) and supernatants resulting from the formulation of PEI–DQA (10 kDa)/pND1 systems (PEI–DQA N/P = 10, 20, 50, 100, 200, and 500) (**A**) and PEI–DQA (25 kDa)/pND1 systems (PEI–DQA N/P = 10, 20, 50, 100, 200, and 500) (**B**).

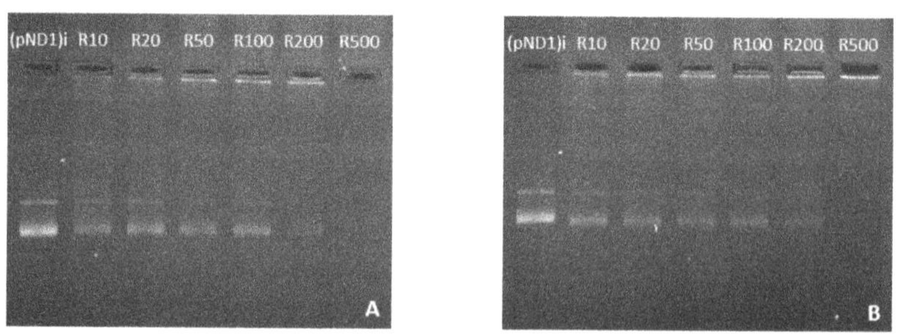

Figure 5. Agarose gel electrophoresis of the initial pND1 solution ((pND1)i) and supernatants resulting from the formulation of PEI–DQA (10 kDa)/TAT/pND1 systems (PEI–DQA N/P = 10, 20, 50, 100, 200, and 500, and TAT N/P = 1) (**A**) and PEI–DQA (25 kDa)/TAT/pND1 systems (PEI–DQA N/P = 10, 20, 50, 100, 200, and 500, and TAT N/P = 1) (**B**).

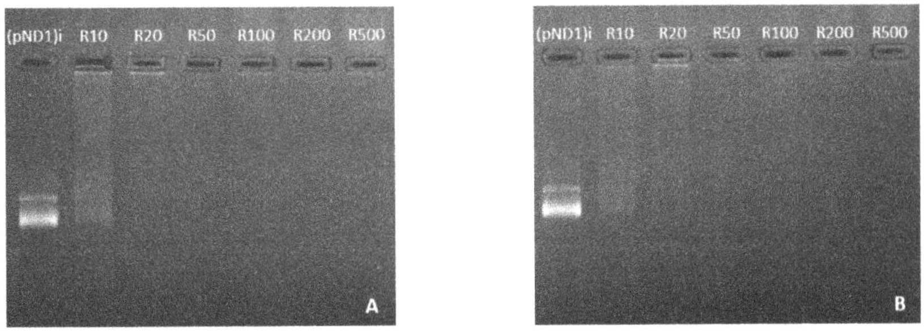

Figure 6. Agarose gel electrophoresis of the initial pND1 solution ((pND1)i) and supernatants resulting from the formulation of PEI–DQA (10 kDa)/TAT/pND1 systems (PEI–DQA N/P = 10, 20, 50, 100, 200, and 500, and TAT N/P = 2) (**A**) and PEI–DQA (25 kDa)/TAT/pND1 systems (PEI–DQA N/P = 10, 20, 50, 100, 200, and 500, and TAT N/P = 2) (**B**).

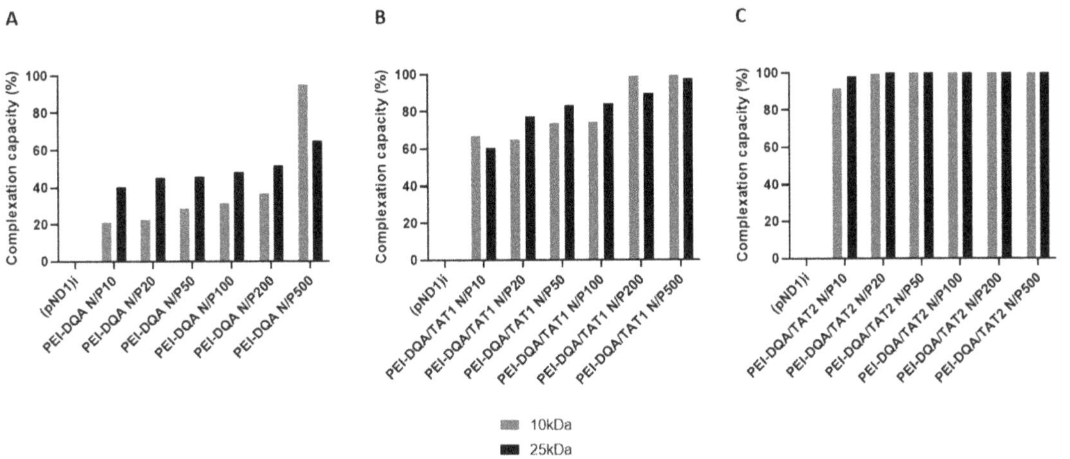

Figure 7. pND1 complexation capacity (CC) of PEI–DQA (10 and 25 kDa)/pND1 (**A**), PEI–DQA (10 and 25 kDa)/TAT(N/P1)/pND1 (**B**), and PEI–DQA (10 and 25 kDa)/TAT(N/P2)/pND1 (**C**) complexes calculated from the band intensity of the agarose gels of Figures 4–6 respectively.

3.3. Characterization of PEI–DQA/TAT/pND1 Complexes

After verifying the complexation capacity of the produced systems, these complexes were characterized by DLS in terms of size and surface charge. The results are shown in Figure 8, corresponding to data from PEI–DQA (10 kDa)/TAT/pND1 and PEI–DQA (25 kDa)/TAT/pND1 complexes. In addition, Table S2 (in SM) presents all of the obtained values of size and zeta potential for both polymer- and peptide-based systems. The results of surface charge measured in zeta potential (mV) for systems with PEI–DQA of 10 kDa revealed a slightly negative overall charge. A decrease in this negative charge with the increase in the N/P ratio was also observed. The complexes prepared with PEI–DQA (25 kDa), for the lowest N/P ratios under study, showed a charge very close to 0 mV, while for N/P ratios of 200:2:1 and 500:2:1 they presented positive zeta potential values (+3.2 and +5.2 mV, respectively). This low global charge can be explained by the presence of DQA, which presents a neutral charge [49], decreasing the influence of positive charges of PEI on the surface of complexes. The main contribution to zeta potential comes from the inclusion of TAT. The surface charge of TAT/pND1 complexes conceived at an N/P ratio of 2 was found to be around +2.3 mV. When compared to these later data, results regarding the physicochemical properties exhibited by the peptide/pND1 complexes [18] presented in the previous publication demonstrated that peptide-based complexes displayed a more positive surface charge, with zeta potential values of up to +20 mV for an N/P = 5, while the highest value for PEI–DQA/TAT/pND1 complexes was +5.2 mV for N/P = 500.

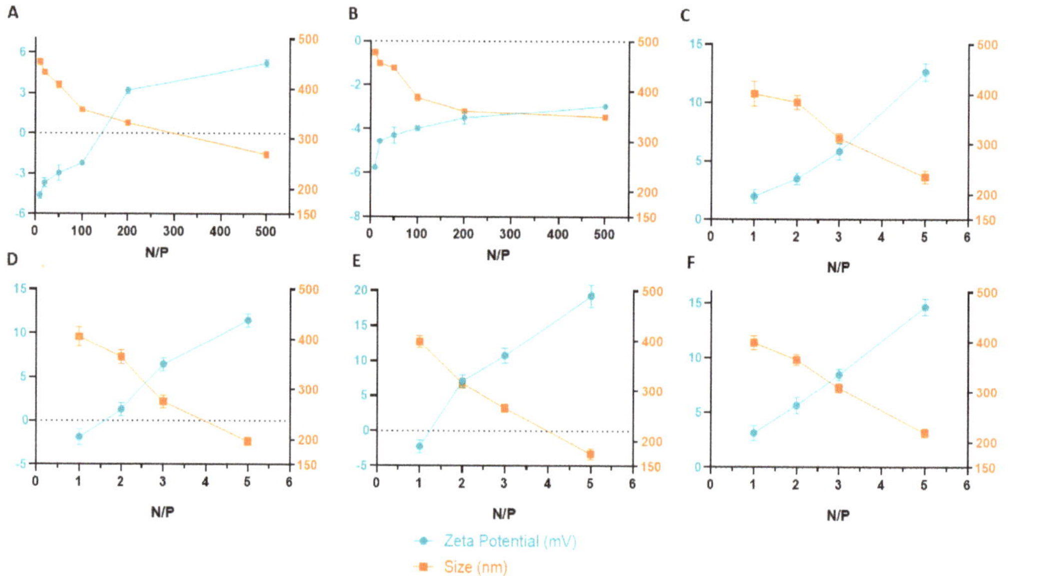

Figure 8. Average zeta potential and mean size displayed by PEI–DQA (10 kDa and 25 kDa)/TAT/pND1 (**A**,**B**), CpMTP/pND1 (**C**), MTS-WRAP1/pND1 (**D**), MTS-WRAP5/pND1 (**E**), and MTS-$(KH)_9$/pND1 (**F**) complexes.

Evaluation of the size of the formulated polyplexes revealed a difference between complexes formulated with PEI of 10 kDa and 25 kDa. For the same N/P ratio, average sizes were smaller in systems prepared with 25 kDa of PEI. The effect of the N/P ratio was also evident within the same type of system—as the ratio increased, there was a decrease in the size of complexes. This evidence was due to the greater capacity of high-molecular-weight PEI to encapsulate pND1, due to high charge density that led to stronger electrostatic interactions. The sizes of all complexes were less than 500 nm, which may have facilitated cell internalization. The sizes evidenced by these systems were superior to the sizes displayed by peptide-based complexes formulated in our previous study [18]. Comparing the results, we previously found that peptide-based complexes had sizes around 150–200 nm for the highest ratios, while in this study PEI–DQA/TAT/pND1 complexes exhibited sizes around 350 nm. Thus, we can infer that peptide/pND1 complexes presented a greater capacity to condense pND1 and formulate smaller complexes. Another relevant characterization parameter is the PdI, which allows us to assess the distribution of sample sizes. PdI values close to 0.01 reveal monodisperse samples; between 0.5 and 0.7 are considered polydisperse particles and a value above 0.7 is indicative of broad particle size distribution [60]. PdI data are presented in Table S2, available in the SM. Analyzing this table, we verified a PdI between 0.4 and 0.6, with an effect on the N/P ratio. Lower ratios corresponded to polydisperse samples, while higher ratio complexes were monodispersed. Compared to PdI values obtained for peptide-based complexes, the difference was minimal. PdI for peptide/pND1 systems was around 0.3, with polydisperse samples found for smaller ratios.

Complexes of MTS-CPP/pND1 and CpMTP/pND1 were previously published [18]. The values were calculated with data obtained from three independent measurements (mean ± SD, $n = 3$).

3.4. Cytotoxic Profile

The cytotoxicity of formulated polymeric complexes was evaluated using the MTT assay in HeLa cells. To verify the safety of the complexes, the viability of HeLa cells was evaluated at 24 h, 48 h, and 72 h after transfection with the various systems under study. PEI–DQA (10 kDa or 25 kDa)/TAT/pND1 was used at N/P ratios of 20, 50, 100, 200, and 500, and TAT/pND1 N/P of 2:1. The results shown in Figure 9A–C correspond to 24, 48, and 72 h transfection, respectively. Non-transfected cells were used as positive controls (100% viability), while cells treated with 70% ethanol were used as negative controls (0% viability). Statistical analysis was performed with the positive controls for comparison. Systems formulated with high ratios demonstrated cytotoxicity when compared to the positive controls (N/P ratios of 100:2:1; 200:2:1, and 500:2:1). Polymeric complexes formulated at low ratios demonstrated biocompatibility (statistically not significant (n.s) versus positive controls). Therefore, these results showed the influence of the N/P ratio on the cytotoxicity of the developed systems. The increment in the N/P ratio led to an increased content of PEI amines, and this high cationic charge corresponded to high cytotoxicity. In this way, it can be pointed out that the N/P ratio can be viewed as a tailoring tool to adjust cellular cytotoxicity levels. At 24 h, the obtained data also demonstrated some differences between complexes formulated with PEI–DQA (10 kDa) and PEI–DQA (25 kDa). At 24 h and for the highest ratios, PEI–DQA (25 kDa) complexes were more cytotoxic than the ones based on PEI–DQA (10 kDa); there was a statistically significant difference (* $p \leq 0.05$) for the comparison between these complexes at N/P ratios of both 200:2:1 and 500:2:1. This tendency, however, was not observed at 48 h. The cytotoxicity evidenced in complexes prepared at high ratios seemed to be more pronounced in the first 24 h when compared to that obtained at 48 h and 72 h after transfection. We can hypothesize that cells suffered an initial shock during the first 24 h of incubation, with a significant loss of their viability, after which they were able to recover over time, showing higher viability. However, a clear explanation for this phenomenon cannot be anticipated at this stage. For instance, at 72 h and at high N/P ratios, the difference from the positive controls was lower (* $p \leq 0.05$ or ** $p < 0.01$ for PEI–DQA (25 kDa)/TAT/pND1 at 500:2:1 versus positive controls) than the difference found at 24 h (for high N/P ratio complexes, **** $p < 0.0001$ versus positive control) or 48 h (for most N/P ratio complexes, *** $p < 0.001$ versus positive control).

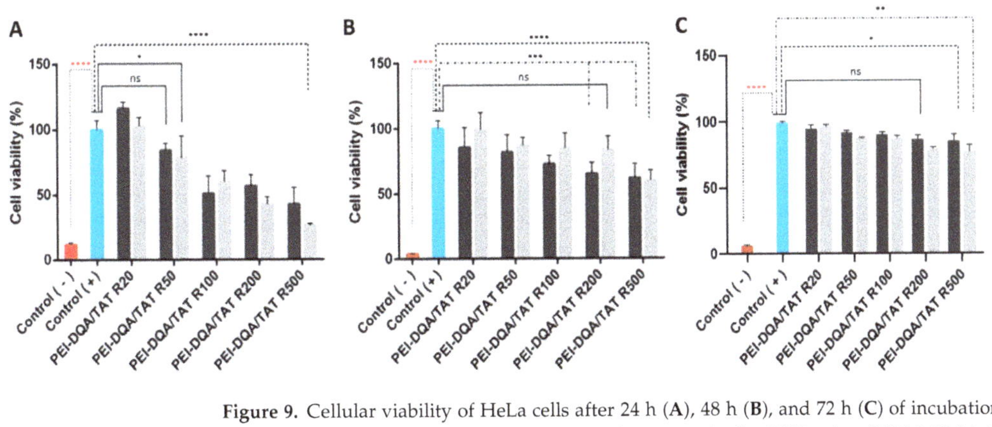

Figure 9. Cellular viability of HeLa cells after 24 h (**A**), 48 h (**B**), and 72 h (**C**) of incubation with PEI–DQA (10 kDa or 25 kDa)/TAT/pND1 complexes conceived at N/P ratios of 20:2:1, 50:2:1, 100:2:1, 200:2:1, and 500:2:1. Non-transfected cells were used as positive controls (100 % viability) and ethanol-treated cells were considered as negative controls (0 % viability). Statistical analysis was carried out using "one-way ANOVA" with data obtained from four independent measurements (mean ± SD, $n = 4$).

For further in vitro studies, and due to their biocompatibility, lower N/P ratio PEI–DQA (10 kDa or 25 kDa)/TAT/pND1 complexes were selected.

The biocompatibility of peptide-based complexes was analyzed and discussed elsewhere [18]. All peptides/pND1 were demonstrated to be non-toxic to HeLa cells at 24 h, with the obtained results being statistically non-significant (n.s) in relation to the positive controls. At 48 h, however, a decrease in biocompatibility was observed for all peptides/pND1 nanoparticles. Furthermore, the presence of the MTS sequence in the peptides seemed to influence the cytotoxic profile displayed by the complexes. MTS-peptide/pND1 carriers exhibited less cellular viability in relation to controls [18]. Comparing cytotoxicity results between polymer and peptide/pND1 complexes, it was verified that the latter complexes presented higher safety in terms of biocompatibility, making them far superior in relation to PEI–DQA/TAT/pND1 complexes. These results can be explained by the greater efficiency of peptides in encapsulating pND1, decreasing the need to use high N/P ratios.

3.5. Mitochondrial Targeting Ability

Over the past few years, researchers have pursued the goal of mitochondrial targeting and mitochondrial transgene expression for the treatment of mitochondrial disorders [52]. This effort has led to the development of efficient delivery systems with the ability of cellular internalization, followed by mitochondrial targeting [52]. Recently, our team has also contributed to this field with the conception of mitochondria-targeted peptide/pND1 complexes using a mitochondrial targeting sequence (MTS) coupled to peptides able to internalize nucleic acids such as MTS-WRAP1, MTS-WRAP5, and MTS-$(KH)_9$ [18]. In this previous work, real live transfection of HeLa cells mediated by these peptide delivery vectors, monitored by fluorescence confocal microscopy, revealed that the complexes were easily taken up by cancer cells and accumulated at the site of the mitochondria [18], unequivocally demonstrating their mitochondrial targeting specificity. Among those peptide complexes, the ones based on the CpMTP peptide seemed to display higher cell penetration ability and targeting performance [18].

In the present study, we investigated the mitochondrial targeting capacity of various synthesized PEI–DQA/TAT/pND1 complexes in HeLa cells. In this sense, the cellular internalization and mitochondrial targeting ability of these polymeric complexes were assessed by confocal microscopy, using appropriate dyes to stain the nuclei and mitochondria, while pND1 was labeled with FITC. In this way, real live transfection of HeLa cells mediated by PEI–DQA (10 kDa or 25 kDa)/TAT/pND1 vectors, conceived at different N/P ratios, was monitored. The obtained cell images, at 6 h of transfection, are visualized in Figures 10 and 11, corresponding to the cellular transfection mediated by PEI–DQA (10 kDa)/TAT/pND1 at an N/P ratio of 50:2:1 and PEI–DQA (25 kDa)/TAT/pND1 at an N/P ratio of 50:2:1, respectively. The images were obtained from a set of consecutive Z planes (Z-stacks; step size of 0.1 μm). In both presented figures, image A represents mitochondria labeled with MitoTracker, image B shows PEI–DQA (10 kDa or 25 kDa)/TAT/pND1 complexes, image C shows nuclei marked with DAPI, and image D corresponds to the merged image. From the green fluorescence visualized in B, we inferred that efficient transfection took place and polymeric complexes were internalized in HeLa cells.

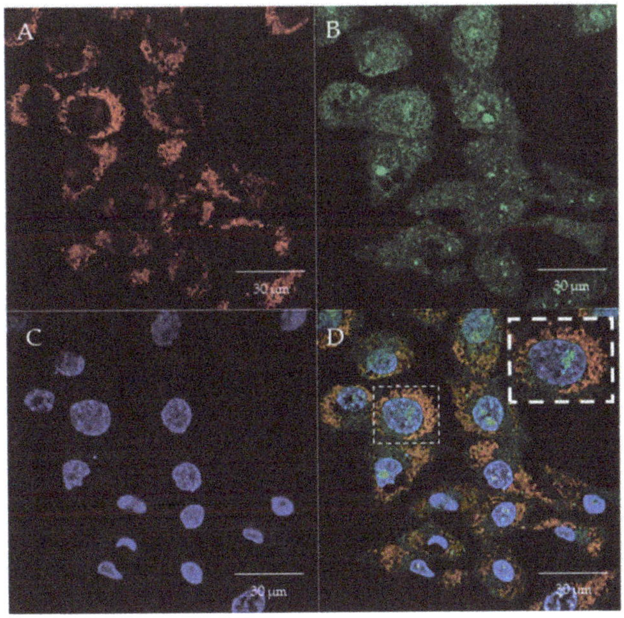

Figure 10. Cellular uptake and intracellular colocalization of PEI–DQA (10 kDa)/TAT/pND1 complexes formulated at an N/P ratio of 50:2:1. Mitochondria stained red by MitoTracker (**A**), green labeled pND1 (**B**), nuclei marked blue by DAPI (**C**), and merged image (**D**).

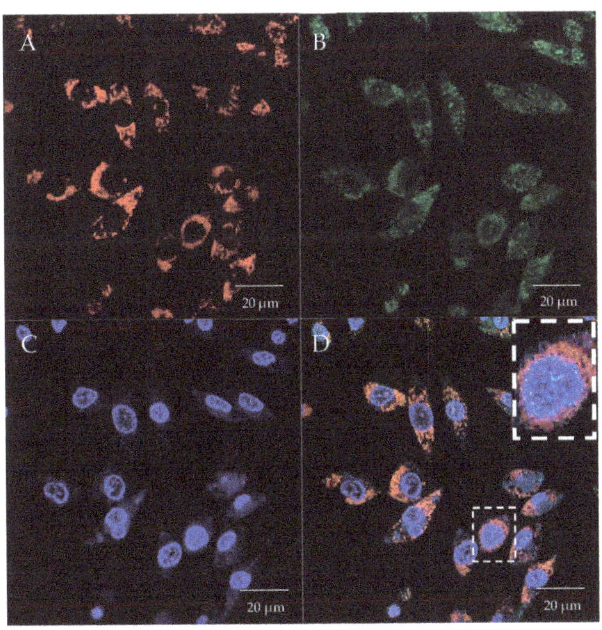

Figure 11. Cellular uptake and intracellular colocalization of PEI–DQA (25 kDa)/TAT/pND1 complexes formulated at an N/P ratio of 50:2:1. Mitochondria stained red by MitoTracker (**A**), green labeled pND1 (**B**), nuclei marked blue by DAPI (**C**), and merged image (**D**).

Moreover, after the uptake, the developed delivery complexes were targeted to the site of the mitochondria. Image D in both figures shows the specific accumulation of PEI–DQA (10 kDa or 25 kDa)/TAT/pND1 complexes in the mitochondria, demonstrating their mitochondria-specific targeting. We also evaluated the effect of the N/P ratio on the targeting capacity by performing the described microscopy study on HeLa cells using PEI–DQA (10 kDa or 25 kDa)/TAT/pND1 formed at a lower N/P ratio of 20:2:1. The collected images can be consulted in the SM (Figures S4 and S5). Very similar observations were made, revealing that both studied N/P ratios were adequate to develop polymeric complexes able to target the mitochondria. This targeting skill, displayed by the novel PEI-based complexes, represents a great asset to further studies on mitochondrial gene delivery and expression. Furthermore, as evidenced in the merged panel of Figures 10 and 11, although the complexes were preferentially located in the mitochondria, some of the complexes were found in the nuclei (especially in the nucleoli).

3.6. Evaluation of Gene Expression

Gene expression was evaluated using recoded mitochondrial pGFP—a pDNA designed for specific expression in the mitochondria [51]. Polymeric and peptide/pGFP delivery systems were prepared at various N/P ratios, following the method described in the Materials and Methods section for pND1. Thereafter, HeLa cells were transfected with the formed complexes, and recoded GFP gene expression was evaluated by RT-PCR. Results obtained for PEI- and peptide-based complexes are shown in Figure 12A,B, respectively. Non-transfected cells were used as controls. The figures show agarose gel electrophoresis of amplification products of the mitochondrial GFP gene, with bands at expected sites. The results obtained for PEI–DQA (10 kDa or 25 kDa)/TAT/pGFP complexes, at N/P ratios of 20:2:1 or 50:2:1 (Figure 12A), demonstrated the efficacy of HeLa cells' transfection and the presence of pGFP transcripts to a higher extent when compared to controls. Although the bands on the agarose gel were faint, it seemed that all considered polymeric complexes led to GFP expression in the mitochondria.

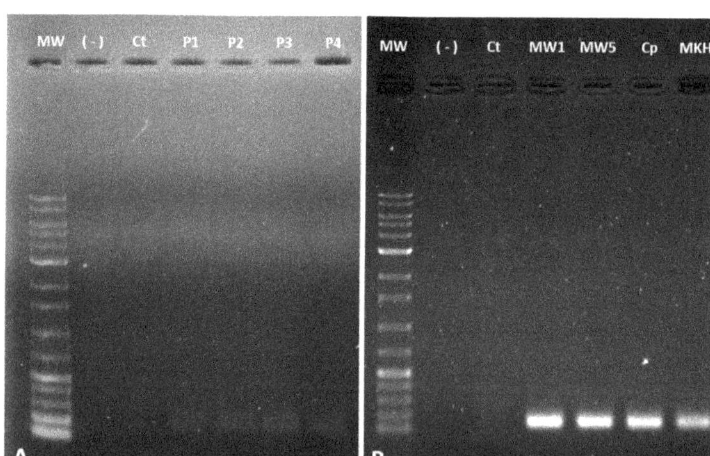

Figure 12. Analysis of GFP gene expression by RT-PCR after transfection of HeLa cells with (**A**) P1—PEI–DQA (10kDa)/TAT/pGFP, N/P of 20:2:1; P2—PEI–DQA (10 kDa)/TAT/pGFP, N/P of 50:2:1; P3—PEI–DQA (25kDa)/TAT/pGFP, N/P of 20:2:1; P4—PEI–DQA (25 kDa)/TAT/pGFP, N/P of 50:2:1; (**B**) MW1—MTS-WRAP1/pGFP, N/P of 5:1; MW5—MTS-WRAP5/pGFP, N/P of 5:1; Cp—CpMTP/pGFP, N/P of 5:1 and MKH—MTS-$(KH)_9$/pGFP, N/P of 5:1. MW—molecular weight; (-)—PCR control; CT—non-transfected cells.

The RT-PCR results for peptide/pGFP systems (Figure 12B) demonstrated a clear detection of GFP mRNA in transfected cells, when compared to mRNA levels detected in control cells. In general, wide and intense bands were obtained for all peptide-based complexes. Since the total amount of RNA transcribed to cDNA was fully comparable in both polymeric and peptide systems, (see the Materials and Methods section for details), it seemed that the extent of gene expression was higher for the latter systems. Apparently, peptide-based systems seemed to be more promising for GFP expression in the mitochondria. We are, however, aware that RT-PCR results are merely qualitative, giving first evidence of gene expression and potentiating further analysis. To deeply infer the content of both gene and protein, more accurate assays are required.

3.7. Quantification of Protein

The evidence of gene delivery and expression in mitochondria instigated us to quantify the produced ND1 protein, after transfection mediated by polymeric or peptide/pND1 complexes. In this way, ND1 protein expression in the mitochondria of HeLa and fibroblast cells was measured by using an ELISA kit. Non-transfected cells served as controls. The results are shown in Figure 13. Control cells presented significant levels of ND1 protein, as ND1 is an endogenous mitochondrial gene. As observed, this amount can be increased by the transfection with the developed mitochondria-targeted/delivery systems. As shown in Figure 13A,B, both fibroblast (A) and HeLa (B) cells transfected with PEI–DQA/TAT/pND1 complexes demonstrated higher levels of ND1 protein when compared to controls (*** $p < 0.001$ or **** $p < 0.0001$). Moreover, it was observed that this increase in the production of ND1 protein was influenced by both the molecular weight of PEI and the N/P ratio. The results demonstrated that PEI (25 kDa) led to a higher amount of ND1 in comparison with the protein levels when transfection was mediated by complexes based on PEI (10 kDa), even at lower N/P ratios. Moreover, high-N/P-ratio systems demonstrated higher ND1 levels compared to lower ratio systems when using PEI of the same molecular weight. The results obtained for fibroblasts and HeLa cells were very similar (Figure 13A,B, respectively), evidencing a pattern independent of the cell type.

Figure 13C showed ND1 content produced from the transfection of HeLa cells with peptide/pND1 complexes. After 48 h of transfection, and similar to polymeric systems, peptide complexes led to high ND1 levels when compared to control cells (**** $p < 0.0001$). Among peptide complexes, the CpMTP peptide was found to induce the production of the lowest amount of ND1. Cells transfected with MTS-WRAP1, MTS-WRAP5, and MTS-(KH)$_9$ systems produced protein levels 3 to 4 times higher than non-transfected cells.

Comparing PEI complexes with peptide-based systems, we found that peptides led to ND1 concentrations above 400 ng/mL (for MTS-WRAP5/pDN1 and MTS-(KH)9/pND1 complexes), while complexes based on PEI did not reach this protein concentration, even when using 10 times higher N/P ratios. Therefore, peptide/pND1 complexes seemed to be more efficient in both gene and protein expression in mitochondria than the polymeric complexes. This assumption also correlated well with previously obtained data regarding the great efficacy of peptide-based complexes for cellular uptake and mitochondrial targeting [52], as well as with the above reported results of GFP gene expression.

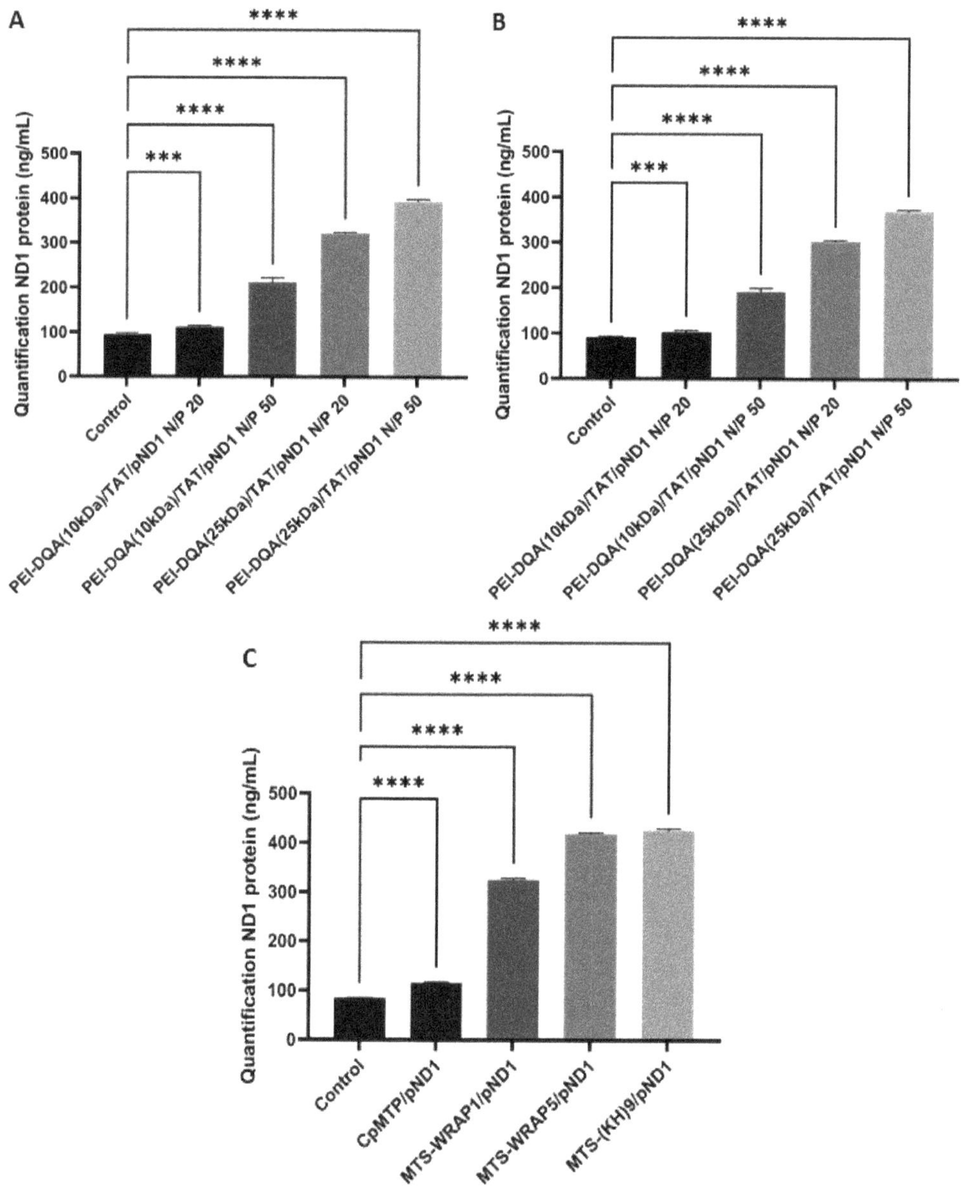

Figure 13. Quantification of ND1 protein 48 h after transfection of fibroblasts (**A**) and HeLa cells (**B**) with PEI–DQA (10 kDa)/TAT/pND1, N/P = 20:2:1; PEI–DQA (10 kDa)/TAT/pND1, N/P = 50:2:1; PEI–DQA (25 kDa)/TAT/pND1, N/P = 20:2:1; and PEI–DQA (25 kDa)/TAT/pND1, N/P = 50:2:1; and 48 h after transfection of HeLa cells (**C**) with CpMTP/pND1, MTS-WRAP1/pND1, MTS-WRAP5/pND1, and MTS-$(KH)_9$/pND1—all complexes prepared at an N/P ratio of 5:1. Non-transfected cells were used as controls. Data were analyzed by one-way ANOVA. (*** $p < 0.001$ or **** $p < 0.0001$).

3.8. Integrity of Mitochondria

After evidence of ND1 protein production in mitochondria was obtained with the reported delivery systems, we were interested to find out whether the transfection with complexes interfered with the normal mitochondrial function—especially with regard to ATP production. To investigate this issue, measurements of ATP in isolated mitochondria of HeLa cells were performed after transfection with peptide/pND1 complexes, as these complexes offered higher performance as pND1 delivery systems to mitochondria in comparison with the complexes based on PEI–DQA. A luminescent ATP detection kit was used (experimental details are available in the SM). Figure S6 (see the SM) shows the luminescence levels found in the mitochondria of cancer cells after transfection with peptide complexes. Although there was a statistically significant difference in luminescence levels in relation to control cells, peptide complexes such as the ones based on MTS-$(KH)_9$ and CpMTP allowed us to maintain the production of high ATP content. This preliminary study demonstrated that, at least, these complexes and the transfection processes they mediated did not greatly interfere with the normal performance of the mitochondria. After transfection, the mitochondria of HeLa cells were able to produce ATP. It seemed that transfection with these peptide/pND1 complexes was an innocuous process for the normal mitochondrial function and, thus, mitochondrial integrity seemed to be preserved. In contrast, the delivery system based on MTS-WRAP1 exhibited a major decrease in ATP levels, in comparison with controls (**** $p < 0.0001$) (Figure S6). The comparison between the MTS-WRAP1/pND1 systems with the other peptide complexes was also statistically significant (**** $p < 0.0001$).

At this stage, we are therefore aware that in order to deeply study the efficiency of the proposed peptide/pND1 complexes for mitochondrial gene therapy, the design/conception of ND1-mutated disease models is imperative. This would bring a realistic platform to further evaluate the capacity of peptide-based delivery vectors for long-term ND1 transgene expression in the mitochondria.

4. Conclusions

Mitochondrial gene therapy has emerged as a potential therapeutic strategy for mitochondrial diseases originating from mtDNA mutations. To make this therapy feasible and clinically viable, the design and conception of a mitochondria-targeted gene delivery system is crucial. To this end, we report in this work the development of novel polymeric complexes able to load/encapsulate a mitochondrial gene, to target mitochondria, and to promote gene delivery and protein expression in this organelle. It was found that their physicochemical properties and cytotoxic profile can be optimized by varying the N/P ratio and PEI molecular weight parameters. In addition to these PEI-based complexes, CPPs with mitochondrial targeting specificity, deeply studied in a previous work for pND1 encapsulation and delivery [18], were also evaluated for mitochondrial gene delivery and protein expression. It was found that both polymeric and peptide/pND1 complexes promoted efficient transfection, with consequent gene expression and ND1 protein production in the mitochondria. Therefore, both nano-platforms can be further explored in the quest for a suitable gene delivery system to mediate mitochondrial gene therapy. Moreover, a comparison between polymeric and peptide/pND1 complexes revealed that the peptide-based ones—mainly due to their greater ability for pND1 complexation—displayed superior performance in terms of cellular uptake, gene delivery, and protein expression. Collectively, our results bring significant and relevant knowledge, instigating progress towards mitochondrial transfection mediated by pDNA complexes, as a powerful tool to fight against mitochondrial DNA diseases.

Supplementary Materials: The following supporting information can be downloaded at https://www.mdpi.com/article/10.3390/pharmaceutics14040757/s1, Table S1: GPC data of synthesized polymers; Table S2: Average zeta potential, mean size, and PdI values for PEI–DQA (10 kDa or 25 kDa)/TAT/pND1, MTS-CPP/pND1, and CpMTP/pND1 complexes; Figure S1: FTIR spectra of PEI–DQA (10 kDa and 25 kDa) and dequalinium chloride; Figure S2: GPC thermogram of PEI (10 kDa), PEI (10 kDa)–SA, PEI (10 kDa)–DQA, PEI (25 kDa), PEI (25 kDa)–SA, and PEI (25 kDa)–DQA; Figure S3: CMC graph of PEI–DQA (25 kDa) and PEI–DQA (10 kDa); Figure S4: Confocal microscopy study of the transfection of HeLa cells with PEI–DQA (10 kDa)/TAT/pND1 complexes at an N/P ratio of 20:2:1; Figure S5: Confocal microscopy study of the transfection of HeLa cells with PEI–DQA (25 kDa)/TAT/pND1 complexes at an N/P ratio of 20:2:1; Figure S6: Detection of ATP in the mitochondria of HeLa cells after transfection mediated by MTS-WRAP1/pND1, MTS-WRAP5/pND1, MTS-$(KH)_9$/pND1, or CpMTP/pND1 complexes at an N/P ratio of 5.

Author Contributions: Conceptualization, R.F., S.B., P.B., Â.S. and D.C.; methodology, R.F., M.P., S.B., E.V., P.B., Â.S. and D.C.; software, R.F., M.P., S.B., P.B., Â.S. and D.C.; validation, S.B., E.V., P.B., Â.S. and D.C.; formal analysis, S.B., E.V., P.B., Â.S. and D.C.; investigation, R.F., M.P., S.B., E.V., P.B., Â.S. and D.C.; resources, S.B., E.V., P.B., Â.S. and D.C.; data curation, R.F., M.P., S.B., E.V., P.B., Â.S. and D.C.; writing—original draft preparation, R.F., S.B. and D.C.; writing—review and editing, R.F., S.B., E.V., P.B., Â.S. and D.C.; visualization, R.F., S.B., E.V., P.B., Â.S. and D.C.; supervision, S.B., P.B., Â.S. and D.C.; project administration, S.B., P.B., Â.S. and D.C.; funding acquisition, S.B., E.V., P.B., Â.S. and D.C. All authors have read and agreed to the published version of the manuscript.

Funding: This research was funded by the Fundação para a Ciência e a Tecnologia (FCT) program contract IF/01459/2015 supported by the European Social Fund (ESF) and the Programa Operacional Potencial Humano, and R.F. (SFRH/BD/148393/2019) was supported by the ESF through the Programa Operacional Regional Centro. This work was developed within the scope of the CICS-UBI projects UIDB/00709/2020 and UIDP/00709/2020, financed by national funds through the Portuguese Foundation for Science and Technology/MCTES. The microscopy facility used in the the development of this work is part of the Portuguese Platform of BioImaging (PPBI), and is partially supported by the project POCI-01-0145-FEDER-022122. P.B. and E.V. were supported by the "Fondation ARC pour la Recherche sur le Cancer" and the "Centre National de la Recherche Scientifique" (CNRS).

Institutional Review Board Statement: Not applicable.

Informed Consent Statement: Not applicable.

Data Availability Statement: Not applicable.

Acknowledgments: D.C. acknowledges the Fundação para a Ciência e a Tecnologia (FCT) program contract IF/01459/2015, supported by the Fundo Social Europeu and the Programa Operacional Potencial Humano; R.F. (SFRH/BD/148393/2019) acknowledges his doctoral fellowship from the FCT, co-financed by the European Social Fund (ESF) through the Programa Operacional Regional Centro. Peptide synthesis was performed with the excellent technical support of the Synbio3 platform, which is supported by the GIS IBISA/ITMO Cancer (Plan Cancer 2014–2019) and Chimie Balard Cirimat Carnot Institute.

Conflicts of Interest: The authors declare no conflict of interest.

References

1. Herst, P.M.; Rowe, M.R.; Carson, G.M.; Berridge, M.V. Functional Mitochondria in Health and Disease. *Front. Endocrinol.* **2017**, *8*, 296. [CrossRef] [PubMed]
2. Filograna, R.; Mennuni, M.; Alsina, D.; Larsson, N. Mitochondrial DNA copy number in human disease: The more the better? *FEBS Lett.* **2021**, *595*, 976–1002. [CrossRef] [PubMed]
3. Zhao, D.; Hong, D.; Zhang, W.; Yao, S.; Qi, X.; Lv, H.; Zheng, R.; Feng, L.; Huang, Y.; Yuan, Y.; et al. Mutations in mitochondrially encoded complex I enzyme as the second common cause in a cohort of Chinese patients with mitochondrial myopathy, encephalopathy, lactic acidosis and stroke-like episodes. *J. Hum. Genet.* **2011**, *56*, 759–764. [CrossRef] [PubMed]
4. Hertweck, K.; Dasgupta, S. The Landscape of mtDNA Modifications in Cancer: A Tale of Two Cities. *Front. Oncol.* **2017**, *7*, 262. [CrossRef] [PubMed]
5. Milenkovic, D.; Blaza, J.; Larsson, N.-G.; Hirst, J. The Enigma of the Respiratory Chain Supercomplex. *Cell Metab.* **2017**, *25*, 765–776. [CrossRef]
6. Filippi, M.-D.; Ghaffari, S. Mitochondria in the maintenance of hematopoietic stem cells: New perspectives and opportunities. *Blood* **2019**, *133*, 1943–1952. [CrossRef]
7. Mehta, M.; Weinberg, S.; Chandel, M.M.M.S.E.W.N.S. Mitochondrial control of immunity: Beyond ATP. *Nat. Rev. Immunol.* **2017**, *17*, 608–620. [CrossRef] [PubMed]

8. Javadov, S.; Kozlov, A.V.; Camara, A.K.S. Mitochondria in Health and Diseases. *Cells* **2020**, *9*, 1177. [CrossRef] [PubMed]
9. Popov, L. Mitochondrial biogenesis: An update. *J. Cell. Mol. Med.* **2020**, *24*, 4892–4899. [CrossRef]
10. Tiwari, M.; Prasad, S.; Tripathi, A.; Pandey, A.N.; Ali, I.; Singh, A.K.; Shrivastav, T.G.; Chaube, S.K. Apoptosis in mammalian oocytes: A review. *Apoptosis* **2015**, *20*, 1019–1025. [CrossRef] [PubMed]
11. Bock, F.J.; Tait, S.W.G. Mitochondria as multifaceted regulators of cell death. *Nat. Rev. Mol. Cell Biol.* **2020**, *21*, 85–100. [CrossRef] [PubMed]
12. Merino, D.; Kelly, G.L.; Lessene, G.; Wei, A.H.; Roberts, A.W.; Strasser, A. BH3-Mimetic Drugs: Blazing the Trail for New Cancer Medicines. *Cancer Cell* **2018**, *34*, 879–891. [CrossRef] [PubMed]
13. Roberts, A.W.; Davids, M.S.; Pagel, J.M.; Kahl, B.S.; Puvvada, S.D.; Gerecitano, J.F.; Kipps, T.J.; Anderson, M.A.; Brown, J.R.; Gressick, L.; et al. Targeting BCL2 with Venetoclax in Relapsed Chronic Lymphocytic Leukemia. *N. Engl. J. Med.* **2016**, *374*, 311–322. [CrossRef] [PubMed]
14. El-Hattab, A.W.; Zarante, A.M.; Almannai, M.; Scaglia, F. Therapies for mitochondrial diseases and current clinical trials. *Mol. Genet. Metab.* **2017**, *122*, 1–9. [CrossRef]
15. Pfeffer, G.; Majamaa, K.; Turnbull, D.M.; Thorburn, D.; Chinnery, P.F. Treatment for mitochondrial disorders. *Cochrane Database Syst. Rev.* **2012**, *2012*, CD004426. [CrossRef] [PubMed]
16. Siva, M.A.; Mahalakshmi, R.; Bhakta-Guha, D.; Guha, G. Gene therapy for the mitochondrial genome: Purging mutations, pacifying ailments. *Mitochondrion* **2019**, *46*, 195–208. [CrossRef] [PubMed]
17. Coutinho, E.; Batista, C.; Sousa, F.; Queiroz, J.; Costa, D. Mitochondrial Gene Therapy: Advances in Mitochondrial Gene Cloning, Plasmid Production, and Nanosystems Targeted to Mitochondria. *Mol. Pharm.* **2017**, *14*, 626–638. [CrossRef] [PubMed]
18. Faria, R.; Vivés, E.; Boisguerin, P.; Sousa, A.; Costa, D. Development of Peptide-Based Nanoparticles for Mitochondrial Plasmid DNA Delivery. *Polymers* **2021**, *13*, 1836. [CrossRef] [PubMed]
19. Yasuzaki, Y.; Yamada, Y.; Ishikawa, T.; Harashima, H. Validation of Mitochondrial Gene Delivery in Liver and Skeletal Muscle via Hydrodynamic Injection Using an Artificial Mitochondrial Reporter DNA Vector. *Mol. Pharm.* **2015**, *12*, 4311–4320. [CrossRef] [PubMed]
20. Vercauteren, D.; Rejman, J.; Martens, T.F.; Demeester, J.; De Smedt, S.C.; Braeckmans, K. On the cellular processing of non-viral nanomedicines for nucleic acid delivery: Mechanisms and methods. *J. Control. Release* **2012**, *161*, 566–581. [CrossRef] [PubMed]
21. Hirano, M.; Emmanuele, V.; Quinzii, C.M. Emerging therapies for mitochondrial diseases. *Essays Biochem.* **2018**, *62*, 467–481. [CrossRef] [PubMed]
22. Neves, A.R.; Sousa, A.; Faria, R.; Albuquerque, T.; Queiroz, J.; Costa, D. Cancer gene therapy mediated by RALA/plasmid DNA vectors: Nitrogen to phosphate groups ratio (N/P) as a tool for tunable transfection efficiency and apoptosis. *Colloids Surf. B Biointerfaces* **2020**, *185*, 110610. [CrossRef]
23. Gómez-Aguado, I.; Rodríguez-Castejón, J.; Vicente-Pascual, M.; Rodríguez-Gascón, A.; Del Pozo-Rodríguez, A.; Aspiazu, M. Ángeles, S. Nucleic Acid Delivery by Solid Lipid Nanoparticles Containing Switchable Lipids: Plasmid DNA vs. Messenger RNA. *Molecules* **2020**, *25*, 5995. [CrossRef] [PubMed]
24. Muhammad, K.; Zhao, J.; Gao, B.; Feng, Y. Polymeric nano-carriers for on-demand delivery of genes via specific responses to stimuli. *J. Mater. Chem. B* **2020**, *8*, 9621–9641. [CrossRef] [PubMed]
25. Wang, H.; Ding, S.; Zhang, Z.; Wang, L.; You, Y. Cationic micelle: A promising nanocarrier for gene delivery with high transfection efficiency. *J. Gene Med.* **2019**, *21*, e3101. [CrossRef] [PubMed]
26. Costa, D.; Albuqerque, T.; Queiroz, J.; Valente, A. A co-delivery platform based on plasmid DNA peptide-surfactant complexes: Formation, characterization and release behavior. *Colloids Surf. B Biointerfaces* **2019**, *178*, 430–438. [CrossRef] [PubMed]
27. Albuquerque, T.; Faria, R.; Sousa, Â.; Neves, A.R.; Queiroz, J.A.; Costa, D. Polymer-peptide ternary systems as a tool to improve the properties of plasmid DNA vectors in gene delivery. *J. Mol. Liq.* **2020**, *309*, 113157. [CrossRef]
28. Neves, A.R.; Albuquerque, T.; Faria, R.; Paul, M.; Biswas, S.; Sousa, Â.; Costa, D. Development of Tailor-Made Dendrimer Ternary Complexes for Drug/Gene Co-Delivery in Cancer. *Pharmaceutics* **2021**, *13*, 1256. [CrossRef]
29. Falanga, A.; Lombardi, L.; Galdiero, E.; Del Genio, V.; Galdiero, S. The world of cell penetrating: The future of medical applications. *Future Med. Chem.* **2020**, *12*, 1431–1446. [CrossRef]
30. Deshayes, S.; Konate, K.; Dussot, M.; Chavey, B.; Vaissière, A.; Van, T.N.N.; Aldrian, G.; Padari, K.; Pooga, M.; Vivès, E.; et al. Deciphering the internalization mechanism of WRAP:siRNA nanoparticles. *Biochim. Biophys. Acta Biomembr.* **2020**, *1862*, 183252. [CrossRef]
31. Sousa, Â.; Almeida, A.M.; Faria, R.; Konate, K.; Boisguerin, P.; Queiroz, J.; Costa, D. Optimization of peptide-plasmid DNA vectors formulation for gene delivery in cancer therapy exploring design of experiments. *Colloids Surf. B Biointerfaces* **2019**, *183*, 110417. [CrossRef] [PubMed]
32. Konate, K.; Dussot, M.; Aldrian, G.; Vaissière, A.; Viguier, V.; Neira, I.F.; Couillaud, F.; Vivès, E.; Boisguerin, P.; Deshayes, S. Peptide-Based Nanoparticles to Rapidly and Efficiently "Wrap 'n Roll" siRNA into Cells. *Bioconjug. Chem.* **2019**, *30*, 592–603. [CrossRef] [PubMed]
33. Khan, M.M.; Filipczak, N.; Torchilin, V.P. Cell penetrating peptides: A versatile vector for co-delivery of drug and genes in cancer. *J. Control. Release* **2021**, *330*, 1220–1228. [CrossRef] [PubMed]
34. Gessner, I.; Neundorf, I. Nanoparticles Modified with Cell-Penetrating Peptides: Conjugation Mechanisms, Physicochemical Properties, and Application in Cancer Diagnosis and Therapy. *Int. J. Mol. Sci.* **2020**, *21*, 2536. [CrossRef] [PubMed]

35. Derakhshankhah, H.; Jafari, S. Cell penetrating peptides: A concise review with emphasis on biomedical applications. *Biomed. Pharmacother.* **2018**, *108*, 1090–1096. [CrossRef] [PubMed]
36. Ruseska, I.; Zimmer, A. Internalization mechanisms of cell-penetrating peptides. *Beilstein J. Nanotechnol.* **2020**, *11*, 101–123. [CrossRef] [PubMed]
37. Bakar, L.M.; Abdullah, M.Z.; Doolaanea, A.A.; Ichwan, S.J.A. PLGA-Chitosan nanoparticle-mediated gene delivery for oral cancer treatment: A brief review. *J. Phys. Conf. Ser.* **2017**, *884*, 012117. [CrossRef]
38. Lellouche, E.; Locatelli, E.; Israel, L.L.; Naddaka, M.; Kurlander, E.; Michaeli, S.; Lellouche, J.-P.; Franchini, M.C. Maghemite-containing PLGA-PEG-based polymeric nanoparticles for siRNA delivery: Toxicity and silencing evaluation. *RSC Adv.* **2017**, *7*, 26912–26920. [CrossRef]
39. Piperno, A.; Sciortino, M.T.; Giusto, E.; Montesi, M.; Panseri, S.; Scala, A. Recent Advances and Challenges in Gene Delivery Mediated by Polyester-Based Nanoparticles. *Int. J. Nanomed.* **2021**, *16*, 5981–6002. [CrossRef] [PubMed]
40. Zhang, F.; Liao, X.; Zhang, G.; Mu, C. Dynamical Analysis of the Generalized Lorenz Systems. *J. Dyn. Control Syst.* **2017**, *23*, 349–362. [CrossRef]
41. Costa, D.; Briscoe, W.H.; Queiroz, J. Polyethylenimine coated plasmid DNA-surfactant complexes as potential gene delivery systems. *Colloids Surf. B Biointerfaces* **2015**, *133*, 156–163. [CrossRef]
42. Faria, R.; Sousa, Â.; Neves, A.R.; Queiroz, J.A.; Costa, D. Methotrexate-plasmid DNA polyplexes for cancer therapy: Characterization, cancer cell targeting ability and tuned in vitro transfection. *J. Mol. Liq.* **2019**, *292*, 292. [CrossRef]
43. Cheraghi, R.; Alipour, M.; Nazari, M.; Hosseinkhani, S. Optimization of conditions for gene delivery system based on PEI. *Nanomedicine* **2017**, *4*, 8–16.
44. Benjaminsen, R.V.; Mattebjerg, M.A.; Henriksen, J.R.; Moghimi, S.M.; Andresen, T.L. The Possible "Proton Sponge" Effect of Polyethylenimine (PEI) Does Not Include Change in Lysosomal pH. *Mol. Ther.* **2013**, *21*, 149–157. [CrossRef] [PubMed]
45. Kaltimbacher, V.; Bonnet, C.; Lecoeuvre, G.; Forster, V.; Sahel, J.-A.; Corral-Debrinski, M. mRNA localization to the mitochondrial surface allows the efficient translocation inside the organelle of a nuclear recoded ATP6 protein. *RNA* **2006**, *12*, 1408–1417. [CrossRef]
46. Bonnet, C.; Kaltimbacher, V.; Ellouze, S.; Augustin, S.; Bénit, P.; Forster, V.; Rustin, P.; Sahel, J.-A.; Corral-Debrinski, M. Allotopic mRNA Localization to the Mitochondrial Surface Rescues Respiratory Chain Defects in Fibroblasts Harboring Mitochondrial DNA Mutations Affecting Complex I or V Subunits. *Rejuvenation Res.* **2007**, *10*, 127–144. [CrossRef]
47. Bonnet, C.; Augustin, S.; Ellouze, S.; Bénit, P.; Bouaita, A.; Rustin, P.; Sahel, J.-A.; Corral-Debrinski, M. The optimized allotopic expression of ND1 or ND4 genes restores respiratory chain complex I activity in fibroblasts harboring mutations in these genes. *Biochim. Biophys. Acta* **2008**, *1783*, 1707–1717. [CrossRef] [PubMed]
48. Ellouze, S.; Augustin, S.; Bouaita, A.; Bonnet, C.; Simonutti, M.; Forster, V.; Picaud, S.; Sahel, J.-A.; Corral-Debrinski, M. Optimized Allotopic Expression of the Human Mitochondrial ND4 Prevents Blindness in a Rat Model of Mitochondrial Dysfunction. *Am. J. Hum. Genet.* **2008**, *83*, 373–387. [CrossRef]
49. Korake, S.; Gajbhiye, K.R. Dequalinium-Derived Nanoconstructs: A Promising Vehicle for Mitochondrial Targeting. *Curr. Drug Deliv.* **2021**, *18*, 1056–1063. [CrossRef]
50. Chuah, J.-A.; Matsugami, A.; Hayashi, F.; Numata, K. Self-Assembled Peptide-Based System for Mitochondrial-Targeted Gene Delivery: Functional and Structural Insights. *Biomacromolecules* **2016**, *17*, 3547–3557. [CrossRef] [PubMed]
51. Lyrawati, D.; Trounson, A.; Cram, D. Expression of GFP in the Mitochondrial Compartment Using DQAsome-Mediated Delivery of an Artificial Mini-mitochondrial Genome. *Pharm. Res.* **2011**, *28*, 2848–2862. [CrossRef] [PubMed]
52. Weissig, V. DQAsomes as the Prototype of Mitochondria-Targeted Pharmaceutical Nanocarriers: Preparation, Characterization, and Use. *Methods Mol. Biol.* **2015**, *1265*, 1–11. [CrossRef]
53. Mallick, S.; Song, S.J.; Bae, Y.; Choi, J.S. Self-assembled nanoparticles composed of glycol chitosan-dequalinium for mitochondria-targeted drug delivery. *Int. J. Biol. Macromol.* **2019**, *132*, 451–460. [CrossRef] [PubMed]
54. Weissig, V.; Torchilin, V.P. Towards Mitochondrial Gene Therapy: DQAsomes as a Strategy. *J. Drug Target.* **2001**, *9*, 1–13. [CrossRef] [PubMed]
55. Zupančič, Š.; Kocbek, P.; Zariwala, M.G.; Renshaw, D.; Gul, M.O.; Elsaid, Z.; Taylor, K.M.G.; Somavarapu, S. Design and Development of Novel Mitochondrial Targeted Nanocarriers, DQAsomes for Curcumin Inhalation. *Mol. Pharm.* **2014**, *11*, 2334–2345. [CrossRef] [PubMed]
56. Wang, H.; Yin, H.; Yan, F.; Sun, M.; Du, L.; Peng, W.; Li, Q.; Feng, Y.; Zhou, Y. Folate-mediated mitochondrial targeting with doxorubicin-polyrotaxane nanoparticles overcomes multidrug resistance. *Oncotarget* **2014**, *6*, 2827–2842. [CrossRef]
57. Wang, Y.; Khan, A.; Liu, Y.; Feng, J.; Dai, L.; Wang, G.; Alam, N.; Tong, L.; Ni, Y. Chitosan oligosaccharide-based dual pH responsive nano-micelles for targeted delivery of hydrophobic drugs. *Carbohydr. Polym.* **2019**, *223*, 115061. [CrossRef]
58. Sun, Y.; Yang, Q.; Xia, X.; Li, X.; Ruan, W.; Zheng, M.; Zou, Y.; Shi, B. Polymeric Nanoparticles for Mitochondria Targeting Mediated Robust Cancer Therapy. *Front. Bioeng. Biotechnol.* **2021**, *9*, 755727. [CrossRef] [PubMed]
59. Faria, R.; Albuquerque, T.; Neves, A.R.; Bhatt, H.; Biswas, S.; Cardoso, A.M.; de Lima, M.C.P.; Jurado, A.S.; Costa, D. Physicochemical characterization and targeting performance of triphenylphosphonium nano-polyplexes. *J. Mol. Liq.* **2020**, *316*, 113873. [CrossRef]

60. Danaei, M.; Dehghankhold, M.; Ataei, S.; Hasanzadeh Davarani, F.; Javanmard, R.; Dokhani, A.; Khorasani, S.; Mozafari, M.R. Impact of Particle Size and Polydispersity Index on the Clinical Applications of Lipidic Nanocarrier Systems. *Pharmaceutics* **2018**, *10*, 57. [CrossRef] [PubMed]

pharmaceutics

MDPI

Article

α/β-Peptides as Nanomolar Triggers of Lipid Raft-Mediated Endocytosis through GM1 Ganglioside Recognition

Anasztázia Hetényi [1,†], Enikő Szabó [2,†], Norbert Imre [1], Kaushik Nath Bhaumik [1], Attila Tököli [1], Tamás Füzesi [1], Réka Hollandi [3], Peter Horvath [3], Ágnes Czibula [2,*], Éva Monostori [2], Mária A. Deli [4] and Tamás A. Martinek [1,*]

1. Department of Medical Chemistry, University of Szeged, Dóm Tér 8, 6720 Szeged, Hungary; hetenyi.anasztazia@med.u-szeged.hu (A.H.); imrenorbert21@gmail.com (N.I.); kaushik.nb@pharm.u-szeged.hu (K.N.B.); attila.tokoli@gmail.com (A.T.); tamasfuzesi99@gmail.com (T.F.)
2. Institute of Genetics, Biological Research Centre, Temesvári krt. 62, 6726 Szeged, Hungary; szabo.eniko@brc.hu (E.S.); monostori.eva@brc.hu (É.M.)
3. Institute of Biophysics, Biological Research Centre, Temesvári krt. 62, 6726 Szeged, Hungary; hollandi.reka@brc.hu (R.H.); horvath.peter@brc.hu (P.H.)
4. Synthetic and Systems Biology Unit, Biological Research Centre, Temesvári krt. 62, 6726 Szeged, Hungary; deli@brc.hu

* Correspondence: czibula.agnes@brc.hu (Á.C.); martinek.tamas@med.u-szeged.hu (T.A.M.)

† These authors contributed equally to this work.

Abstract: Cell delivery of therapeutic macromolecules and nanoparticles is a critical drug development challenge. Translocation through lipid raft-mediated endocytic mechanisms is being sought, as it can avoid rapid lysosomal degradation. Here, we present a set of short α/β-peptide tags with high affinity to the lipid raft-associated ganglioside GM1. These sequences induce effective internalization of the attached immunoglobulin cargo. The structural requirements of the GM1-peptide interaction are presented, and the importance of the membrane components are shown. The results contribute to the development of a receptor-based cell delivery platform.

Keywords: cell delivery; glycan recognition; alpha-beta peptide; endocytosis; immunoglobulin

Citation: Hetényi, A.; Szabó, E.; Imre, N.; Bhaumik, K.N.; Tököli, A.; Füzesi, T.; Hollandi, R.; Horvath, P.; Czibula, Á.; Monostori, É.; et al. α/β-Peptides as Nanomolar Triggers of Lipid Raft-Mediated Endocytosis through GM1 Ganglioside Recognition. *Pharmaceutics* **2022**, *14*, 580. https://doi.org/10.3390/pharmaceutics14030580

Academic Editors: Prisca Boisguérin and Sébastien Deshayes

Received: 27 January 2022
Accepted: 4 March 2022
Published: 7 March 2022

Copyright: © 2022 by the authors. Licensee MDPI, Basel, Switzerland. This article is an open access article distributed under the terms and conditions of the Creative Commons Attribution (CC BY) license (https://creativecommons.org/licenses/by/4.0/).

1. Introduction

The efficient translocation of macromolecular cargoes and nanoparticles through the mammalian cell membrane is an important challenge in modern drug development [1–3]. Endocytic routes avoiding rapid lysosomal degradation are of interest because these facilitate endosomal escape and delivery to organelles or cytosol with limited cargo decomposition [4,5]. Triggering internalization at the lipid rafts is an emerging approach because this gateway generates endosomes that rarely fuse with lysosomes [6–8]. Lipid raft-mediated endocytosis can facilitate endosomal escape for the internalized cargoes [9,10]. Often, an abundance of different ganglioside molecules decorates the extracellular surface of the lipid rafts, on which pattern-specific recognition induces the desired endocytosis [11,12]. Ganglioside GM1 is the major receptor for the Cholera toxin, which utilizes this pathway [13,14]. In a recent study, we showed that a pentapeptide tag (WYKYW) could bind ganglioside GM1 with high affinity and thereby internalize an IgG complex (580 kDa) at low nanomolar extracellular concentrations [15]. The endocytotic pathways were tested with inhibitor experiments, demonstrating that lipid raft-mediated internalization is the major pathway for this process. The involvement of GM1 in the mechanism was supported by Cholera toxin co-localization and an inhibition experiment with the GM1 binder lectin galectin-1. Cell lines expressing different amounts of cell surface GM1 displayed internalization efficiency in accordance with their GM1 content. Inducing endocytosis with WYKYW avoided lysosomes and retained the functional structure of the cargo, which justifies further investigations.

Using various foldamers as cell-penetrating agents can offer efficient translocation of cargoes while being less susceptible to hydrolysis and the immune system [16–20]. Specifically, incorporating β-amino acids can lead to tuned interactions with the target molecule, which is beneficial to bioavailability [21]. Prior work determined the required sugar moieties on gangliosides for high-affinity binding with the WYKYW sequence [15]. Removal of the sialic acid or the terminal disaccharide moieties caused substantial affinity loss. In this work, our goal was to explore the structural requirements of the WYKYW–GM1 ganglioside interaction for the peptidic partner to help rational designs for expanding this family of sequences for cell delivery. Establishing a pharmacophore model of the ganglioside–peptide interaction appeared feasible in a structure-affinity relationship approach. To meet this challenge, we aimed to determine the effects of various amino acid replacements in the parent sequence, including side-chain removal (Ala-scan), stereochemical inversion (D-scan), and backbone homologation (β^3-scan). In addition, the role of terminal tryptophan residues was investigated with the help of spectroscopic methods. The structure-affinity relationship analysis yielded improved unnatural peptides, efficiently internalizing large protein cargoes.

2. Materials and Methods

Peptide Synthesis and Purification: Peptide amides were synthesized with (7-azabenzotriazol-1-yl)tetramethyluronium hexafluorophosphate as a coupling agent (HATU) on TentaGel R RAM resin. (Sigma-Aldrich, Budapest, Hungary, product code: 86359) Coupling was carried out with a three times-equivalent excess of amino acid at room temperature for 3 h. The peptide cleavage from the resin was performed with TFA/water/D,L-dithiothreitol/triisopropylsilane (90:5:2.5:2.5), which was then precipitated in ice-cold diethyl ether. The resin was washed with acetic acid and water. The raw products were subsequently filtered and lyophilized. Peptide purification was performed using RP–HPLC with a C18 column. The HPLC eluents were 0.1% TFA in water and 0.1% TFA in ACN. Analytical RP–HPLC and ESI–MS measurements confirmed the purity of the peptides [22].

Isothermal Titration Calorimetry: ITC was performed using a MicroCal VP-ITC microcalorimeter at 35 °C. The monosialoganglioside GM1 was obtained from Biosynth-Carbosynth (Bratislava, Slovakia, product code: OG03918, $C_{73}H_{131}N_3O_{31}\cdot xNa$, HPLC purity > 95%). The n-dodecylphosphocholine (DPC) was obtained from Avanti (Alabaster, Alabama, product code: 850336, $C_{17}H_{38}NO_4P$, purity > 99%). In individual titrations, 15 μL of solution containing GM1:DPC 1:5 was injected into the ligand solution in the cell from a computer-controlled 300 μL microsyringe at time intervals of 300 s. The GM1:DPC micelle mixture was prepared in the same pH 7.2 phosphate buffer as the ligand in the cell. The concentration of the ligand in the cell was 15 μM. The concentration of GM1 in the syringe was 300 μM. In control experiments, GM1:DPC was titrated into the cell containing buffer without ligand. To rule out lipid clustering, control measurements were performed with pure GM1 in the syringe, which yielded a marked decrease in ΔH values, indicating the GM1:DPC interaction in the micelles (Figure S6). No aggregation of the micelles was observed after titration, which rules out any secondary inter-particle effects induced by ligand binding. We note that such an uncontrolled aggregation phenomenon would have been detrimental to the very sensitive ITC detection. Our finding is in agreement with the literature results that mixed micelles of GM1 and DPC are readily formed [23]. The experiments were repeated twice. The experimental data were fitted to one-binding-site or two-independent-site models with adjustable parameters of ΔH_{b1}, K_{d1}, n_1, and ΔH_{b2}, K_{d2}, n_2, respectively. Background subtraction and spline baseline correction was applied prior to the application of a generalized reduced gradient nonlinear least-squares procedure. The inert counter ions and residual solvent in the peptide, protein, and lipid samples normally caused residual heat during mixing, which was corrected with a constant term as an additional parameter in the model. Errors were estimated by using jack-knife resampling [22].

Tryptophan fluorescence blue-shift measurements: Fluorescence experiments were carried out at room temperature, in 20 mM PBS (pH 7.4) with a Hitachi F-2500 fluorescence spectrophotometer (PMT voltage: 700 V, response time: 0.08 s). Tryptophan was excited at the wavelength of 295 nm. Emission spectra were recorded in the range 300 to 400 nm. The excitation and emission bandwidths were set to 5 nm. Peptides were measured alone at a concentration of 2.5 μM and with the addition of either 250 μM DPC micelles or 50 μM GM1:250 μM DPC bicelles. Control measurements in the absence of the peptide were subtracted from the corresponding data [22].

Circular dichroism: Circular dichroism measurements were carried out with a Jasco J-815 CD Spectrometer. Spectra were recorded using a thermally jacketed 1 mm quartz cuvette in the wavelength range 260 to 195 nm. The scan speed was 50 nm min^{-1} with 5 accumulations. Peptide concentration was 200 μM in 10 mM PBS (pH 7.4). The effect of the binding on the CD curve was measured after adding 100:500 μM GM1:DPC bicelles. A Julabo water thermostat controlled the temperature with an equilibration time of 10 min at each temperature. Solvent baseline subtraction was applied [22].

Cell Culture: HeLa cells were cultured in advanced MEM (Gibco®/Invitrogen Corporation, New York, NY, USA) supplemented with 10% fetal bovine serum (FBS, PAN-Biotech, Aidenbach, Germany). Penicillin–streptomycin (100 U mL^{-1}, Gibco®/Invitrogen Corporation, New York, NY, USA) and 2 mM L-glutamine (Gibco®/Invitrogen Corporation, New York, NY, USA) was added to the medium. The cells were grown at 37 °C in a humidified incubator containing 5% CO_2 [22].

Preparation of Carrier–Protein Complexes: A solution of the biotinylated peptide, biotinylated monoclonal mouse anti-human [24] antibody and unlabelled Neutravidin (NA) (ThermoFisher Scientific, Watham, MA, USA, product code: 31000) was mixed at a molar ratio of 3:1:1. The secondary Alexa Fluor 647-conjugated F(ab')2-goat anti-mouse IgG (Gibco®/Invitrogen Corporation, New York, NY, USA, product code: 31006) was added to the solution subsequently at a molar ratio of 1:1 relative to the primary antibody. The dilution of this complex was set before adding to HeLa cells [22].

Live Confocal Laser Scanning Microscopy: HeLa cells were plated on six-chamber μ-Slides VI 0.4 (ibidi, Gräfelfing, Germany) for overnight culturing in MEM + 10% FBS at 1.25×10^4 cells per cm^2 (or 1.5×10^4 cells per channel). After washing with PBS, the cells were incubated at 37 °C with the complexes in MEM + 1% FBS medium for the required time lengths, which was followed by washing with PBS. The cells were stained with 100 ng mL^{-1} Hoechst 33342 (Sigma-Aldrich St. Louis, MO, USA) in MEM medium for 30 min at 37 °C. Cell membranes were visualized after a 5 min treatment with FITC-labelled WGA lectin at 0.2 μg mL^{-1} at room temperature. The cells were incubated in Leibovitz's L-15 medium (Gibco®/Invitrogen Corporation, New York, NY, USA) during microscopic analysis. Cell fluorescence was analyzed to observe the localization of the cargo, using a Leica SP5 AOBS confocal laser scanning microscope with a 405 nm UV diode (for Hoechst staining), a 488 nm argon laser line (for FITC staining), and a 633 HeNe laser line (for Alexa Fluor 647 staining). An appropriate spectral filter was used for each channel for the detection of the emissions [22].

Image Analysis: We used Mask R-CNN, a deep learning-based image segmentation platform for identifying cells and extracting their properties; U-Net, another deep learning approach; and CellProfiler software for feature extraction. Cell nuclei were identified on the basis of the Hoechst signal using a very heavily augmented training set of The Data Science Bowl 2018 competition. The augmentation was performed by learning image styles and generating synthetic images of similar types with Pix2Pix, a generative adversarial network (GAN) deep network. Subsequently, a Mask R-CNN network was trained, and individual nuclei were inferred. A similarly augmented image set was used to train a U-Net deep convolutional neural network, using the FITC-WGA lectin channel images and binary masks marking the cytoplasm as foreground. The trained U-Net network predicted the foreground pixels corresponding to cytoplasm. We approximated the cytoplasm with a watershed region propagation algorithm on the weighted sum image of the U-Net prediction and the FITC-WGA lectin channel. With the help of the detected objects (nucleus and cytoplasm) as masks, cellular features such as Alexa 647 intensity values, textural properties, and morphological descriptors were extracted. The integrated intensities of individual cells were used for the final statistical analysis [22].

Protease assays: Peptides were dissolved in 100 mM TRIS–HCl buffer (pH 8.0) containing 10 mM $CaCl_2$ at 100 mM concentration. Chymotrypsin (Sigma-Aldrich, Budapest, Hungary) stock solution was 0.5 mg/mL in the same buffer. A total 500 mL peptide solution was measured in a reaction vessel and 50 mL protease stock was added. Samples of 50 mL were taken after 2, 6, 25 and 60 min. The samples were diluted into 500 mL 5% TFA aliquots. The samples were then injected into HPLC–MS using a Phenomenex C18-XB Peptide column (250 × 4.6 mm, 3.6 mm). For the experiments with trypsin, the same method was used, except for the buffer, which did not contain any $CaCl_2$.

3. Results

First, we performed an Ala-scan on the WYKYW sequence, substituting each of the five amino acids one by one, and measured their affinity to GM1:DPC bicelles using isothermal titration calorimetry (ITC) [25]. Dissociation constants (Table 1, Figures S1–S3) revealed that any side chain removal from the original sequence was detrimental to binding. This finding suggests that WYKYW is a minimal sequence because all side chains were essential in stabilizing the GM1–WYKYW complex. Modification of the central Lys led to the largest affinity decrease, indicating the role of the cationic function. Removal of the neighboring Tyr residues had the second largest effect on the affinity. From the Ala-scan, we can conclude that the core tripeptidic sequence (YKY) has a dominant contribution to stabilizing the GM1–WYKYW complex.

Table 1. Binding affinities (K_D [nM]) of the WYKYW analogues to ganglioside GM1. The binding stoichiometry (n_1) was 0.5 in all cases. Superscripts indicate the corresponding peptide termini.

	^{N}W	Y	K	Y	W^{C}
original			23.8		
Ala-scan	n.f. [a]	5755	10,467	1694	1060
D-scan	881	892	4523	3243	3926
β-scan	4.3	60	332	40	86

[a] not fittable.

High-affinity WYKYW–GM1 interactions project a precise fit between the two partners, which requires a specific stereochemical pattern along the peptidic chain. We tested this with a systematic D-amino acid substitution. Again, we found that any configuration change resulted in a marked decrease in the affinities. The central Lys residue was the most sensitive. Interestingly, inversion of Tyr4 and Trp5 yielded lower affinities than the Ala replacement. This observation suggests that stereochemical inversion not only disrupted the interaction, but that the modified side-chain orientation exerted an extra destabilization effect. We concluded that no change in the amino acid configuration is tolerable, which strongly supports that the molecular recognition between GM1 and WYKYW requires a specific peptide geometry. To further test the sensitivity of the interaction to the backbone homologation, we carried out β^3-amino acid replacements in the sequence. Strikingly, the incorporation of the β^3-h-Trp at the N-terminal yielded improved affinity. The GM1–WYKYW complex displayed good tolerance to the backbone homologation, except for the central β^3-h-Lys residue. This pattern in the structure-affinity relationship corroborates that the central YKY motif and its fine geometry play crucial roles in ganglioside recognition. The protease resistance of the homologated sequences was also measured. As expected, the single β^3-amino acid replacements at the central residues provided increased stability against proteases (Table S1).

While the structure-affinity relationship study revealed the importance of the central YKY motif, the Trp residues at the termini contribute a large interaction surface, pointing to the role of hydrophobic interactions. We hypothesized that these residues immerse into the fatty acid region of the membrane [25] and thereby stabilize the WYKYW–GM1 interaction over the membrane surface. The Trp fluorescence maximum wavelength depends on the environment; relocation of the aromatic side chain from an aqueous solvent to a hydrophobic environment causes a blue shift [26]. The Trp fluorescence was measured in the presence of DPC micelles and GM1:dodecylphosphocholine (DPC) bicelles (Figure 1). WYKYW alone had an emission maximum of 350 nm. The addition of DPC micelles caused a minor blue shift and intensity increase. In contrast, GM1:DPC bicelles induced a blue shift of the emission maximum to 341 nm, accompanied by a marked intensity increase. This observation suggests that Trp moved to a hydrophobic environment in the membrane and became partially shielded from the water. The extent of the shift was consistent with a relative permittivity of 7, which suggests that the Trp side chains were neither in the aqueous medium nor completely buried in the hydrophobic interior near the headgroups of the amphipathic lipids [27]. This location is in good accordance with the requirement that the segment YKY is close to the carbohydrate moiety of GM1.

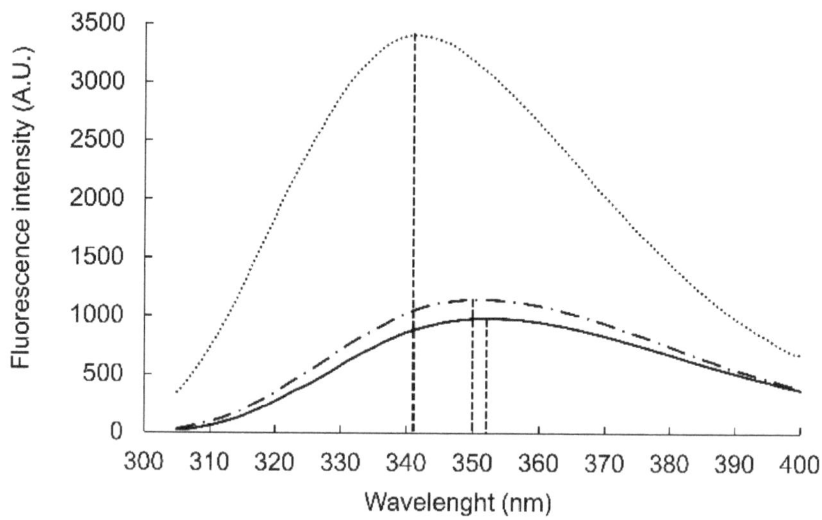

Figure 1. The blue-shift of the tryptophan fluorescence emission in interaction with GM1:DPC micelles. Emission spectra obtained for peptide WYKYW alone (black, 2.5 μM), WYKYW + DPC (DPC 250 μM, dash-dotted), and WYKYW + GM1:DPC (+GM1:DPC 50:250 μM, dotted).

The aromatic side chains dominated the circular dichroism spectrum, preventing direct conclusions about the backbone geometry. However, intensity loss at 198 nm was observed upon adding the GM1:DPC bicelles (Figure 2). To test the contribution to this region from the aromatic rings, WAKAW and AYKYA were synthesized and measured (Figure S4 long-dashed). Thus, the intensive peak at 198 nm could be assigned to Trp–Trp face-to-edge interactions in solution [28,29], which is lost in the GM1-bound form.

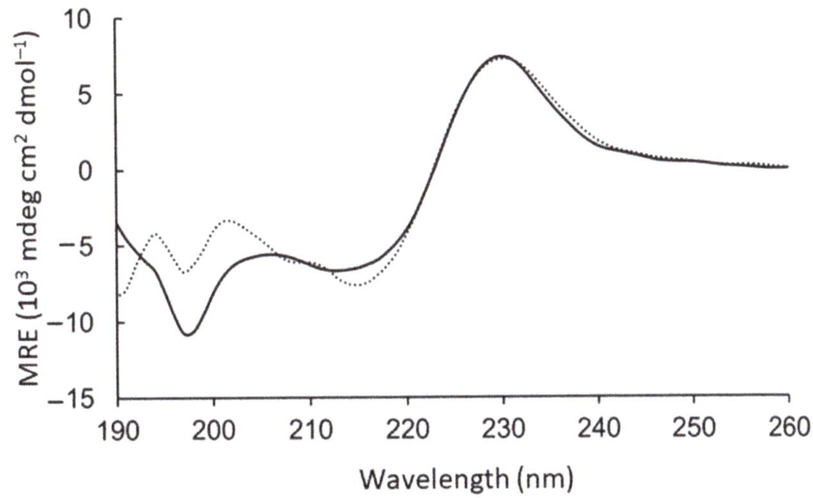

Figure 2. CD spectra measured for peptide WYKYW alone (black, 200 μM) and after adding 100:500 μM GM1:DPC bicelles (dotted).

As previously shown, binding did not occur with the pentasaccharide moiety of GM1 without the hydrophobic sphingosine and fatty acid parts in the membrane [15]. Removal of the sialic acid also strongly decreased affinity. Together with the Trp residues' membrane insertion, these findings suggested an interaction model for WYKYW–GM1 interactions (Figure 3).

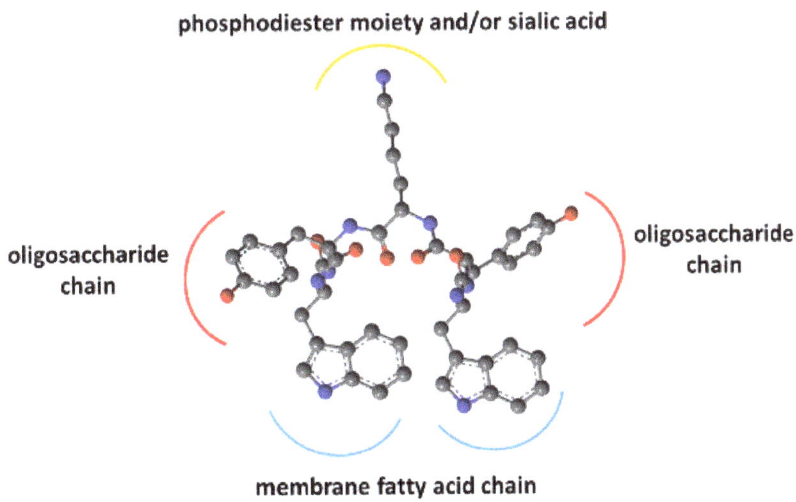

Figure 3. Pharmacophore hypothesis generated from the SAR and Trp-fluorescence data.

The model rationalizes the importance of the stereochemical configuration. It also explains the backbone homologation tolerance at the termini. The position of Trp side chains in the dynamic membrane is adaptive, whereas the side chain distances have to be more specific in the YKY segment to recognize the oligosaccharide moiety.

The nanomolar affinity to GM1 of the α/β-peptidic sequences projected their ability to trigger endocytosis. Therefore, we tested these carriers for protein internalization as previously described [30]. The peptides were attached to a Neutravidin hub through a biotinylated linker (Figure 4). The cargo was a protein complex containing a primary and a fluorescent secondary antibody. We chose penetratin as a linker sequence because it is unable to internalize the protein cargo by itself, but its cationic nature improves the solubility of the construct. Control experiments were run with the Neutravidin-antibody complex without carrier peptides.

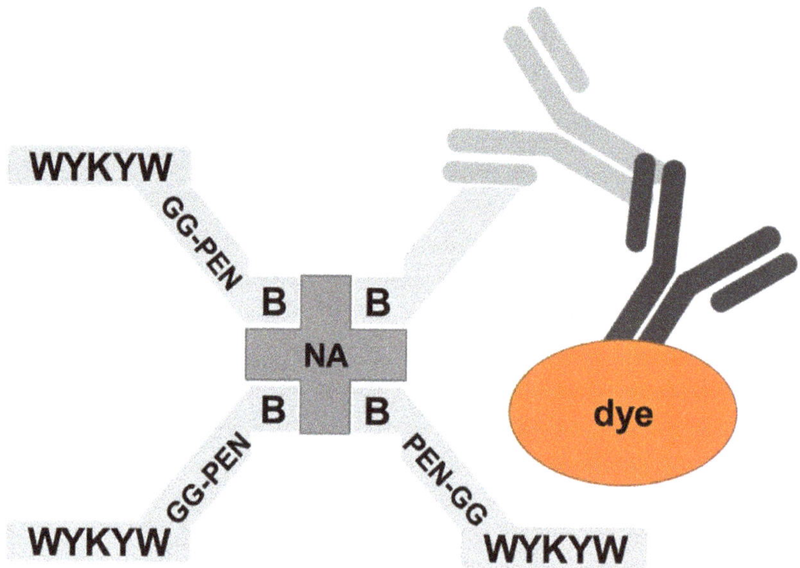

Figure 4. Schematic representation of the bottom-up designed modular carrier–hub–antibody cargo–secondary antibody–fluorescent-dye construct.

Their internalization efficiency was monitored with CLSM (Figure 5). HeLa cells were incubated with the carrier–cargo complexes for 1 or 4 h at an extracellular concentration of 80 nM. The α/β-peptidic tags were able to induce endocytosis. We carried out an AI-based quantitation with the different analogues (Figure 6). Strikingly, no correlation was found with the GM1 binding affinities measured for the single carrier sequences. We hypothesized that the multiple presentations of the carrier on the complex increases avidity, thereby facilitating the endocytosis. To this end, we prepared the NA(biotinyl-Penetratin-WYK$^{\beta}$YW)$_4$ constructs, where the K_D of the monomers were 332 nM. ITC measurements showed a K_D of 38 nM for the multivalent construct (Figure S5), explaining the efficient internalization. The stoichiometry of this interaction was 1:1 relative to the concentration of the hub Neutravidin, in contrast to the 2:1 carrier:GM1 ratio. We note that Neutravidin is a tetrameric protein that can display two carrier peptides simultaneously toward the membrane. These findings strongly support the avidity-increasing effect of the multivalent presentation of the carrier.

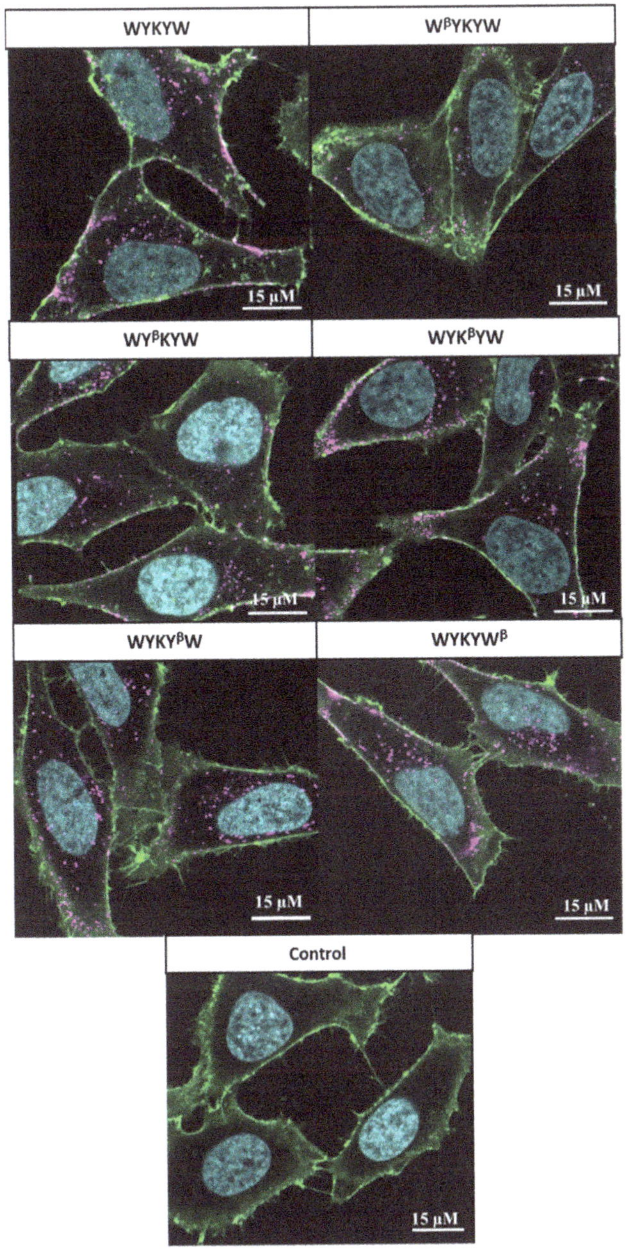

Figure 5. Delivery of the IgG cargo into HeLa cells, using the carrier WYKYW and its α,β-peptidic derivatives. The schematic representation of the carrier–cargo complex is given in Figure 4. Images are tagged by the carrier sequence measured (β denotes the corresponding $β^3$-amino acid). Alexa Fluor 647-conjugated secondary antibody is indicated in magenta; green staining defines cell membranes (WGA-FITC). Nuclei are indicated in cyan. The carrier–cargo complexes we applied at a concentration of 80 nM, and cells were incubated for 1 h. In the control measurement, cells were treated with IgG complex without the ganglioside-binding carrier peptides attached.

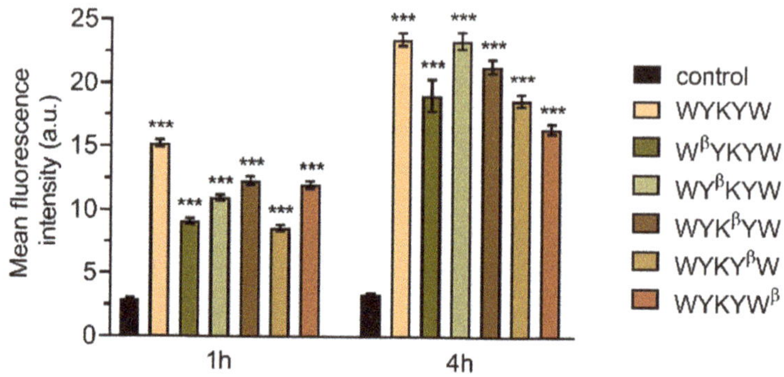

Figure 6. Artificial intelligence-aided quantitative analysis of the live CLSM images. HeLa cells were incubated for 1 or 4 h with the carrier–cargo complexes at 80 nM. At least 150 representative cells were then analyzed at each setup. The graph shows mean fluorescence intensity values ± standard error of the mean (SEM). Statistical analysis was performed using one-way analysis of variance (ANOVA) with Dunnett's multiple comparison test, where each sample was compared to the control sample of the matching time point. *** $p < 0.001$.

4. Discussion

In summary, we found a sequence-dependent affinity for the WYKYW derivatives to the GM1 ganglioside. Contribution to the binding of each residue was demonstrated. We found that the stereochemical configuration pattern is crucial. The central YKY segment recognizes the oligosaccharide moiety, while the terminal Trp residues stabilize the complex by partial membrane insertion. We identified backbone-homologated analogues, which have affinities comparable with the parent sequence. We showed that the α/β-peptidic sequences internalize the multivalent carrier–IgG cargo complex with good efficiency with the help of an avidity increase.

5. Patents

A patent application has been accepted overlapping with the findings presented in this manuscript: WO2020245617.

Supplementary Materials: The following supporting information can be downloaded at: https://www.mdpi.com/article/10.3390/pharmaceutics14030580/s1, Figure S1: ITC detection of interactions of GM1 with Ala derivatives, Figure S2: ITC detection of interactions of GM1 with d-derivatives, Figure S3: ITC detection of interactions of GM1 with β-derivatives, Figure S4: CD spectrum of peptides, Figure S5: ITC detection of interactions of GM1 with derivatives, Figure S6: ITC enthalpograms of WYKYW–GM1, Table S1: Half-lives of compounds in protease assay with chymotrypsin, Table S2: Statistical Analysis, and Peptide characterization data.

Author Contributions: A.H. designed experiments, carried out data analysis and ITC measurements. E.S., N.I. and T.F. performed cell delivery experiments. N.I. carried out CD and Trp-fluorescence measurements. K.N.B. and A.T. synthesized all studied materials. R.H. and P.H. carried out image analysis. É.M. and Á.C. designed and analyzed cell delivery experiments, M.A.D. and T.A.M. wrote the paper with input from all co-authors. All authors have read and agreed to the published version of the manuscript.

Funding: This research was funded by the National Research, Development and Innovation Office of Hungary, grant number GINOP-2.2.1-15-2016-00007, the Hungarian Ministry of Innovation and Technology, TUDFO/47138-1/2019-ITM, and the Hungarian National Brain Research Program (MTA-SE-NAP B-BIOMAG). T.A.M. acknowledges support from the Hungarian Academy of Sciences LENDULET-Foldamer, and NKFI K134754. P.H. acknowledges support from the Finnish TEKES FiDiPro Fellow Grant 40294/13, and the Hungarian Academy of Sciences LENDULET-Biomag.

Support by the Ministry of Innovation and Technology of Hungary through the National Research, Development and Innovation Fund (TKP2021-EGA-32) is acknowledged.

Institutional Review Board Statement: Not applicable.

Informed Consent Statement: Not applicable.

Data Availability Statement: All data supporting the findings of this study are available within the article and its supplemental information, or from the corresponding author upon reasonable request.

Acknowledgments: Viktor Farkas (Eötvös Lorand University, Budapest) is acknowledged for their valuable help in the evaluation of the CD curves.

Conflicts of Interest: The authors declare no conflict of interest.

References

1. Sánchez-Navarro, M.; Teixidó, M.; Giralt, E. Just passing through. *Nat. Chem.* **2017**, *9*, 727–728. [CrossRef] [PubMed]
2. Fosgerau, K.; Hoffmann, T. Peptide therapeutics: Current status and future directions. *Drug Discov. Today* **2015**, *20*, 122–128. [CrossRef] [PubMed]
3. Xie, J.; Bi, Y.; Zhang, H.; Dong, S.Y.; Teng, L.H.; Lee, R.B.J.; Yang, Z.G. Cell-Penetrating Peptides in Diagnosis and Treatment of Human Diseases: From Preclinical Research to Clinical Application. *Front. Pharmacol.* **2020**, *11*, 697. [CrossRef] [PubMed]
4. Varkouhi, A.K.; Scholte, M.; Storm, G.; Haisma, H.J. Endosomal escape pathways for delivery of biologicals. *J. Control. Release* **2011**, *151*, 220–228. [CrossRef] [PubMed]
5. Rennick, J.J.; Johnston, A.P.R.; Parton, R.G. Key principles and methods for studying the endocytosis of biological and nanoparticle therapeutics. *Nat. Nanotechnol.* **2021**, *16*, 266–276. [CrossRef] [PubMed]
6. Kiss, A.L.; Botos, E. Endocytosis via caveolae: Alternative pathway with distinct cellular compartments to avoid lysosomal degradation? *J. Cell. Mol. Med.* **2009**, *13*, 1228–1237. [CrossRef]
7. Matsubara, T.; Otani, R.; Yamashita, M.; Maeno, H.; Nodono, H.; Sato, T. Selective intracellular delivery of ganglioside GM3-binding peptide through Caveolae/raft-mediated endocytosis. *Biomacromolecules* **2017**, *18*, 355–362. [CrossRef]
8. Li, Y.H.; Gao, L.; Tan, X.; Li, F.Y.; Zhao, M.; Peng, S.Q. Lipid rafts-mediated endocytosis and physiology-based cell membrane traffic models of doxorubicin liposomes. *BBA-Biomembranes* **2016**, *1858*, 1801–1811. [CrossRef]
9. Ruseska, I.; Zimmer, A. Internalization mechanisms of cell-penetrating peptides. *Beilstein J. Nanotechnol.* **2020**, *11*, 101–123. [CrossRef]
10. Pietiainen, V.; Marjomaki, V.; Upla, P.; Pelkmans, L.; Helenius, A.; Hyypia, T. Echovirus 1 endocytosis into caveosomes requires lipid rafts, dynamin II, and signaling events. *Mol. Biol. Cell* **2004**, *15*, 4911–4925. [CrossRef]
11. Ewers, H.; Helenius, A. Lipid-mediated endocytosis. *Cold Spring Harb. Perspect. Biol.* **2011**, *3*, a004721. [CrossRef] [PubMed]
12. Kabbani, A.M.; Raghunathan, K.; Lencer, W.I.; Kenworthy, A.K.; Kelly, C.V. Structured clustering of the glycosphingolipid GM1 is required for membrane curvature induced by cholera toxin. *Proc. Natl. Acad. Sci. USA* **2020**, *117*, 14978–14986. [CrossRef] [PubMed]
13. Fujinaga, Y.; Wolf, A.A.; Rodighiero, C.; Wheeler, H.; Tsai, B.; Allen, L.; Jobling, M.G.; Rapoport, T.; Holmes, R.K.; Lencer, W.I. Gangliosides that associate with lipid rafts mediate transport of cholera and related toxins from the plasma membrane to endoplasmic reticulm. *Mol. Biol. Cell* **2003**, *14*, 4783–4793. [CrossRef] [PubMed]
14. Montesano, R.; Roth, J.; Robert, A.; Orci, L. Non-coated membrane invaginations are involved in binding and internalization of cholera and tetanus toxins. *Nature* **1982**, *296*, 651–653. [CrossRef]
15. Imre, N.; Hetényi, A.; Szabó, E.; Bodnár, B.; Szkalisity, A.; Gróf, I.; Bocsik, A.; Deli, M.A.; Horvath, P.; Czibula, Á. Routing Nanomolar Protein Cargoes to Lipid Raft-Mediated/Caveolar Endocytosis through a Ganglioside GM1-Specific Recognition Tag. *Adv. Sci.* **2020**, *7*, 1902621. [CrossRef]
16. Vezenkov, L.L.; Martin, V.; Bettache, N.; Simon, M.; Messerschmitt, A.; Legrand, B.; Bantignies, J.L.; Subra, G.; Maynadier, M.; Bellet, V. Ribbon-like Foldamers for Cellular Uptake and Drug Delivery. *ChemBioChem* **2017**, *18*, 2110–2114. [CrossRef]
17. Oba, M. Cell-Penetrating Peptide Foldamers: Drug-Delivery Tools. *ChemBioChem* **2019**, *20*, 2041–2045. [CrossRef]
18. Misawa, T.; Ohoka, N.; Oba, M.; Yamashita, H.; Tanaka, M.; Naito, M.; Demizu, Y. Development of 2-aminoisobutyric acid (Aib)-rich cell-penetrating foldamers for efficient siRNA delivery. *Chem. Commun.* **2019**, *55*, 7792–7795. [CrossRef]
19. Peraro, L.; Kritzer, J.A. Emerging methods and design principles for cell-penetrant peptides. *Angew. Chem. Int. Ed.* **2018**, *57*, 11868–11881. [CrossRef]
20. Monreal, I.A.; Contreras, E.M.; Wayman, G.A.; Aguilar, H.C.; Saludes, J.P. SialoPen peptides are new cationic foldamers with remarkable cell permeability. *Heliyon* **2020**, *6*, e05780. [CrossRef]
21. Quartararo, A.J.; Gates, Z.P.; Somsen, B.A.; Hartrampf, N.; Ye, X.; Shimada, A.; Kajihara, Y.; Ottmann, C.; Pentelute, B.L. Ultra-large chemical libraries for the discovery of high-affinity peptide binders. *Nat. Commun.* **2020**, *11*, 3183. [CrossRef] [PubMed]
22. Imre, N. Intracellular Protein Delivery with the Use of Endocytosis Routing Sequences. Ph.D. Thesis, University of Szeged, Szeged, Hungary, 2020.

23. Cantu, L.; Corti, M.; Del Favero, E.; Raudino, A. Phase transition at the surface of mixed micelles of the ganglioside GM1 and dodecylphosphocholine. *J. Phys. Condens. Matter* **2000**, *12*, A321–A325. [CrossRef]
24. Oravecz, T.; Monostori, E.; Kurucz, E.; Takacs, L.; Ando, I. Cd3-Induced T-Cell Proliferation and Interleukin-2 Secretion Is Modulated by the Cd45 Antigen. *Scand. J. Immunol.* **1991**, *34*, 531–537. [CrossRef] [PubMed]
25. de Jesus, A.J.; Allen, T.W. The role of tryptophan side chains in membrane protein anchoring and hydrophobic mismatch. *Biochim. Biophys. Acta (BBA)-Biomembr.* **2013**, *1828*, 864–876. [CrossRef]
26. Vivian, J.T.; Callis, P.R. Mechanisms of tryptophan fluorescence shifts in proteins. *Biophys. J.* **2001**, *80*, 2093–2109. [CrossRef]
27. Munishkina, L.A.; Fink, A.L. Fluorescence as a method to reveal structures and membrane-interactions of amyloidogenic proteins. *Biochim. Biophys. Acta (BBA)-Biomembr.* **2007**, *1768*, 1862–1885. [CrossRef]
28. Anderson, J.M.; Kier, B.L.; Jurban, B.; Byrne, A.; Shu, I.; Eidenschink, L.A.; Shcherbakov, A.A.; Hudson, M.; Fesinmeyer, R.M.; Andersen, N.H. Aryl-Aryl Interactions in Designed Peptide Folds: Spectroscopic Characteristics and Optimal Placement for Structure Stabilization. *Biopolymers* **2016**, *105*, 337–356. [CrossRef]
29. Guvench, O.; Brooks, C.L. Tryptophan side chain electrostatic interactions determine edge-to-face vs parallel-displaced tryptophan side chain geometries in the designed beta-hairpin "trpzip2". *J. Am. Chem. Soc.* **2005**, *127*, 4668–4674. [CrossRef]
30. Doorley, G.W.; Payne, C.K. Nanoparticles act as protein carriers during cellular internalization. *Chem. Commun.* **2012**, *48*, 2961–2963. [CrossRef]

pharmaceutics

Article

Improving Membrane Activity and Cargo Delivery Efficacy of a Cell-Penetrating Peptide by Loading with Carboranes

Tamara Lützenburg [1], Nele Burdina [1], Matthias S. Scholz [2] and Ines Neundorf [1,*]

[1] Institute for Biochemistry, Department of Chemistry, University of Cologne, Zülpicher Str. 47a, 50674 Cologne, Germany; tluetze3@uni-koeln.de (T.L.); nburdina@smail.uni-koeln.de (N.B.)
[2] Pharmaceutical Chemistry I & II, Pharmaceutical Institute, Faculty of Mathematics and Natural Sciences, University of Bonn, An der Immenburg 4, 53121 Bonn, Germany; scholz@uni-bonn.de
* Correspondence: ines.neundorf@uni-koeln.de

Abstract: Cell-penetrating peptides (CPPs) have emerged as versatile tools to increase the intracellular accumulation of different kinds of cargoes. For an efficient cellular uptake and drug delivery, their organization into a distinct and stable secondary structure at the outer surface of the plasma membrane is a hallmark and supports optimal lipid–peptide interactions. Incorporation of hydrophobic moieties, such as carboranes (CBs), has the potential to increase the lipophilicity of peptides, and thus, to facilitate the formation of secondary structures. Herein, we present synthesis and biophysical as well as biological characterization of carborane-CPP conjugates having incorporated one or more CB clusters. Our results highlight the possibility to modulate the secondary structure of CPPs by the addition of CB's leading to constructs with altered membrane activity and promising use in terms of nucleic acid delivery.

Keywords: alpha-helix stabilization; secondary structure; cell-penetrating peptides; lipid–peptide interaction; peptide engineering

Citation: Lützenburg, T.; Burdina, N.; Scholz, M.S.; Neundorf, I. Improving Membrane Activity and Cargo Delivery Efficacy of a Cell-Penetrating Peptide by Loading with Carboranes. *Pharmaceutics* **2021**, *13*, 2075. https://doi.org/10.3390/pharmaceutics13122075

Academic Editors: Prisca Boisguérin and Sébastien Deshayes

Received: 27 October 2021
Accepted: 1 December 2021
Published: 3 December 2021

Publisher's Note: MDPI stays neutral with regard to jurisdictional claims in published maps and institutional affiliations.

Copyright: © 2021 by the authors. Licensee MDPI, Basel, Switzerland. This article is an open access article distributed under the terms and conditions of the Creative Commons Attribution (CC BY) license (https://creativecommons.org/licenses/by/4.0/).

1. Introduction

Biologically active peptides often develop distinct and stable conformations when in proximity to their binding partners. In case of cell-penetrating peptides (CPPs), structural organization is essential for efficient interaction with components of the plasma membrane [1]. This structuring readily occurs when CPPs are in close proximity to the lipid bilayer [2,3] and induces the following internalization process [4,5]. Consequently, chemical modifications that evoke and stabilize such secondary structures are of paramount importance to enhance the intracellular accumulation of CPPs.

During the past years, CPPs have emerged as promising delivery systems for small drug molecules, nanoparticles, nucleic acids or proteins [6–11], and have been widely applied for, e.g., anticancer treatments [12]. The formation and stability of their secondary structure is influenced by numerous factors including sequence length and amino acid composition [13,14]. For instance, strategies to increase the internalization efficiency of CPPs include the introduction of modifications or substitution of distinct amino acids within the peptide sequence [15,16]. Interestingly, it has been found out that the presence of hydrophobic amino acids, or the attachment of other hydrophobic moieties, e.g., lipidation, might further induce superior interaction with lipid bilayers [15,17,18]. In this respect, so-called car(ba)boranes (CBs) have emerged as interesting lipophilic substitutes and were already integrated as pharmacophores into different small molecule drugs. In addition, the incorporation of carboranes into therapeutic peptides was already described [19–21]. In this regard, CB-peptide conjugates turned out to be promising for application in tumor-selective boron neutron capture therapy (BNCT), and also CPPs were used to create such boron delivery systems [22,23]. However, to our knowledge, it is not clear and has not been elucidated yet how carboranes influence the physicochemical and biological characteristics of a CPP, and if one can use CBs to modulate the potential and activity of CPPs. Recently, we reported the straight-forward synthesis of peptide conjugates containing one CB cluster [24]. In this present study, we extended our efforts to engineer and modulate CPPs having one or more boron clusters incorporated. As a model CPP, we used sC18, which was developed in our group and turned out as a useful transporter for many different cargoes [25–29]. We studied the physicochemical properties of the new CB_x-CPP conjugates, as well as their biological activity when in contact with different membrane models. In a preliminary proof-of-principle study, we confirmed the cargo delivery potential of CB_4-sC18, which proved to be highly promising as a transporter for pDNA.

2. Materials and Methods

2.1. Carborane Synthesis

The carborane building block (1,7-dicarba-closo-dodecarbaborane) was purchased from Katchem Ltd. (Prague, Czech Republic). Additional chemicals used were bought from Acros (Geel, Belgium), Alfa Aesar (Haverhill, MA, USA), Fluka (Taufkirchen, Germany), Merck (Darmstadt, Germany), Sigma-Aldrich (Merck group: Darmstadt, Germany), and Carbolution (St. Ingbert, Germany). *Meta*-carborane-1-carboxylic acid was synthesized according to published procedures [24].

2.2. Peptide Synthesis

All used amino acids were purchased from IRIS Biotech (Marktredwitz, Germany) and were bought as Nα-Fmoc-protected versions. Additional chemicals and consumables required for synthesis were derived from Fluka (Taufkirchen, Germany), Merck (Darmstadt, Germany), Sarstedt (Nümbrecht, Germany), Sigma-Aldrich (Merck group: Darmstadt, Germany), and VWR (Darmstadt, Germany). This includes ethyl cyano(hydroxyimino)acetate (Oxyma), *N,N′*-diisopropylcarbodiimide (DIC), 1-[bis(dimethylamino)methylene]-1H-1,2,3-triazolo[4,5-b]pyridinium 3-oxid hexafluorophosphate (HATU), *N,N*-diisopropylethylamine (DIPEA), acetonitrile (ACN), and trifluoroacetic acid (TFA). For peptide synthesis, a combination of standard Fmoc/*t*-Bu solid-phase peptide synthesis (SPPS) on a Syro I peptide synthesizer (MultiSynTech, Bochum, Germany) and manual coupling protocols were used [25]. Peptides were generated as *C*-terminally amidated molecules on a Rink amide resin. The identity of peptides was confirmed via HPLC-ESI mass spectrometry (LTQ XL, Thermo Scientific, Waltham, MA, USA).

2.3. Carboranyl-Peptide Synthesis

Synthesis of carborane-peptide conjugates was carried out on solid support (loading 0.5 mmol/gram resin). The carborane cluster was introduced using 2 eq activated *meta*-carborane-1-carboxylic acid with 5 eq Oxyma/DIC overnight at room temperature. Carboranyl-peptides were cleaved from the solid support using trifluoroacetic acid/triisopropylsilane/water (95/2.5/2.5, *v*/*v*/*v*). Identification was performed via HPLC-ESI (Chromolith® Performance RP-18e, 100-4.6 mm, Merck, Darmstadt, Germany; 10–60% ACN in water (incl. 0.1% formic acid) over 15 min; 0.6 mL/min flow rate) mass spectrometry. Purification was carried out using preparative HPLC (Nucleodur C18ec; 100-5; Macherey-Nagel, Düren, Germany), with a flow rate of 1.5 mL/min over 45 min. Different gradients were used, according to the hydrophobicity of the conjugate; CB_1-sC18: 10–60% ACN in water; CB_2-sC18: 20–70% ACN in water; CB_3-sC18, CB_4-sC18, CB_5-sC18: 30–90% ACN in water. Water was supplemented with 0.1% TFA, ACN with 0.08% TFA. Final chromatograms were recorded using a linear gradient from 20–80% ACN in water (incl. 0.1% TFA) using a Nucleodur C18ec; 100-5 (Macherey-Nagel, Düren, Germany) column (1 mL/min flow rate).

2.4. Circular Dichroism Spectroscopy

Conjugates were dissolved in 10 mM phosphate buffer (pH 7.0), either alone or supplemented with 50% TFE (*v*/*v*), to a final concentration of 20 μM. CD spectra of all conjugates were recorded from 180–260 nm (0.5 nm intervals) on a J-715 spectropolarimeter (JASCO, Pfungstadt, Germany) in an N_2 atmosphere.

2.5. Cell Viability Assay

In total, 17.000 HeLa cells per well were seeded onto 96-well plates (Sarstedt, Nümbrecht, Germany). After confluency of 80–90% was reached, the growth medium was removed, and cells were incubated with 100 µL of peptide solution of different concentrations (125 nM–100 µM in serum-free medium) for 24 h at 37 °C. For a positive control, cells were treated with 70% EtOH for 10 min. Cells were washed with serum-free medium and incubated with 10% (*v*/*v*) resazurin (Sigma-Aldrich, Merck group: Darmstadt, Germany) in serum-free medium for 1.5 h at 37 °C. Cell viability was determined by measurement of the resorufin product at 595 nm (λ_{ex} = 550 nm) on a Tecan infinite M200 plate reader (Tecan Group AG, Männedorf, Switzerland). Viability was calculated from the fluorescence values of treated cells in relation to untreated cells.

2.6. CF-Leakage Assay

Large unilamellar vesicles (LUVs) were prepared by dissolving different lipids in $CHCl_3$, to reach the desired compositions (lipids were purchased from Avanti Polar Lipids, Inc., Alabaster, AL, USA). To remove the solvent, the mixture was placed into a round-bottomed flask on ice and the solvent was evaporated under reduced pressure for 20 min at 40 °C and afterwards under vacuum for another 5 min. The dried lipid film was then hydrated with HKS buffer (25 mM HEPES, 150 mM KCl, pH 7.4, 10% sucrose (*w*/*v*)), supplemented with 100 mM 5(6)-carboxyfluorescein (CF), and homogenized for 5 min at 45 °C to form liposomes (final concentration: 8 mM). The resulting multilamellar vesicle dispersion underwent 10 freeze/thawing cycles and was passed 21 times through a mini extruder (0.4 µm polycarbonate treck-etch membrane; Avanti Polar Lipids, Inc., Alabaster, AL, USA). External CF was removed by size exclusion chromatography (PD10 column, GE Healthcare, Chicago, IL, USA). Peptides were added to the LUVs and the CF release was monitored by measuring the increased fluorescence intensity on a plate reader (Biotek, Winooski, VT, USA). After 90 min, Triton-x-100 was added to a final concentration of 0.4% (*v*/*v*) to measure the maximal leakage and normalize data.

2.7. Hemolysis Assay

Human red blood cells were washed five times with PBS (4 °C, 3000× *g*, 5 min) and diluted to 5% (*v*/*v* in PBS). In total, 100 µL of this solution were transferred into a fresh 96-well plate. In total, 50 µL of peptide solution were added and gently resuspended. Red blood cells were incubated for either 1 or 24 h at 37 °C and 5% CO_2, respectively. In total, 50 µL of 10% Triton-x-100 were added 10 min prior to the end of the incubation time and served as a positive control. After incubation, cells were centrifuged (r.t., 2500× *g*, 3 min). The supernatant was carefully transferred into a fresh 96-well plate and the absorption at 560 nm was measured on a Tecan infinite M200 plate reader (Tecan Group AG, Männedorf, Switzerland). The measured absorption referred to the hemoglobin concentration.

2.8. Flow Cytometry

HeLa cells were seeded onto 24-well plates (Sarstedt, Nümbrecht, Germany, 100,000 cells per well, respectively). After reaching 80–90% confluency, medium was removed, and cells were incubated with 1 µM CF-labeled peptide solution in serum-free medium for either 30 min or 2 h at 37 °C. The peptide solution was removed, cells were washed twice with PBS, and trypsinized for 5 min. Detached cells were resuspended in full medium (phenol red free) and afterwards analyzed using a guava easycyte HT flow cytometer (Merck, Darmstadt, Germany).

2.9. Confocal Laser Scanning Microscopy (CLSM)

HeLa cells were seeded into 8-well plates (ibidi, Gräfelfing, Germany, 40,000 cells per well, respectively) and grown to 80–90% confluency. Cells were treated with 1 µM CF-labeled peptide in serum-free medium for 30 min or 2 h at 37 °C. Ten minutes prior to the end of the incubation time, nuclei were stained with Hoechst33342. The peptide solution was removed, and cells were treated with trypan blue (150 mM in 0.1 M acetate buffer, pH 4.15) for 30 s. Cells were washed twice with PBS and covered with fresh medium supplemented with FBS. Microscopic analysis was performed using an inverse confocal TCS SP8 microscope (Leica Microsystems, Wetzlar, Germany), equipped with a 63× oil-immersion objective. Images were recorded with LAS X software (Leica Microsystems, Wetzlar, Germany) and adjusted with Fiji.

2.10. Critical Micelle Concentration (CMC) Determination

Peptide conjugates were dissolved in water and different concentrations (1–200 µM) were prepared. UV-VIS spectra ranging from 200–700 nm were recorded using a 1 cm path length quartz cuvette in a Spectroquant Pharo 300 photometer (Merck, Darmstadt, Germany). The absorbance at 255 nm was plotted against the logarithm of the peptide concentration. The intersection point of two lines was determined as CMC.

2.11. Proliferation Assay

In total, 3000 HeLa cells per well were seeded onto 96-well plates (Sarstedt, Nümbrecht, Germany). When a confluency of 30–40% was reached, medium was removed and cells were incubated with peptides and Daunorubicin at different concentrations (1 µM peptide concentration and 100 nM or 1 µM Daunorubicin in serum-free medium) for 2 h at 37 °C. Afterwards, a washout was performed, and cells were washed once with 100 µL of serum-free medium. Then, 200 µL of medium with serum were added and cells were grown for 72 h at 37 °C. As a positive control, cells were treated with 70% EtOH for 10 min. Afterwards, cells were covered with 10% (*v*/*v*) resazurin (Sigma-Aldrich, Merck group: Darmstadt, Germany) in serum-free medium for 1.5 h at 37 °C. The cell viability of treated cells was determined relative to that of untreated cells by measurement of the resorufin product at 595 nm (λ_{ex} = 550 nm) on a Tecan infinite M200 plate reader (Tecan Group AG, Männedorf, Switzerland). Viability was calculated from the fluorescence values of treated cells in relation to untreated cells.

2.12. Electrophoretic Mobility Shift Assay (EMSA)

Peptides were co-incubated with mCherry plasmid at various ratios for 1 h at 37 °C in nuclease-free water (total volume: 15 µL). The following mass ratios were chosen (peptide:plasmid): 0.25:1, 0.5:1, 1:1, and 3:1 (mCherry plasmid: 250 ng). After incubation, 3 µL of 6x Loading Dye were added into the reaction mixture and the mixture was loaded onto a 1% agarose gel. The agarose gel was run for 1.5 h at 100 V and the gel analyzed in a ChemiDoc XRS (Bio-Rad, Hercules, CA, USA).

2.13. Transfection with mCherry Plasmid

HeLa cells were seeded into 8-well plates (ibidi, Gräfelfing, Germany, 8000 cells per well, respectively) and grown to 30–40% confluency. Peptides and mCherry plasmid were co-incubated for 1 h at 37 °C at a ratio of 3:1 in serum-free medium (mCherry plasmid: 250 ng). Afterwards, the medium was removed, and the solution was added to the cells. Additional serum-free medium was added to ensure a final peptide concentration of 1 µM. Cells were incubated for 2 h with the peptides and peptide/plasmid solution at 37 °C, respectively. As a positive control, Lipofectamine 2000 (Invitrogen, Thermo Scientific, Waltham, MA, USA) was used and as a negative control, cells were treated with plasmid only. Afterwards, a washout was performed, and cells were washed once with 150 µL of serum-free medium. In total, 300 µL of medium with serum were added and cells were grown for 72 h at 37 °C. Ten minutes prior to microscopic analysis, 0.6 µL of Hoechst33342 were added. Then, cells were washed twice with 150 µL of serum-free medium. In total, 300 µL of serum-supplemented medium were added and cells analyzed using a BZ-X800E microscope (Keyence, Osaka, Japan).

3. Results

3.1. Synthesis of Carborane-Peptide Conjugates

First, we synthesized a series of CB-CPP conjugates bearing up to five *meta*-CBs by using a combination of automated and manual solid-phase peptide synthesis (SPPS). In order to couple CB stepwise to sC18, additional lysines, e.g., Dde-Lys(Fmoc)-OH, were introduced (Scheme 1).

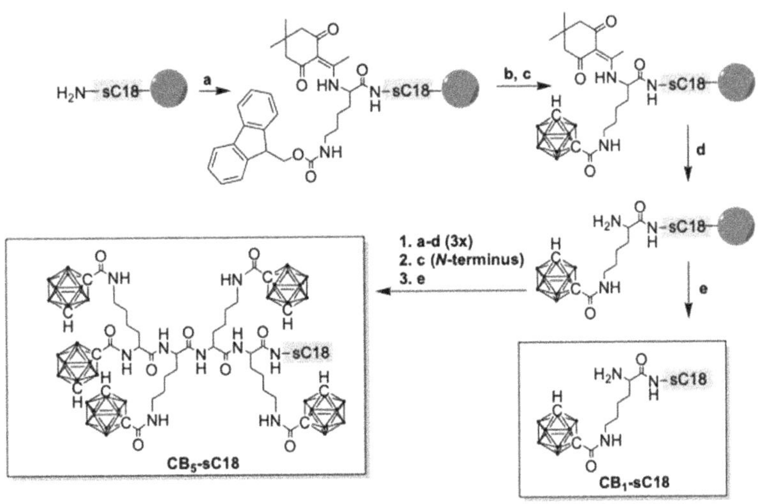

Scheme 1. Synthesis of CB-sC18 conjugates; a: Dde-Lys(Fmoc)-OH, HATU, DIPEA; b: 30% piperidine/DMF; c: *meta*-carborane-1-carboxylic acid, Oxyma, DIC; d: 3% hydrazine/DMF; e: TFA/TIS/water (95/2.5/2.5, *v*/*v*/*v*).

Moreover, a second fluorescent series was synthesized with all conjugates labeled with 5(6)-carboxyfluorescein (CF) at the ε-amino group of lysine at position 12 of sC18. All products were obtained in high purities and confirmed via liquid chromatography mass spectrometry analysis (Table 1, Table S1 and Figures S1–S12).

Table 1. Names, sequences, and analytical data of synthesized peptides. Peptides are *C*-terminally amidated. ***K***: lysine with CB attached at the ε-amino group. K: labeling position for CF.

Peptide	Sequence	$MW_{calc.}$ [Da]	$MW_{exp.}$ [Da]
sC18	[1]GLRKRLRKFRNKIKEK[16]	2069.55	2070.34
CB_1-sC18	*K*-GLRKRLRKFRNKIKEK	2367.95	2368.74
CB_2-sC18	*KK*-GLRKRLRKFRNKIKEK	2666.37	2666.97
CB_3-sC18	*KKK*-GLRKRLRKFRNKIKEK	2965.79	2965.99
CB_4-sC18	*KKKK*-GLRKRLRKFRNKIKEK	3264.21	3263.82
CB_5-sC18	CB-*KKKK*-GLRKRLRKFRNKIKEK	3433.44	3434.29

3.2. Physicochemical Properties of CB-CPP Conjugates

Comparing the HPLC retention profiles of all CB-sC18 conjugates revealed a stepwise shift towards higher retention times dependent on the number of CBs coupled (Figure S13). Since this effect was obviously dependent on the hydrophobic nature of the conjugates, we were then interested in whether they would also self-assemble and form nanostructures, such as micelles or fibrils. An easy to apply and fast method makes use of changes in spectroscopic properties associated with micelle formation [30]. For this, one measures the UV absorbance at a specific wavelength as a function of the concentration. Because aggregates like micelles and free molecule species contribute to a different extent to the absorbed UV light, an abrupt change of the slope occurs above a certain point. The corresponding concentration is defined as the critical micelle concentration (CMC). We measured the UV-VIS spectra of all CB-sC18 conjugates and observed for all an additional concentration-dependent maximum at 255 nm (Figure 1A,B and Table 2), whereas this maximum was not detected for meta-carborane-1-carboxylic acid (Figure S14) or sC18 alone (Figure S15).

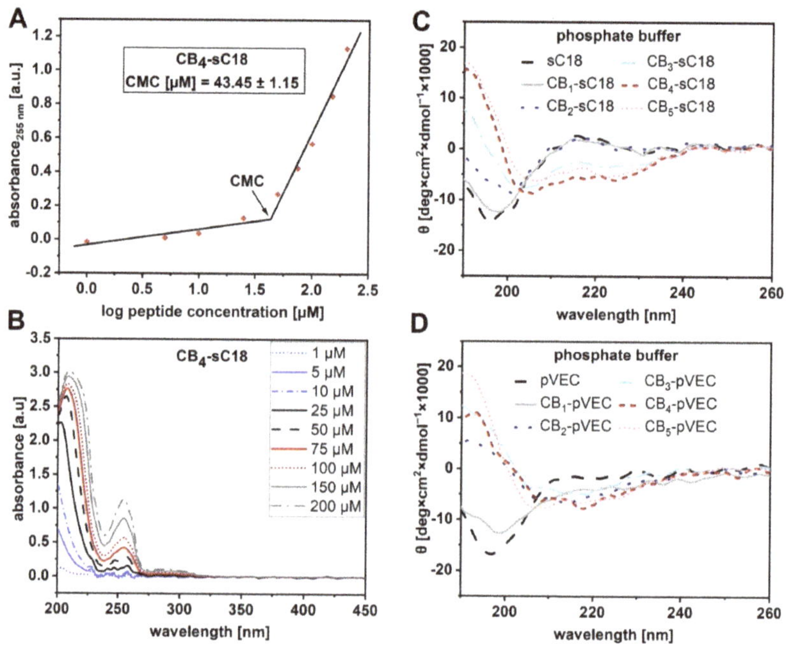

Figure 1. Determination of CMC and CD spectra of CB-CPP conjugates. (**A**) The determination of CMC for CB_4-sC18 is shown by measuring the UV-VIS spectra at various concentrations (**B**). (**C**) CD spectra of CB-sC18 conjugates in phosphate buffer. (**D**) CD spectra of CB-pVEC conjugates in phosphate buffer.

Table 2. Percentage of α-helical content [31], and determined R-values of CB-sC18 and CB-pVEC conjugates calculated as the ratio between the molar ellipticity values at 222 and 208 nm [32].

Peptide	Phosphate Buffer [1]	50% TFE [1]	Phosphate Buffer [2]	50% TFE [2]	CMC [μM]
sC18	3.45	71.22	/	0.89	-
CB_1-sC18	5.02	50.68	/	0.85	-
CB_2-sC18	5.45	60.49	/	0.89	-
CB_3-sC18	16.68	56.10	0.56	0.88	44.46 ± 1.14
CB_4-sC18	21.68	56.30	0.73	0.98	43.45 ± 1.15
CB_5-sC18	21.09	50.15	0.87	0.92	44.37 ± 1.16
pVEC	12.18	48.60	0.37	0.73	-
CB_1-pVEC	17.72	44.28	0.57	0.82	-
CB_2-pVEC	20.41	51.36	0.84	0.82	-
CB_3-pVEC	16.98	65.76	0.99	0.82	-
CB_4-pVEC	22.59	59.84	0.92	0.84	-
CB_5-pVEC	22.96	45.48	0.79	0.79	-

[1] % of alpha helix, [2] *R*-value.

Thus, we hypothesized this maximum as evidence for a possible self-assembly, and by plotting the absorbance at 255 nm against the logarithm of the peptide concentration, we assigned the intersection point as CMC. Interestingly, determination of a CMC was only feasible for $CB_{3–5}$-sC18 variants (Table 2, Figures S16–S19), letting us assume that self-assembly accompanied by a change in secondary structure would only be observed when three or more CBs were attached.

To find out more, we performed CD spectroscopy for all peptides solved either in phosphate buffer or in phosphate buffer supplemented with trifluoroethanol (TFE), respectively. All conjugates and sC18 adopted an alpha-helix, when present in phosphate buffer with TFE (Figure S20). On the other hand, when CB-conjugates were solved in phosphate buffer only (Figure 1C), some of them directly formed distinct secondary structures. For instance, CB_2-sC18 and more clearly CB_3-sC18 did not display a clear random coil structure, but the shapes of their CD curves changed to ones with a more alpha-helical character, and for CB_4- and CB_5-sC18, the alpha-helical structure was even more evident (Table 2). Taking this observation into account, we concluded that the number of CB moieties actually promoted the formation of alpha-helical structures. This effect might be further explained by the helical wheel projection of each conjugate showing that the hydrophobic face of the peptide was clearly enlarged when three or more CB units were attached (see Figure S21).

In order to verify the observed influence of CBs, we chose the CPP pVEC (LLIILRRRIRKQAHAHSK) [33] and synthesized the analogous series of $CB_{1–5}$-pVEC conjugates by introducing CBs in the same way as described before (Figures S22–S27). Again, we observed that the elution profile of the different conjugates was shifted towards higher retention times, similarly as we have seen it before for the CB-sC18 series (Figure S28). More interestingly, CD spectroscopy measurements indicated a change in the secondary structure when two or more CBs were attached, while a more distinct alpha-helical structure was recognized for CB_4- and CB_5-pVEC (Figure 1D and Figure S29, Table 2). In conclusion, the attached CBs contribute to the formation of defined secondary structures, which is important in light of the membrane activity of CPPs.

3.3. Interaction with Artificial and Biological Membranes

Our results thus far inspired us to investigate how the novel CB-conjugates would perform in a lipid environment. Therefore, we first used large unilamellar vesicles (LUVs) composed of different phospholipids as suitable model systems for mimicking different types of cell membranes. We prepared negatively charged (DOPC/DOPE/DOPG; 40/30/30) LUVs, representing cancer cell membranes, and zwitterionic (DOPC/DOPE; 50/50) ones, representing healthy cell membranes, and incubated them at 25 °C for 90 min with 1 µM peptide solutions (Figure 2). Both lipid systems were additionally loaded with CF in order to monitor membrane leakage via fluorescence dye outflow.

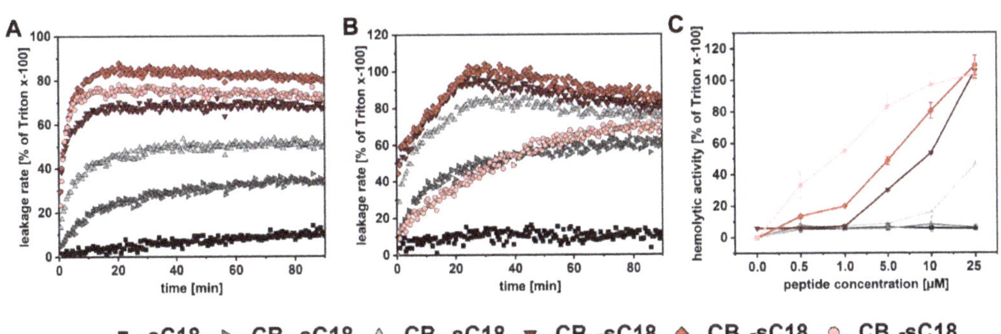

Figure 2. Time-dependent release of CF from negatively charged (**A**) or zwitterionic LUVs (**B**). Data were normalized to Triton x-100 control. (**C**) Hemolytic activity of CB-sC18 conjugates towards RBCs. Data were normalized to the control. Experiments were conducted in duplicate (**A**,**B**) or triplicate (**C**) with $n = 2$.

As depicted in Figure 2A, sC18 did not cause any leakage to negatively charged LUVs, while all conjugates bearing carboranes led to an increased outflow of CF, particularly CB_3-, CB_4-sC18, and CB_5-sC18. The most potent conjugate was CB_4-sC18, exhibiting not only the highest leakage rate but also the most rapid release of CF. Interestingly, this effect also appeared when CB_4-sC18 was incubated with zwitterionic LUVs (Figure 2B), followed by CB_3-sC18 and CB_2-sC18. Notably, CB_5-sC18 demonstrated the same leakage profile as CB_1-sC18, and was, therefore, not as potent as it was towards negatively charged LUVs. This finding might be allocated to the formation of micelles and/or a higher tendency to tightly interact with the lipid phase, also leading to slight precipitation over longer time incubations. In summary, CB_4-sC18 exhibited the highest leakage rate towards both tested model systems, while interaction with negatively charged membranes seemed to be stronger and more rapid.

Next, we incubated human red blood cells (RBCs) with different concentrations of CB-sC18 conjugates at 37 °C for 1 or 24 h, respectively. Since their cell membranes are mainly composed of neutral lipids, RBCs serve as an excellent biological model for the lytic activity of a peptide. As depicted in Figure 2C, sC18 and CB_1-sC18 did not exhibit any effect on RBCs after 24 h of incubation even at the highest concentration used (25 μM). For sC18, this agreed with the detected low activity when in contact with zwitterionic LUVs. Compared to this, CB_2-sC18 and all other conjugates were able to lyse RBCs, exhibiting up to 100% hemolytic activity. Interestingly, the efficacy of the conjugates to lyse RBCs increased with the number of CB attached (CB_2-sC18 < CB_3-sC18 < CB_4-sC18 < CB_5-sC18). The same activity profile was observed when incubating peptide conjugates with RBCs for only one hour (Figure S30), whereas the effects started at concentrations higher than 5 μM. Altogether, we concluded all CB-sC18 conjugates had high membrane activities. However, this property is impressively dependent on the number of CBs attached to the CPP.

3.4. *Cytotoxicity and Cellular Uptake Studies*

Following, CB-sC18 conjugates' interaction with living cells was of interest. We chose HeLa cells, since we have already used them in many studies including sC18 [25,34]. We first assessed the cytotoxicity of the novel peptides towards this cell line (Figure 3A). As expected from our previous studies, sC18 exhibited no cytotoxicity when incubated for 24 h with the cells [35]. In contrast, a clear decrease in cell viability was visible for all novel CB-sC18 conjugates (Figure 3A). While the cytotoxic activity increased from CB_1-sC18 to CB_2-sC18, conjugates CB_3-, CB_4-, and CB_5-sC18 showed nearly the same activity profile, concluding that there is no direct correlation between the number of CBs coupled to the peptide and their cytotoxic properties. Nevertheless, the observed cytotoxic profiles agreed with the LUV experiments using negatively charged vesicles, where we measured a very fast dye release for all three conjugates CB_3- to CB_5-sC18, presumably due to membrane distortion. Since for both controls alone (*meta*-carborane-1-carboxylic acid as well as sC18) nearly no cytotoxicity was detected (Figure S31), we allocated this high membrane activity to the higher alpha-helical content in combination with the enlarged hydrophobic face within the CB-conjugates that was present after introducing CBs to sC18.

Next, we investigated the cellular uptake of the different CB-sC18 conjugates in HeLa cells. We performed microscopy studies using confocal laser scanning microscopy (CLSM), and quantified the uptake by flow cytometry (Figure 3B,C). Since it has been previously demonstrated that CPP uptake might be influenced by the choice of fluorophore and its position within the CPP sequence [36,37], we first assessed whether it would make any difference if sC18 was labeled with CF at the *N*-terminal or at position 12. As depicted in Figure S32, we did not determine any impact on the internalization ability of the peptide, verifying position 12 of sC18 as suitable for CF labeling. However, all conjugates were able to internalize into HeLa cells already after 30 min of incubation time and at non-toxic concentrations of 1 μM (Figure S33). The control peptide sC18 as well as CB_1-sC18 were internalized nearly to the same extent. Again, cellular uptake increased when more CBs were attached to the peptide, whereas it was highest for CB_4-sC18 and decreased again for the most hydrophobic conjugate CB_5-sC18. Additionally, all conjugates showed a vesicular distribution within the cytosol of the cells, which was even more visible, when incubating the cells for 2 h (Figure 3B). This observation might be a hint of an endocytotic internalization pathway. Moreover, as illustrated in Figure 3C, CB_4-sC18 was again internalized the most, and the cellular uptake significantly increased for all conjugates when incubating cells for 2 h, besides for CB_1-sC18. Particularly, CB_{3-5}-sC18 were internalized to a much higher extent when incubating the cells for 2 h. This might be explained by their highly increased hydrophobic character, and therefore, stronger and more intense interaction with the hydrophobic membrane core. The general decreased uptake of CB_5-sC18 might be a result of formed self-assemblies, which are presumably taken up to a lesser extent or need more time for internalization. All in all, these results emphasized again that the attachment of CBs highly supported and adjusted the cellular uptake of all conjugates, highlighting CB_4-sC18 as the most active candidate.

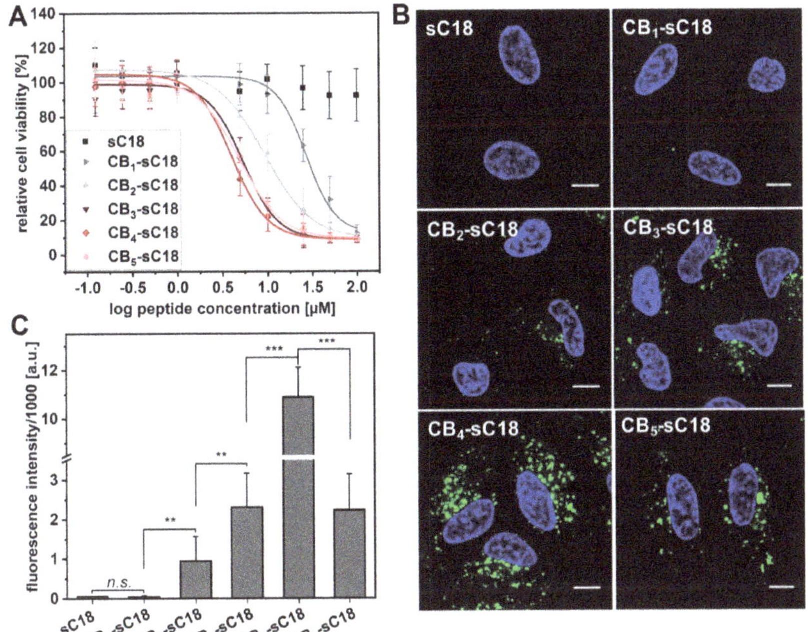

Figure 3. Cytotoxicity and cellular uptake into HeLa cells. (**A**) Cytotoxicity profile of CB-peptide conjugates towards HeLa cells. (**B**) CLSM analysis of 1 µM CF-labeled peptide conjugates after 2 h of incubation in HeLa cells; scale bar: 10 µm. (**C**) Flow cytometry analysis of 1 µM CF-labeled peptide conjugates after 2 h of incubation in HeLa cells. Significances were determined by Student's *t*-test (** $p < 0.005$; *** $p < 0.0005$). CB_4-sC18 is significant (*** $p < 0.0005$) to all conjugates.

3.5. Cargo Delivery Studies

Lastly, we were interested in whether the CB-sC18 conjugates would be useful for intracellular cargo delivery. We chose two different cargoes, daunorubicin (Dau), which has a reasonably good cell membrane permeability by itself, as well as plasmid DNA encoding for the red fluorescent protein mCherry, in order to analyze if CB-sC18 conjugates were able to support the intracellular accumulation in HeLa cells of otherwise non-cell-permeable molecules. Since CB_4-sC18 performed the best, but CB_1-sC18 exhibited the lowest cytotoxic effect, we used both conjugates for our studies. Furthermore, we were interested in whether the conjugates could be applied as delivery enhancers by just co-incubating the cargo with transporter, e.g., the CB-CPP conjugate.

Our results demonstrated that both CB_1-sC18 as well as CB_4-sC18 did not alter the cytotoxicity of Dau. Moreover, daunorubicin's activity was slightly increased using 100 nM Dau in combination with CB_4-sC18 (Figure 4A). Of more interest was that CB_4-sC18 proved to be versatile to support the internalization of pDNA encoding for mCherry. As depicted in Figure 4B, when cells were inspected 72 h after transfection, we detected no fluorescent signal for the control cells (pDNA alone) and only a weak red signal when the cells were incubated with a mixture of CB_1-sC18 and mCherry pDNA. Instead, cells that were transfected with a solution of CB_4-sC18 and mCherry pDNA demonstrated strong red intracellular signals derived from a successful plasmid transfection. Having a closer look at the intensity values of fluorescent cells, we observed an almost similar intensity of lipofectamine-transfected cells compared to cells that were covered with CB_4-sC18 and mCherry pDNA (see the line profiles in Figure 4B and Figure S35).

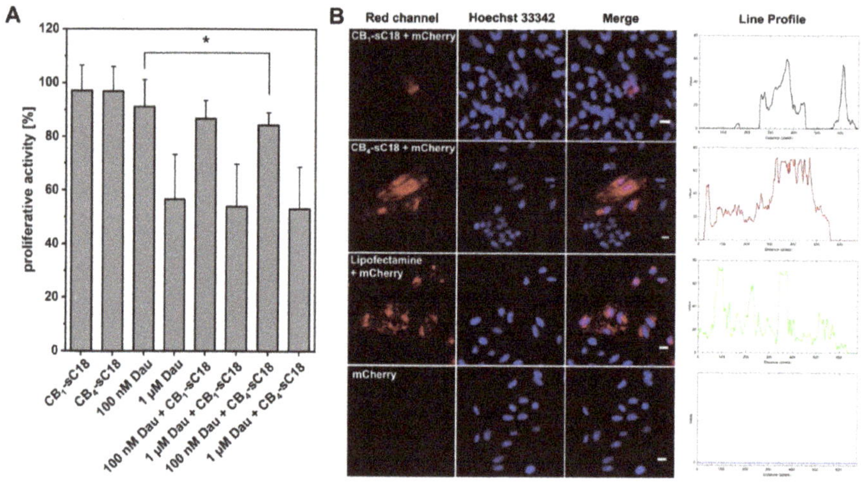

Figure 4. Cargo delivery properties of CB-sC18 conjugates. (**A**) Proliferative activity of CB-sC18 conjugates, Dau and CB-sC18/Dau mixture towards HeLa cells. Significances were determined by Student's *t*-test (* $p < 0.05$). (**B**) Microscopic analysis of HeLa cells transfected with mCherry encoding pDNA using either CB_1-sC18, CB4-sC18, or lipofectamine 2000, scale bar: 20 µm.

4. Discussion

Our findings presented in this work prove that introducing CBs to cell-penetrating peptides modulates and enhances their membrane activity. We could demonstrate that the increased hydrophobicity caused by the attached CBs combined with the higher alpha-helical content significantly improved lipid phase interaction and thus, cellular uptake into HeLa cells for the CPP sC18. Beside their cellular uptake mediated by endocytosis, we assume that CB-CPPs intensively interact with lipid environments. More precisely, we hypothesize that CB-peptides, particularly those having more than two CBs included, tightly interact with lipid membranes in a way that they, in a first step, approach the membrane surface attracted by electrostatic interaction and facilitated by their amphipathic helix. Second and probably simultaneously, the CB tail is likely be inserted into the lipid core, tightly interacting with the fatty acid alkyl chains, and presumably sticking to it, inducing the following cell entry processes by formation of transient pores or inverted micelles. We cannot exclude that these interactions also lead to the recruitment of membrane components important for endocytotic uptake pathways. Since the CMC measured was above the concentration that was used in these experiments, we assume that CB-peptides contact the membrane surface in a singular state. However, we cannot completely exclude that self-assembled structures also play a role in membrane interaction, as displayed in

Figure 5. In addition, we observed an intracellular vesicular distribution of the conjugates and therefore, we suppose that mainly endocytosis took place as the entry pathway at the concentrations applied. However, which pathway exactly is the favored one has to be elucidated in more detail in future studies.

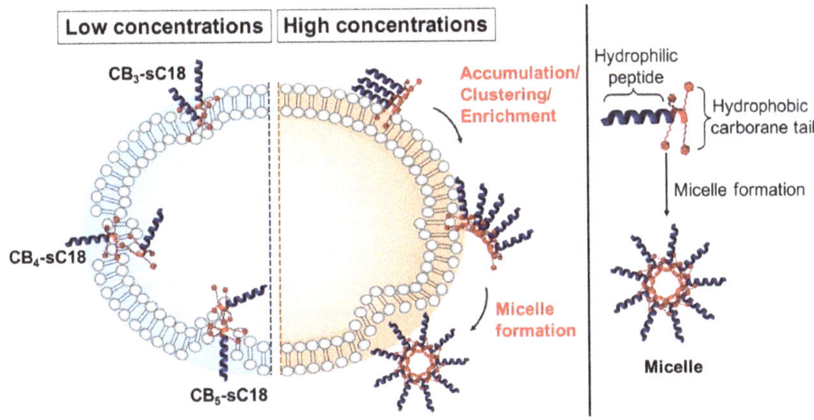

Figure 5. The hypothetical mechanism of interaction of CB-peptide conjugates with cellular membranes is demonstrated. Not shown is endocytosis, which we observed as one of the preferred internalization pathways (see Figure 3). Left: low concentrations of CB-CPP conjugates interact with the outer surface of the plasma membrane, leading to pore formation events and potential membrane distortion. Right: high concentrations of CB-peptide conjugates might result in potential self-assembly processes.

Interestingly, our findings were transferable to the CPP pVEC, which was affected in the same manner as sC18 (Figure 1), concluding that CB coupling indeed alters the conformation of CPPs. Thus, the herein reported CB-CPP conjugates might present an interesting and valuable toolbox to study the first steps of CPP cellular uptake at the cellular membrane. More importantly, with the attachment of CB, one might further improve the overall performance of CPPs in terms of lipid–peptide interaction and internalization ability. Our results fit recent reports in which it was shown that peptides comprising high hydrophobic/amphipathic character induce membrane penetration and disrupting events when they insert deeply into the hydrophobic core of phospholipid membranes [38–40]. Additionally, we propose that in our case, the developed secondary structure played a crucial role for CB-CPP interplay and insertion within the lipid bilayer [41].

In line with these observations were the results obtained from our cellular studies. When increasing the applied conjugate concentrations, we observed increased cytotoxicity in HeLa cells as well as increased hemolytic activity in RBCs dependent on the number of CBs attached. However, all these effects were observed beyond a concentration threshold of 1 μM. Of note is that the new CB-conjugates were internalized highly efficiently and exhibited no cytotoxic effects when applied at lower concentrations of about 1 μM or below (Figure S33), which could make them interesting candidates for drug delivery applications.

Therefore, in preliminary studies, we investigated their performance as transport enhancers by simple co-incubation experiments. When incubating the anticancer drug daunorubicin with CB-sC18 conjugates, the activity of Dau was principally not diminished. Furthermore, CB_4-sC18 showed slightly better activities compared to CB_1-sC18, since it could slightly increase the cytotoxic activity of Dau (when co-incubated with 100 nM Dau). The fact that the preferred uptake of daunorubicin was not decreased in the presence of the peptides was possibly due to very fast uptake of the drug itself. Notably, we have seen a similar effect in previous studies, in which we used a cyclic peptidomimetic version of sC18 for enhancing the intracellular accumulation of daunorubicin. In this case, no improvement was detected when both substances were co-incubated and added to the cells [29]. On the other side, we investigated the ability of CB_1-sC18 and CB_4-sC18 in terms of nucleic acid delivery. Obviously, the transfection efficiency of the CB_1-sC18/pDNA solution was only low, since only less cells showed a relatively weak signal after 72 h. Compared to this, we detected a clear strong red signal within cells that were transfected with the corresponding CB_4-sC18/pDNA solution. Since the pDNA alone did not show any signal, we assume a successful delivery of the pDNA facilitated by CB_4-sC18. In addition, we performed electro mobility shift assays (EMSA, see Figure S34) and observed that at the used pDNA/CB-CPP ratio, the pDNA was retarded. Thus, we conclude that a successful complex-like formation between both compounds is a prerequisite for successful transfection of the pDNA. Moreover, the intensity of the signal is comparable to the signal detected by the lipofectamine control (see Figure 4B and Figure S35), highlighting the high potential of CB-sC18 conjugates, particularly CB_4-sC18, to act as transfection agent.

In conclusion, we designed and synthesized a series of novel CB-CPP variants and analyzed how CB clusters influence the biophysical characteristics of CPPs. We found out that CBs affect secondary structure formation, probably leading to enhanced membrane activities. Therefore, the herein created new conjugates proved to be versatile tools for biological investigations, including, e.g., lipid–peptide interaction, as well as cargo delivery. In terms of further possible future studies, we highlight them as interesting candidates for application in BNCT therapy, where they might be ideal to transport high boron loads to the site of interest. In fact, as we have seen mainly endosomal accumulation of our CB-sC18 conjugates, the boron delivery to the perinuclear site might be significantly increased by making use of the endocytotic transport machinery.

Supplementary Materials: The following are available online at https://www.mdpi.com/1999-4923/13/12/2075/s1, Figure S1: UV-chromatogram and corresponding ESI-MS spectrum of sC18, Figure S2: UV-chromatogram and corresponding ESI-MS spectrum of CB_1-sC18, Figure S3: UV-chromatogram and corresponding ESI-MS spectrum of CB_2-sC18, Figure S4: UV-chromatogram and corresponding ESI-MS spectrum of CB_3-sC18, Figure S5: UV-chromatogram and corresponding ESI-MS spectrum of CB_4-sC18, Figure S6: UV-chromatogram and corresponding ESI-MS spectrum of CB_5-sC18, Figure S7: UV-chromatogram and corresponding ESI-MS spectrum of CF-sC18 (pos. 12), Figure S8: UV-chromatogram and corresponding ESI-MS spectrum of CB_1-CF-sC18, Figure S9: UV-chromatogram and corresponding ESI-MS spectrum of CB_2-CF-sC18, Figure S10: UV-chromatogram and corresponding ESI-MS spectrum of CB_3-CF-sC18, Figure S11: UV-chromatogram and corresponding ESI-MS spectrum of CB_4-CF-sC18, Figure S12: UV-chromatogram and corresponding ESI-MS spectrum of CB_5-CF-sC18, Figure S13: UV-chromatograms of all conjugates recorded using a linear gradient from 20–80% ACN in water (incl. 0.1% trifluoroacetic acid) within 15 min, Figure S14: Determination of critical micelle concentration (CMC) of *meta*-carboranyl-carboxylic acid, Figure S15: Determination of critical micelle concentration (CMC) of sC18, Figure S16: Determination of critical micelle concentration (CMC) of CB_1-sC18, Figure S17: Determination of critical micelle concentration (CMC) of CB_2-sC18, Figure S18: Determination of critical micelle concentration (CMC) of CB_3-sC18, Figure S19: Determination of critical micelle concentration (CMC) of CB_5-sC18, Figure S20: CD spectra of CB-sC18 conjugates (20 μM peptide concentration) in 10 mM phosphate buffer (pH 7.0) with the addition of 50% TFE, Figure S21: Helical wheel projections of CB-sC18 conjugates, Figure S22: (A) UV-chromatogram and (B) corresponding ESI-MS spectrum of pVEC after purification, Figure S23: (A) UV-chromatogram and (B) corresponding ESI-MS spectrum of CB_1-pVEC after purification. Figure S24: (A) UV-chromatogram and (B) corresponding ESI-MS spectrum of CB_2-pVEC after purification, Figure S25: (A) UV-chromatogram and (B) corresponding ESI-MS spectrum of CB_3-pVEC after purification Figure S26: (A) UV-chromatogram and (B) corresponding ESI-MS spectrum of CB_4-pVEC after purification, Figure S27: (A) UV-chromatogram and (B) corresponding ESI-MS spectrum of CB_5-pVEC after purification, Figure S28: UV-chromatogram overlay of CB-pVEC conjugates, Figure S29: CD spectra of CB-pVEC conjugates (20 μM peptide concentration) in 10 mM phosphate buffer (pH 7.0) with the addition of 50% TFE, Figure S30: Hemolytic activity of carborane-peptide conjugates towards human red blood cells (RBCs), Figure S31: Cytotoxicity profile of *meta*-1-carborane-carboxylic acid towards HeLa cells, Figure S32: Cellular uptake of sC18 into HeLa cells via flow cytometry, Figure S33: Internalization of CB-sC18 conjugates after 30 min,

Figure S34: Electrophoretic mobility shift assay (EMSA), Figure S35: Overlay of intensity profiles, Table S1: Names, sequences, analytical data and critical micelle concentration (CMC) of the CB-sC18 series, the CF-labeled CB-sC18 series and the CB-pVEC series.

Author Contributions: Conceptualization, T.L., M.S.S. and I.N.; methodology, T.L. and N.B.; investigation, T.L. and N.B.; writing—original draft preparation, T.L., N.B., M.S.S. and I.N.; writing—review and editing, T.L., M.S.S. and I.N.; visualization, M.S.S. and I.N.; funding acquisition, M.S.S. and I.N. All authors have read and agreed to the published version of the manuscript.

Funding: This research was funded by the JÜRGEN-MANCHOT STIFTUNG (T.L.) and the DEUTSCHE FORSCHUNGSGEMEINSCHAFT (DFG), grant number SCHO 1593/1-1 (M.S.).

Institutional Review Board Statement: Not applicable.

Informed Consent Statement: Not applicable.

Acknowledgments: We thank the CECAD Imaging Facility for their support in microscopy.

Conflicts of Interest: The authors declare no conflict of interest.

References

1. Wagstaff, K.; Jans, D. Protein Transduction: Cell Penetrating Peptides and Their Therapeutic Applications. *Curr. Med. Chem.* **2006**, *13*, 1371–1387. [CrossRef] [PubMed]
2. Di Pisa, M.; Chassaing, G.; Swiecicki, J.-M. Translocation Mechanism(s) of Cell-Penetrating Peptides: Biophysical Studies Using Artificial Membrane Bilayers. *Biochemistry* **2015**, *54*, 194–207. [CrossRef]
3. Kardani, K.; Milani, A.; Shabani, S.H.; Bolhassani, A. Cell Penetrating Peptides: The Potent Multi-Cargo Intracellular Carriers. *Expert Opin. Drug Deliv.* **2019**, *16*, 1227–1258. [CrossRef]
4. Yamashita, H.; Demizu, Y.; Shoda, T.; Sato, Y.; Oba, M.; Tanaka, M.; Kurihara, M. Amphipathic Short Helix-Stabilized Peptides with Cell-Membrane Penetrating Ability. *Bioorg. Med. Chem.* **2014**, *22*, 2403–2408. [CrossRef]
5. Copolovici, D.M.; Langel, K.; Eriste, E.; Langel, Ü. Cell-Penetrating Peptides: Design, Synthesis, and Applications. *ACS Nano* **2014**, *8*, 1972–1994. [CrossRef]
6. Aroui, S.; Brahim, S.; De Waard, M.; Bréard, J.; Kenani, A. Efficient Induction of Apoptosis by Doxorubicin Coupled to Cell-Penetrating Peptides Compared to Unconjugated Doxorubicin in the Human Breast Cancer Cell Line MDA-MB 231. *Cancer Lett.* **2009**, *285*, 28–38. [CrossRef] [PubMed]
7. Gessner, I.; Neundorf, I. Nanoparticles Modified with Cell-Penetrating Peptides: Conjugation Mechanisms, Physicochemical Properties, and Application in Cancer Diagnosis and Therapy. *Int. J. Mol. Sci.* **2020**, *21*, 2536. [CrossRef]
8. de Figueiredo, I.R.; Freire, J.M.; Flores, L.; Veiga, A.S.; Castanho, M.A.R.B. Cell-Penetrating Peptides: A Tool for Effective Delivery in Gene-Targeted Therapies. *IUBMB Life* **2014**, *66*, 182–194. [CrossRef]
9. Künnapuu, K.; Veiman, K.-L.; Porosk, L.; Rammul, E.; Kiisholts, K.; Langel, Ü.; Kurrikoff, K. Tumor Gene Therapy by Systemic Delivery of Plasmid DNA with Cell-Penetrating Peptides. *FASEB Bioadvances* **2018**, *1*, 105–114. [CrossRef] [PubMed]
10. Kurrikoff, K.; Langel, Ü. Recent CPP-Based Applications in Medicine. *Expert Opin. Drug Deliv.* **2019**, *16*, 1183–1191. [CrossRef] [PubMed]
11. Sebbage, V. Cell-Penetrating Peptides and Their Therapeutic Applications. *Biosci. Horiz. Int. J. Stud. Res.* **2009**, *2*, 64–72. [CrossRef]
12. Klimpel, A.; Lützenburg, T.; Neundorf, I. Recent Advances of Anti-Cancer Therapies Including the Use of Cell-Penetrating Peptides. *Curr. Opin. Pharmacol.* **2019**, *47*, 8–13. [CrossRef] [PubMed]
13. Tang, H.; Yin, L.; Kim, K.H.; Cheng, J. Helical Poly(Arginine) Mimics with Superior Cell-Penetrating and Molecular Transporting Properties. *Chem. Sci.* **2013**, *4*, 3839–3844. [CrossRef] [PubMed]
14. Karle, I.L. Controls Exerted by the Aib Residue: Helix Formation and Helix Reversal. *Pept. Sci.* **2001**, *60*, 351–365. [CrossRef]
15. Jobin, M.-L.; Blanchet, M.; Henry, S.; Chaignepain, S.; Manigand, C.; Castano, S.; Lecomte, S.; Burlina, F.; Sagan, S.; Alves, I.D. The Role of Tryptophans on the Cellular Uptake and Membrane Interaction of Arginine-Rich Cell Penetrating Peptides. *Biochim. Biophys. Acta (BBA)-Biomembr.* **2015**, *1848*, 593–602. [CrossRef] [PubMed]
16. Fernández-Carneado, J.; Kogan, M.J.; Van Mau, N.; Pujals, S.; López-Iglesias, C.; Heitz, F.; Giralt, E. Fatty Acyl Moieties: Improving Pro-Rich Peptide Uptake inside HeLa Cells. *J. Pept. Res.* **2005**, *65*, 580–590. [CrossRef]
17. Som, A.; Reuter, A.; Tew, G.N. Protein Transduction Domain Mimics: The Role of Aromatic Functionality. *Angew. Chem. Int. Ed.* **2012**, *51*, 980–983. [CrossRef] [PubMed]
18. Carrigan, C.N.; Imperiali, B. The Engineering of Membrane-Permeable Peptides. *Anal. Biochem.* **2005**, *341*, 290–298. [CrossRef] [PubMed]
19. Scholz, M.S.; Wingen, L.M.; Brunst, S.; Wittmann, S.K.; Cardoso, I.L.A.; Weizel, L.; Proschak, E. Soluble Epoxide Hydrolase Inhibitors with Carboranes as Non-Natural 3-D Pharmacophores. *Eur. J. Med. Chem.* **2020**, *185*, 111766. [CrossRef] [PubMed]
20. Worm, D.J.; Hoppenz, P.; Els-Heindl, S.; Kellert, M.; Kuhnert, R.; Saretz, S.; Köbberling, J.; Riedl, B.; Hey-Hawkins, E.; Beck-Sickinger, A.G. Selective Neuropeptide Y Conjugates with Maximized Carborane Loading as Promising Boron Delivery Agents for Boron Neutron Capture Therapy. *J. Med. Chem.* **2019**, *63*, 2358–2371. [CrossRef] [PubMed]

21. Scholz, M.; Hey-Hawkins, E. Carbaboranes as Pharmacophores: Properties, Synthesis, and Application Strategies. *Chem. Rev.* **2011**, *111*, 7035–7062. [CrossRef]
22. Hoppenz, P.; Els-Heindl, S.; Kellert, M.; Kuhnert, R.; Saretz, S.; Lerchen, H.-G.; Köbberling, J.; Riedl, B.; Hey-Hawkins, E.; Beck-Sickinger, A.G. A Selective Carborane-Functionalized Gastrin-Releasing Peptide Receptor Agonist as Boron Delivery Agent for Boron Neutron Capture Therapy. *J. Org. Chem.* **2020**, *85*, 1446–1457. [CrossRef]
23. Michiue, H.; Sakurai, Y.; Kondo, N.; Kitamatsu, M.; Bin, F.; Nakajima, K.; Hirota, Y.; Kawabata, S.; Nishiki, T.I.; Ohmori, I.; et al. The Acceleration of Boron Neutron Capture Therapy Using Multi-Linked Mercaptoundecahydrododecaborate (BSH) Fused Cell-Penetrating Peptide. *Biomaterials* **2014**, *35*, 3396–3405. [CrossRef]
24. Lützenburg, T.; Neundorf, I.; Scholz, M. Direct Carborane-Peptide Conjugates: Synthesis and Evaluation as Non-Natural Lipopeptide Mimetics. *Chem. Phys. Lipids* **2018**, *213*, 62–67. [CrossRef]
25. Neundorf, I.; Rennert, R.; Hoyer, J.; Schramm, F.; Löbner, K.; Kitanovic, I.; Wölfl, S. Fusion of a Short HA2-Derived Peptide Sequence to Cell-Penetrating Peptides Improves Cytosolic Uptake, but Enhances Cytotoxic Activity. *Pharmaceuticals* **2009**, *2*, 49–65. [CrossRef]
26. Negrete-Hurtado, A.; Overhoff, M.; Bera, S.; De Bruyckere, E.; Schätzmüller, K.; Kye, M.J.; Qin, C.; Lammers, M.; Kondylis, V.; Neundorf, I.; et al. Autophagy Lipidation Machinery Regulates Axonal Microtubule Dynamics but Is Dispensable for Survival of Mammalian Neurons. *Nat. Commun.* **2020**, *11*, 1535. [CrossRef]
27. Feni, L.; Parente, S.; Robert, C.; Gazzola, S.; Arosio, D.; Piarulli, U.; Neundorf, I. Kiss and Run: Promoting Effective and Targeted Cellular Uptake of a Drug Delivery Vehicle Composed of an Integrin-Targeting Diketopiperazine Peptidomimetic and a Cell-Penetrating Peptide. *Bioconjugate Chem.* **2019**, *30*, 2011–2022. [CrossRef]
28. Klimpel, A.; Stillger, K.; Wiederstein, J.L.; Krüger, M.; Neundorf, I. Cell-Permeable CaaX-Peptides Affect K-Ras Downstream Signaling and Promote Cell Death in Cancer Cells. *FEBS J.* **2021**, *288*, 2911–2929. [CrossRef]
29. Feni, L.; Jütten, L.; Parente, S.; Piarulli, U.; Neundorf, I.; Diaz, D. Cell-Penetrating Peptides Containing 2,5-Diketopiperazine (DKP) Scaffolds as Shuttles for Anti-Cancer Drugs: Conformational Studies and Biological Activity. *Chem. Commun.* **2020**, *56*, 5685–5688. [CrossRef]
30. Ihsan, A.B.; Nargis, M.; Koyama, Y. Effects of the Hydrophilic-Lipophilic Balance of Alternating Peptides on Self-Assembly and Thermo-Responsive Behaviors. *Int. J. Mol. Sci.* **2019**, *20*, 4604. [CrossRef]
31. Sparks, D.L.; Lund-Katz, S.; Phillips, M.C. The Charge and Structural Stability of Apolipoprotein A-I in Discoidal and Spherical Recombinant High Density Lipoprotein Particles. *J. Biol. Chem.* **1992**, *267*, 25839–25847. [CrossRef]
32. Lakshminarayanan, R.; Fan, D.; Du, C.; Moradian-Oldak, J. The Role of Secondary Structure in the Entropically Driven Amelogenin Self-Assembly. *Biophys. J.* **2007**, *93*, 3664–3674. [CrossRef]
33. Elmquist, A.; Lindgren, M.; Bartfai, T.; Langel, U. VE-Cadherin-Derived Cell-Penetrating Peptide, PVEC, with Carrier Functions. *Exp. Cell Res.* **2001**, *269*, 237–244. [CrossRef] [PubMed]
34. Gessner, I.; Klimpel, A.; Klußmann, M.; Neundorf, I.; Mathur, S. Interdependence of Charge and Secondary Structure on Cellular Uptake of Cell Penetrating Peptide Functionalized Silica Nanoparticles. *Nanoscale Adv.* **2020**, *2*, 453–462. [CrossRef]
35. Gronewold, A.; Horn, M.; Ranđelović, I.; Tóvári, J.; Muñoz Vázquez, S.; Schomäcker, K.; Neundorf, I. Characterization of a Cell-Penetrating Peptide with Potential Anticancer Activity. *ChemMedChem* **2017**, *12*, 42–49. [CrossRef]
36. Hedegaard, S.F.; Derbas, M.S.; Lind, T.K.; Kasimova, M.R.; Christensen, M.V.; Michaelsen, M.H.; Campbell, R.A.; Jorgensen, L.; Franzyk, H.; Cárdenas, M.; et al. Fluorophore Labeling of a Cell-Penetrating Peptide Significantly Alters the Mode and Degree of Biomembrane Interaction. *Sci. Rep.* **2018**, *8*, 6327. [CrossRef] [PubMed]
37. Birch, D.; Christensen, M.V.; Staerk, D.; Franzyk, H.; Nielsen, H.M. Fluorophore Labeling of a Cell-Penetrating Peptide Induces Differential Effects on Its Cellular Distribution and Affects Cell Viability. *Biochim. Biophys. Acta (BBA)-Biomembr.* **2017**, *1859*, 2483–2494. [CrossRef]
38. Nair, P.; Dey, D.; Borkotoky, S.; Shukla, A.; Banerjee, M. Hydrophobicity and Oligomerization Are Essential Parameters for Membrane Penetration Activity of the VP4 Peptide from Hepatitis A Virus (HAV). *Arch. Biochem. Biophys.* **2019**, *678*, 108188. [CrossRef] [PubMed]
39. Sani, M.-A.; Separovic, F. How Membrane-Active Peptides Get into Lipid Membranes. *Acc. Chem. Res.* **2016**, *49*, 1130–1138. [CrossRef]
40. Takayama, K.; Nakase, I.; Michiue, H.; Takeuchi, T.; Tomizawa, K.; Matsui, H.; Futaki, S. Enhanced Intracellular Delivery Using Arginine-Rich Peptides by the Addition of Penetration Accelerating Sequences (Pas). *J. Control. Release* **2009**, *138*, 128–133. [CrossRef] [PubMed]
41. Eiríksdóttir, E.; Konate, K.; Langel, Ü.; Divita, G.; Deshayes, S. Secondary Structure of Cell-Penetrating Peptides Controls Membrane Interaction and Insertion. *Biochim. Biophys. Acta (BBA)-Biomembr.* **2010**, *1798*, 1119–1128. [CrossRef] [PubMed]

pharmaceutics

MDPI

Article

Bi-Functional Peptides as a New Therapeutic Tool for Hepatocellular Carcinoma

Eric Savier [1,2], Lorena Simon-Gracia [3], Frederic Charlotte [4], Pierre Tuffery [5], Tambet Teesalu [3,6], Olivier Scatton [1,2] and Angelita Rebollo [7,*]

1 Department of Hepatobiliary and Liver Transplantation Surgery, AP-HP, Pitié–Salpêtrière Hospital, Sorbonne Université, 75006 Paris, France; eric.savier@aphp.fr (E.S.); olivier.scatton@gmail.com (O.S.)
2 Sant Antoine Research Center (CRSA), Institut Nationale de la Santé et la Recherche Médicale (Inserm), Institute of Cardiometabolism and Nutrition (ICAN), Sorbonne Université, 75006 Paris, France
3 Laboratory of Precision and Nanomedicine, Institute of Biomedicine and Translational Medicine, University of Tartu, 50090 Tartu, Estonia; lorenasimongracia@gmail.com (L.S.-G.); tambet.teesalu@ut.ee (T.T.)
4 Department of Pathology, AP-HP, Pitié–Salpêtrière Hospital, 75006 Paris, France; frederic.charlotte@aphp.fr
5 Biologie Fontionelle Adaptative (BFA), Unité Mixte de Recherche (UMR) 8251, Centre National de la Recherche Scientifique (CNRS) ERL U1133, Inserm, Université de Paris, 75006 Paris, France; pierre.tuffery@univ-paris-diderot.fr
6 Center for Nanomedicine and Department of Cell, Molecular and Developmental Biology, University of California, Santa Barbara, CA 93106, USA
7 Faculté de Pharmacie, Unité des Technologies Chimiques et Biologiques pour la Santé (UTCBS), Inserm U1267, Centre National de la Recherche Scientifique CNRS UMR8258, Université de Paris, 75006 Paris, France
* Correspondence: angelita.rebollo@parisdescartes.fr

Citation: Savier, E.; Simon-Gracia, L.; Charlotte, F.; Tuffery, P.; Teesalu, T.; Scatton, O.; Rebollo, A. Bi-Functional Peptides as a New Therapeutic Tool for Hepatocellular Carcinoma. *Pharmaceutics* **2021**, *13*, 1631. https://doi.org/10.3390/pharmaceutics13101631

Academic Editors: Prisca Boisguérin and Sébastien Deshayes

Received: 6 September 2021
Accepted: 30 September 2021
Published: 6 October 2021

Publisher's Note: MDPI stays neutral with regard to jurisdictional claims in published maps and institutional affiliations.

Copyright: © 2021 by the authors. Licensee MDPI, Basel, Switzerland. This article is an open access article distributed under the terms and conditions of the Creative Commons Attribution (CC BY) license (https://creativecommons.org/licenses/by/4.0/).

Abstract: Background: The interfering peptides that block protein–protein interactions have been receiving increasing attention as potential therapeutic tools. Methods: We measured the internalization and biological effect of four bi-functional tumor-penetrating and interfering peptides into primary hepatocytes isolated from three non-malignant and 11 hepatocellular carcinomas. Results: These peptides are internalized in malignant hepatocytes but not in non-malignant cells. Furthermore, the degree of peptide internalization correlated with receptor expression level and tumor aggressiveness levels. Importantly, penetration of the peptides iRGD-IP, LinTT1-IP, TT1-IP, and RPARPAR-IP induced apoptosis of the malignant hepatocytes without effect on non-malignant cells. Conclusion: Receptor expression levels correlated with the level of peptide internalization and aggressiveness of the tumor. This study highlights the potential to exploit the expression of tumor-penetrating peptide receptors as a predictive marker of liver tumor aggressiveness. These bi-functional peptides could be developed for personalized tumor treatment.

Keywords: hepatocellular carcinoma; tumor-penetrating peptides; interfering peptides

1. Introduction

Despite significant progress in translational cancer research, advances in the design of targeted anti-cancer therapies have remained disappointingly slow [1–6]. The two most important issues with current cancer therapies are the lack of tumoral specificity and the lack of selectivity. Treatments thus induce off-target effects and adverse side effects, and the amount of drug that actually reaches its target remains relatively low. Consequently, there is a real need for selective anti-cancer drugs.

Various targeted delivery strategies have been developed in an effort to overcome these limitations. One strategy is the use of Tumor-Penetrating Peptides (TPP), which are recognized as tumor-specific drug delivery vehicles that can penetrate into tumor cells to deliver cargo. TPPs are internalized via specific receptors expressed on tumor cells and vasculature [7,8], and are characterized by the presence of the C-end Rule (CendR) motif with the consensus sequence R/KXXR/K [9,10]. This motif has to be C-terminally exposed to allow tumor-specific binding and penetration via the Neuropilin-1 (NRP-1) receptor.

RPARPAR is a prototypic CendR peptide that binds and internalizes via the NRP-1 receptor [11,12]. Another widely used TPP is iRGD (CRGDKGPDC), which is recruited to tumors via interaction with the integrins $\alpha v\beta 3/5$ through the RGD motif. On the cell surface, the tumor proteases cleave iRGD to C-terminally expose the CRGDK CendR motif that triggers internalization with the NRP-1 receptor [13]. The cyclic TT1 and its linear version, LinTT1, bind first to the p32 protein expressed on the tumor cell surface [14]. Both are also cleaved by tumoral proteases, thus exposing the CendR sequence that can then bind to NRP-1 [14–19].

TPPs are widely used as tumor-homing affinity ligands in targeted therapies, as they combine tumor specificity with cargo to reduce toxicity and increase efficacy [17,20–25]. In previous work, we paired TPPs with an Interfering Peptide (IP) that blocks the interaction between the phosphatase PP2A and its physiological inhibitor, the oncoprotein SET. This leads to bi-functional peptides that are able to specifically target tumoral cells where, once internalized, they dissociate the PP2A/SET interaction [26].

Hepatocellular Carcinoma (HCC) is a primary liver cancer that originates from hepatocytes [27]. HCC is the sixth most frequent cancer and the fourth leading cause of cancer-related mortality worldwide. Risk factors for HCC include viral infection, alcohol abuse, non-alcoholic fatty liver disease, certain toxins, and genetic diseases. These factors are responsible for chronic liver inflammation, fibrosis, and ultimately cellular transformation and liver function impairment [28]. Multimodal lines of therapy against HCC include surgical resection, chemotherapy, or radiotherapy. Tyrosine kinase inhibitors have some efficacy but are contraindicated in cases that involve altered liver function. Immunotherapy treatments hold promise for treating HCC [28], but more efficient targeted therapies are needed. An alternative way to treat HCC could be to use a dual-peptide strategy combining a TPP with an IP. Here, we show that the fused TPP-IPs can selectively internalize into primary tumoral hepatocytes isolated from HCC patients. Our results show that level of TPP-IP receptor expression by HCC tumor cells correlates with degree of peptide internalization and tumor aggressiveness, which raises prospects for a selective liver tumor-targeting approach.

2. Materials and Methods

2.1. Patients

Samples of benign and tumoral liver were collected from 14 patients. All patients gave informed consent. Samples 1 to 3 correspond to non-malignant tumors or necrotic HCC (following sorafenib treatment), and samples 4 to 14 correspond to HCC (Table 1). A tumor aggressiveness score was calculated based on histological or biological factors known to be associated with poor prognosis. This tumor aggressiveness score included tumor encapsulation [29], tumor differentiation [30], presence of satellite nodules, vascular invasion, macrotrabecular type [31], and $\log_{10}$ of the preoperative Alpha Feto Protein (AFP) value [32]. Tumors that scored 0 had zero aggressiveness, tumors that scored <6 were considered moderately aggressive, and tumors that scored >6 were considered highly aggressive.

Table 1. Clinical characteristics of the patients.

Patient	Sex	Age	Type Tumor	AFP	Log_{10} AFP	Partially Encapsulated (0/1)	Satellite Nodule (0/1)	Vascular Invasion (0/1)	Differentiation (1/2/3) [1]	Macrotrabecular (0/1)	Aggressiveness [2]	Aggression Class
1	F	48.5	Hepatocellular adenoma	1.6	0	0	0	0	0	0	0	Null
2	M	65.3	Necrotic lymph node	2.6	0	0	0	0	0	0	0	Null
3	F	53.5	Angiomyolipoma	4.9	0	0	0	0	0	0	0	Null
4	M	56.5	Microtrabecular and pseudoglandular, Nuclear grade 2	7.7	0	1	0	0	2	0	3	Moderate
5	F	47.4	Microtrabecular	1010	3	1	1	0	2	0	7	High
6	F	59.9	Trabecular, Edmondson grade 2, nuclear grade 2	28,000	4	1	0	0	2	0	7	High
7	M	76.4	Microtrabecular and pseudoglandular	6662	3	1	0	0	2	1	7	High

Table 1. *Cont.*

Patient	Sex	Age	Type Tumor	AFP	Log_{10} AFP	Partially En-capsulated (0/1)	Satellite Nodule (0/1)	Vascular Invasion (0/1)	Differentiation (1/2/3) [1]	Macrotrabecular (0/1)	Aggressiveness [2]	Aggression Class
8	M	47.1	Macro-trabecular, Edmonson grade 3, nuclear grade 3	6	0	1	1	1	2	1	6	High
9	M	73.4	Edmondson grade 3, nuclear grade 3	1.4	0	1	0	0	2	1	4	Moderate
10	M	67.3	Edmonson grade 2, nuclear grade 2	6.4	0	1	0	0	2	0	3	Moderate
11	M	57.5	Macrotrabecular	5.1	0	1	1	1	2	1	6	High
12	M	69.3	Trabecular, Edmonson grade 2, nuclear grade 2	343	2	1	0	2	2	0	6	High
13	M	68.7	Edmonson grade 2 HCC, nuclear grade 2	2.5	0	1	0	2	2	0	3	Moderate
14	M	78.8	Trabecular	341	2	0	0	2	2	0	5	Moderate

[1] Well differentiated HCC = 1, moderately differentiated HCC = 2, undifferentiated HCC = 3. [2] Sum of Log_{10} AFP, partially encapsulated, satellite nodule, vascular invasion, differentiation, macrotabecular HCC (min = 0, max: 7).

2.2. Peptide Synthesis and Sequences

The peptides were synthesized in an automated multiple peptide synthesizer with solid-phase and standard Fmoc chemistry (GL Biochem, Shanghai, China). The characterization was performed by High-Performance Liquid Chromatography (HPLC, Shimadzu France, Marne-la-Valle) and Mass Spectrometry (MS, Bruker, Wissembourg, France). For internalization experiments, the peptides were synthesized with a fluorochrome (FITC, Sigma-Aldrich, Saint Quentin, France). The peptides and sequences used are shown in Table 2.

Table 2. Sequence of the peptides used in this study.

Peptide ID	Sequence
iRGD-IP	FITC -Ahx-ETVTLLVALKVRYRERIT-Ahx-CRGDKGPDC-$CONH_2$ (C-C disulfide bond)
RPARPAR-IP	FITC -Ahx-ETVTLLVALKVRYRERIT-Ahx-RPARPAR-OH
LinTT1-IP	FITC -Ahx-ETVTLLVALKVRYRERIT-Ahx-AKRGARSTA-$CONH_2$
TT1-IP	FITC -Ahx-ETVTLLVALKVRYRERIT-Ahx-CKRGARSTC-$CONH_2$ (C-C disulfide bond)

Ahx: aminohexanoic acid.

2.3. Isolation and Culture of Primary Human Tumoral Hepatocytes

Healthy hepatocytes were isolated from patient samples following a protocol previously described [26]. Human tumoral hepatocytes were isolated from tumoral liver samples collected from adult patients undergoing surgery. Samples were cut into small pieces and treated with 4 mL of dispase (Gibco, Ref 17105-041, Thermo Fisher, France; 10 mg/mL in PSA buffer) (NaCl 8 g/L, KCl 0.2 g/L, glucose 1 g/L, $NaHCO_3$ 0.35 g/L, phenol red 1 mL/L; Thermo Fisher, France) and 2 mL of collagenase type I (Gibco Ref 17100-017; 5 mg/mL in PSA buffer). Samples were incubated at 37 °C under agitation for a maximum of 1 h, and then the solution was filtered and passed through needles of different diameters. The volume was filled up to 50 mL with culture medium and centrifuged at 177× g for 5 min. Arginase expression confirmed that the isolated cells were hepatocytes. The supernatant was discarded, and the cells were cultured in DMEM medium (Thermo Fisher, France) supplemented with 10% Fetal Calf Serum (FCS, Gibco, Thermo Fisher, France) and antibiotics until treatment with peptides. The hepatocytes were maintained in culture for no more than 36 h to ensure that they did not enter differentiation.

2.4. Quantification of Cellular Internalization

Primary human hepatocytes were seeded overnight on 24-well plates and then incubated for 4 h with FITC-labeled peptides. After treatment, cells were detached, treated with trypsin (Gibco, Thermo Fisher, France) to remove non-internalized peptides, washed twice with PBS (Gibco, Thermo Fisher, France) to remove free peptides, and resuspended in 200 µL of PBS. FITC fluorescence intensity of internalized peptides was measured using a FACSCanto II flow cytometry system (Beckton Dickinson, Franklin Lakes, NJ, USA). Data were analyzed using FACSDiva 6.1.3 software (DB Biosciences, Franklin Lakes, NJ, USA). Healthy primary hepatocytes were used as control. For detection of TPP receptors on the cell surface, anti-p32 (Sigma Aldrich, St. Louis, MO, USA, AB2991), anti-NRP-1 (antibody generated in house, prepared by immunizing rabbits with human recombinant NRP1, followed by affinity purification) [24], and anti-integrin v/β3 (Abcam, Cambridge, UK, ab203123) antibodies were incubated with the cells for 30 min at room temperature. Cells were then washed and incubated with fluorophore-labeled secondary goat anti-mouse antibody (Alexa Fluor 647 goat anti-mouse, Thermo Fisher, Waltham, MA, USA, A-21238) or goat anti-rabbit antibody (Thermo Fisher, Waltham, MA, USA, A48285) for 10 min. Cells were then washed again, and receptor expression was analyzed by flow cytometry as described above.

2.5. *Immunohistochemistry*

The immunostaining procedure was performed on formalin-fixed, deparaffinized, 3μm-thick sections using a Ventana Benchmark Ultra platform (Roche Diagnostics, Basel, Switzerland) and the Ultraview visualization system (Roche Diagnostics, Basel, Switzerland) according to the manufacturer's instructions. The following primary antibodies were used: mouse monoclonal anti-CK19 antibody (dilution 1/100; clone RCK108; ref. M088801-2, Agilent, Santa Clara, CA, USA) followed by CC1 antigen retrieval buffer (36 min, 95 °C) and an antibody incubation time of 20 min at 20 °C; mouse monoclonal anti-human hepatocyte (HepPar) (dilution 3/100; clone OCH1E5; ref. M715801-2, Agilent, Santa Clara, CA, USA) followed by CC1 antigen retrieval buffer (64 min, 95 °C) and an antibody incubation time of 32 min at 20 °C; mouse monoclonal anti-human Glypican-3 (prediluted; clone 1G12; ref. F/261M-98, MM, Brignais, France) followed by CC1 antigen retrieval buffer (64 min, 95 °C) and an antibody incubation time of 32 min at 37 °C; mouse monoclonal anti-human b-catenin (prediluted; clone 14; ref. 05269016001, Roche Diagnostics, Basel, Switzerland) followed by CC1 antigen retrieval buffer (64 min, 95 °C) and an antibody incubation time of 32 min at 37 °C, and mouse monoclonal anti-human glutamine synthetase (prediluted; clone GS6; ref. 07107757001, Roche Diagnostics, Basel, Switzerland) followed by antigen retrieval protease (4 min, 20 °C) and an antibody incubation time of 40 min at 20 °C.

2.6. *Detection of Apoptosis by Annexin-V Staining*

The degree of apoptosis induced by the four TPP-IPs on primary benign and tumoral treated hepatocytes was measured by flow cytometry on cells stained with annexin-V FITC (Biosciences, Fischer Scientific, Hampton, NH, USA). The primary cells were incubated with the peptides for 12 h at 37 °C in DMEM supplemented with 10% FCS (Gibco, Thermo Fisher, France), then washed and treated according to the manufacturer's protocol. Level of apoptosis was measured using FACSCanto II flow cytometry system (Becton Dickinson Biosciences, Franklin Lakes, NJ, USA).

2.7. *Immunoprecipitation and Western Blotting*

MDA-MB231 cells (ATCC, HTB-26) (5×10^6) were lysed for 20 min at 4 °C in lysis buffer (50 mM Tris pH8, 1% NP40, 137 mM NaCl, 1 mM $MgCl_2$, 1 mM $CaCl_2$, 10% glycerol and protease inhibitor mixture, Sigma Aldrich, St. Louis, MO, USA). Lysates (500 μg) were immunoprecipitated with the appropriate antibody overnight at 4 °C, and protein A/G Sepharose (Santa Cruz, Dallas, TX, USA) was added for 1 h at 4 °C. After washing with TBST (20 mM Tris-HCl pH7.5, 150 mM NaCl, 0.05% Tween 20; Gibco, Thermo Fisher, France), the PP2A/SET interaction was competed using 1 mM of PP2A/SET or Ras/Raf IP (GL Biochem, Shanghai, China) for 30 min at room temperature. After several washing steps, immunoprecipitates were separated by SDS-PAGE, transferred to nitrocellulose, and blotted with anti-PP2A antibody (Sigma Aldrich, St. Louis, MO, USA). The membrane was washed and incubated with HRP-conjugated secondary antibody (Dako, Hamburg, Germany, 1:1000 dilution). Protein detection was performed using the ECL system (Bio-

Rad, Hercules, CA, USA). The blot was also hybridized with anti-SET antibody as internal control (Thermo Fischer, Waltham, MA, USA, MA5-34662).

2.8. Statistical Analysis

The data were analyzed using SigmaPlot version 12.0, Systat Software, Inc. (D-40699, Erkrath, Germany) and logarithmic regression with StatView version 5.0 for Windows SAS Institute Inc. Statistical tests used include *t*-tests and a Mann-Whitney rank sum test or Pearson correlations, as appropriate. Values of $p < 0.05$ were considered statistically significant.

3. Results

3.1. Clinical Characteristics of the Patients and Tumor Aggressiveness Classification

Samples from 14 patients were analyzed. Of these, three were from non-malignant tumors and 11 from HCC. The patient population had a median age of 62 years (range: 47–78 years) with a large predominance of males (71%). Clinical aggressiveness was calculated according to six parameters: AFP, non-encapsulation, satellite nodules, vascular embolization, differentiation, and macrotrabecular type (Table 1). Samples 1 to 3 corresponded to zero aggressiveness hepatocellular adenoma, necrotic tissue and angiomyolipoma, respectively. Samples 4 to 14 corresponded to HCC of moderate or high aggressiveness and were classified using the following parameters: for encapsulation, non-encapsulated = 0, partially encapsulated = 1; for differentiation, well-differentiated = 1, moderately differentiated = 2, undifferentiated = 3; for satellite nodes, positive = 1, negative = 0; for vascular invasion, positive = 1, negative = 0; for macrotrabecular type, positive = 1, negative = 0.

3.2. Immunohistochemical Characteristics of the Patients

The immunohistochemical markers that were analyzed in the patient samples were: CK19, to differentiate HCC from cholangiocarcinoma; HepPar, a marker that differentiates HCC from metastatic carcinoma [33]; GPC3, a member of the glypican family involved in progression of HCC [34]; β-catenin, a marker of development and progression of HCC [35]; and glutamine synthetase [36], which may enhance metastatic potential in HCC. Absence of CK19 expression confirmed that the patient samples corresponded to HCC but not to cholangiocarcinoma. Note that samples from patients #7, #8, #11 and #12, which were classified as highly aggressive HCC, showed the highest levels of HepPar marker expression. Similarly, samples from patients #6 and #7, which were also classified as highly aggressive HCCs, expressed the highest levels of glutamine synthase (Table 3).

Table 3. Immunohistochemical characteristics of the patients.

Patient	CK19	HepPar	GPC3	Nuclear β-Catenin	Glutamine Synthetase
1	–	+	–	+	–
2	necrosis	necrosis	necrosis	necrosis	necrosis
3	–	–	–	–	–
4	–	+	+	10–20%	+++
5	–	–	–	0	0
6	–	+	–	–	++
7	–	+++	–	+	+++
8	+	+++	–	–	–
9	–	+	+	–	–
10	–	+	–	–	+
11	–	+++	+	–	–
12	–	+++	+	–	–
13	–	+++	–	–	–
14	–	+	+++	+	–

0: no material was available.

3.3. *In Vitro Competition against PP2A/SET Interaction*

In vitro competition testing was performed to confirm that the IP targeted the PP2A/SET interaction. Lysates from MDA-MB231 cells were immunoprecipitated with anti-PP2A antibody, and the interaction with SET was competed using IP PP2A/SET (Figure 1). SET was detected in the control immunoprecipitates and in immunoprecipitates after competition with the IP disrupting the Ras/Raf interaction (peptide sequence: MEHIQGAWKTISGFGLK), whereas the levels detected were much lower after competition with 1 mM of the IP blocking the PP2A/SET interaction. PP2A was used as internal control of protein loading.

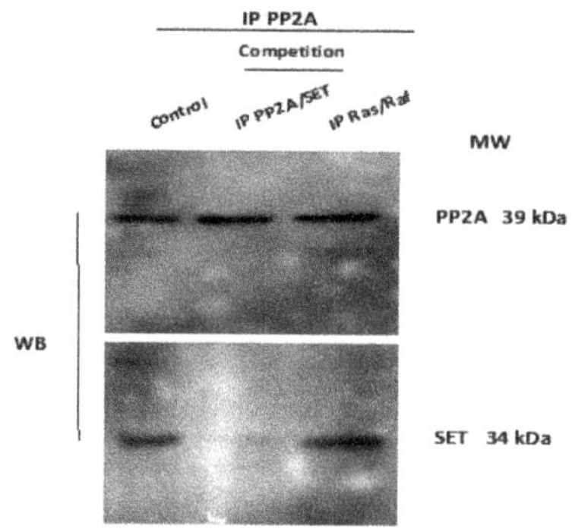

Figure 1. The IP disrupted PP2A/SET interaction in vitro. Lysates were immunoprecipitated with anti PP2A antibody. The PP2A/SET interaction was competed in vitro with 1 mM of the PP2A/SET IP and with an irrelevant Ras/Raf IP used as a negative control. The total amount of PP2A was used as internal control.

3.4. *Internalization of Tumor-Penetrating and Interfering Peptides (TPP-IP) into Primary Tumoral Hepatocytes via Specific Receptors*

We generated four bi-functional peptides composed of a TPP (iRGD, RPARPAR, LinTT1, or TT1) paired with the IP blocking the interaction between the phosphatase PP2A and its physiological inhibitor, the oncoprotein SET. These peptides penetrated specifically into tumoral B-cells [26]. We analyzed the intracellular penetration of these TPP-IP in a group of 14 samples of non-malignant tumors (samples #1 to #3) or HCC (samples #4 to #14), graded according to histological tumor type.

Figure 2A shows that none of the four TPP-IPs penetrated non-malignant tumors. In HCC (samples #4 to #14), iRGD-IP showed the lowest level of internalization but with a significant difference between benign and aggressive tumors ($p = 0.02$). Bi-functional peptides RPARPAR-IP ($p = 0.005$), LinTT1-IP ($p = 0.002$), and TT1-IP ($p = 0.005$) showed higher levels of penetration in tumoral hepatocytes, again with a significant difference compared to non-malignant samples (Figure 2A). Interestingly, RPARPAR-IP showed the highest level of internalization in HCC samples, ahead of LinTT1-IP and TT1-IPs which showed very similar levels of internalization (Figure 2A). Crucially, none of the TPP-IPs internalized into healthy hepatocytes (control in Figure 2A).

Given that these peptides are internalized by tumoral hepatocytes via specific receptors on tumoral cells, we analyzed the expression of integrin v/β3, p32 and NRP-1. Figure 2B shows that samples #1 to #3 (non-malignant tumors) and healthy control hepatocytes all showed very low levels of cell surface receptor expression, whereas samples #4 to #14 (HCC tumors) showed significantly higher receptor expression levels compared to non-malignant tumors ($p = 0.05$ for integrin v/β3; $p = 0.05$ for p32; $p = 0.05$ for NRP-1). We previously showed that the IP without TPP failed to internalize into malignant B cells and tumoral hepatocytes, whereas a non-tumoral-specific cell-penetrating peptide alone or combined with the IP effectively internalized in both malignant and healthy B cells and hepatocytes. The new results reported here confirm that the specific internalization of the TPP-IPs into tumor cells is due to internalization via specific receptors.

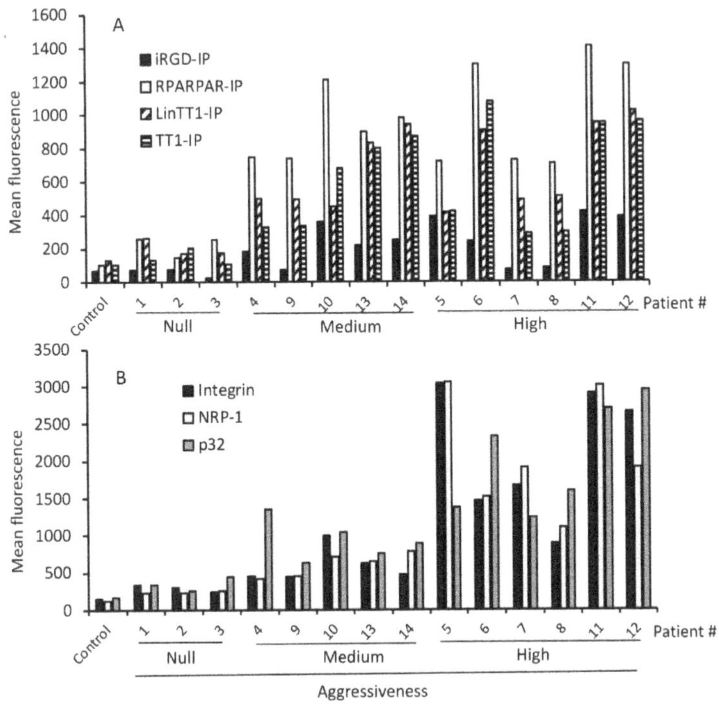

Figure 2. Selective internalization of TPP-IPs and receptor expression in malignant hepatocytes. (**A**) TPP-IP internalization. Hepatocytes isolated from benign or malignant liver patient samples were incubated for 4 h with 25 µM of FITC-labeled peptides. The mean fluorescence of internalized peptides was analyzed by flow cytometry and compared to control healthy hepatocytes. Internalization of the TPP-IPs was significantly higher in HCC compared to non-malignant tumors (HCC versus non-malignant tumors or normal hepatocytes respectively, $n = 4$ to 11 per group, mean ± standard error, iRGD-IP: 61 ± 12 vs. 244 ± 40, $p = 0.022$; RPARPAR-IP: 191 ± 38 vs. 977 ± 83, $p = 0.005$; LinTT1-IP: 184 ± 27 vs. 682 ± 73, $p = 0.002$; TT1-IP: 137 ± 23 vs. 637 ± 93, $p = 0.005$). (**B**) Receptor expression. Hepatocytes isolated from non-malignant or tumoral liver samples were incubated with antibodies against NRP-1, p32, and integrin v/b3, followed by an APC-labeled secondary antibody. Samples were analyzed by flow cytometry. Healthy hepatocytes were used as control. Receptor expression levels were significantly higher in tumoral hepatocytes than in non-malignant tumors (HCCs vs. non-malignant tumors or normal hepatocytes, respectively, $n = 4$ to 11 per group, Integrin v/b3: 254 ± 40 vs. 1416 ± 306, $p = 0.005$; NRP-1: 211 ± 31 vs. 1410 ± 290, $p = 0.005$; p32: 303 ± 58 vs. 1530 ± 238, $p = 0.01$).

3.5. TPP Internalization and Receptor Expression Correlated with Tumor Aggressiveness

iRGD is recruited via interaction with integrins and then cleaved by tumoral proteases, thus allowing interaction with the NRP-1 receptor. Similarly, LinTT1 and TT1 first bind to p32, a mitochondrial protein aberrantly expressed on the cell surface of tumoral cells and are then cleaved by proteases expressed by the tumor cells, allowing them to interact with NRP-1. Finally, the RPARPAR peptide binds directly to tumoral cells expressing NRP-1.

We tested whether there was a correlation between the level of primary receptor expression on the tumoral cells and level of peptide internalization. Figure 3A shows a low level of iRGD-IP internalization (compared with RPARPAR-IP, LinTT1-IP, and TT1-IP) and variable expression of its receptor, integrin v/β3. The highest levels of integrin v/β3 expression were found in samples from patients #5, #6, #7, #11, and #12, which matched to the samples with the high tumor aggressiveness scores. Moreover, samples #5, #11, and #12 showed a higher degree of internalized iRGD-IP, which also matched with high tumor aggressiveness. A similar pattern was found for NRP-1 receptor expression and RPARPAR-IP internalization (Figure 3B), where the highest level of NRP-1 expression was found in samples from patients #5, #6, #7, #11, and #12, and the highest RPARPAR-IP internalization was found in samples #6, #11, and #12 that also corresponded to the most aggressive tumors. Finally, there was a different pattern of p32 receptor expression and LinTT1-IP/TT1-IP internalization, with the highest expression of the receptor in samples from patients #6, #11, and #12 that were classified as aggressive tumors, and these same samples also showed the highest peptide internalization (Figure 3C).

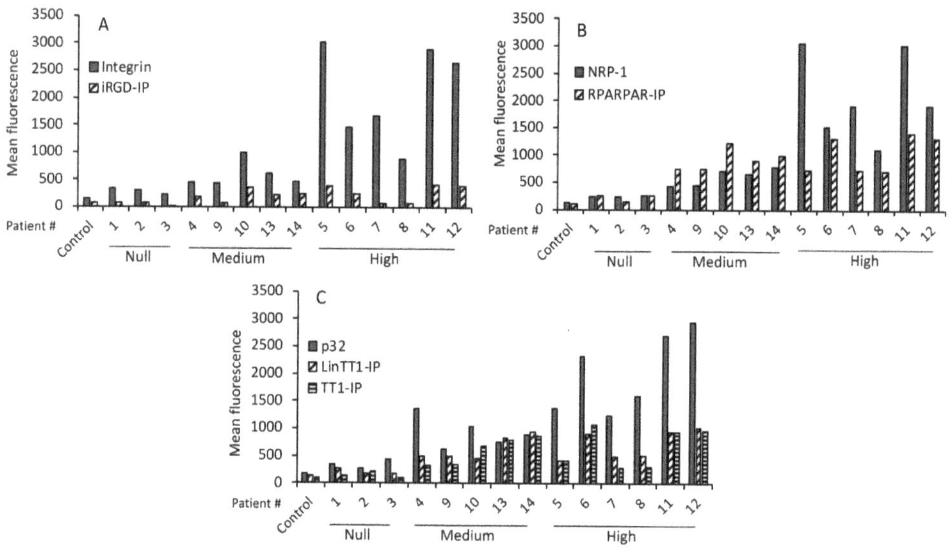

Figure 3. Primary receptor expression and internalization of TPP-IPs. (**A**) Expression of primary receptor integrin and iRGD-IP internalization. (**B**) Expression of NRP-1 receptor and RPARPAR-IP internalization. (**C**) Expression of the primary p32 receptor and LinTT-1-IP and TT1-IP internalization. Data are from Figure 2.

Analysis of NRP-1 expression levels in comparison to internalization of iRGD-IP (Figure 4A), LinTT1-IP and TT1-IP (Figure 4B), or RPARPAR-IP (Figure 3B) found that samples with the highest receptor expression also had high tumor aggressiveness scores and showed prominent TPP-IP internalization.

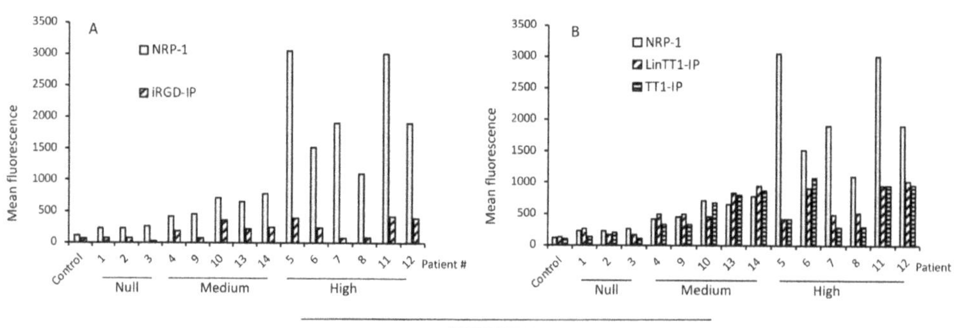

Figure 4. Secondary receptor expression and internalization of TPP-IPs. (**A**) Expression of secondary receptor (NRP-1) and iRGD-IP internalization. (**B**) Expression of secondary receptor (NRP-1) and internalization of LinTT1-IP and TT1-IP. Data are from Figure 2.

Figure 5A–C shows that there was a significant correlation between the levels of primary receptor expression (integrin vβ3, NRP-1 and p32) and peptide internalization (p = 0.010 for iRGD-IP; p = 0.045 for RPARPAR-IP; p = 0.02 for LinTT1-IP; p = 0.03 for TT1-IP). Figure 5C also shows that tumor aggressiveness score correlates with TPP-IP internalization (p = 0.02).

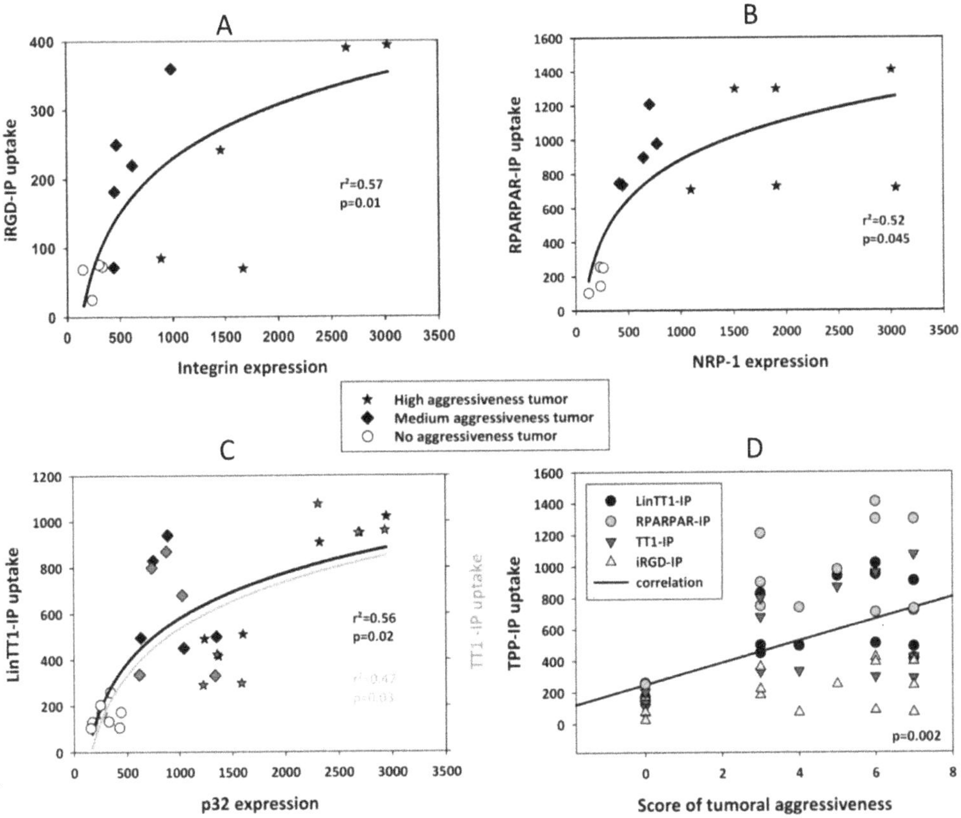

Figure 5. Correlation between receptor expression, peptide internalization, and aggressiveness of HCC. Correlation between integrin expression and internalization of iRGD-IP followed the equation: y = b0 + b1 × log(x). White circle: non-malignant tumor or normal hepatocytes; black squares: samples from medium aggressiveness tumors; black stars: samples from high aggressiveness tumors. (**A**) Integrin expression and iRGD-IP internalization, (b0 = −610 ± 207, b1 = 121 ± 31) (**B**); NRP-1 expression and RPARPAR-IP internalization, (b0 = −1344 ± 600, b1 = 322 ± 89) (**C**), p32 expression and internalization of LinTT1-IP (b0 = −1456 ± 529, b1 = 294 ± 76) and TT1-IP (b0 = −1636 ± 665, b1 = 313 ± 96) (**D**) Correlation between score of aggressiveness and TTP-IP internalization (r = 0.479, standard error of estimate = 333.2, f = y0 + a × x with y0 = 249 ± 83, a = 69.7 ± 17).

3.6. Apoptotic Effect of TPP-IPs on Tumoral Hepatocytes

We have previously demonstrated that TPP-IPs induced apoptosis in tumoral B-cells [26]. Figure 6 shows that iRGD-IP, RPARPAR-IP, LinTT1-IP, and TT1-IP induced apoptosis in HCC (sample #7 here) but not in non-malignant samples (sample #1). Apoptotic effect was stronger for LinTT1-IP and TT1-IP peptides, suggesting a tumor-specific induction of apoptosis.

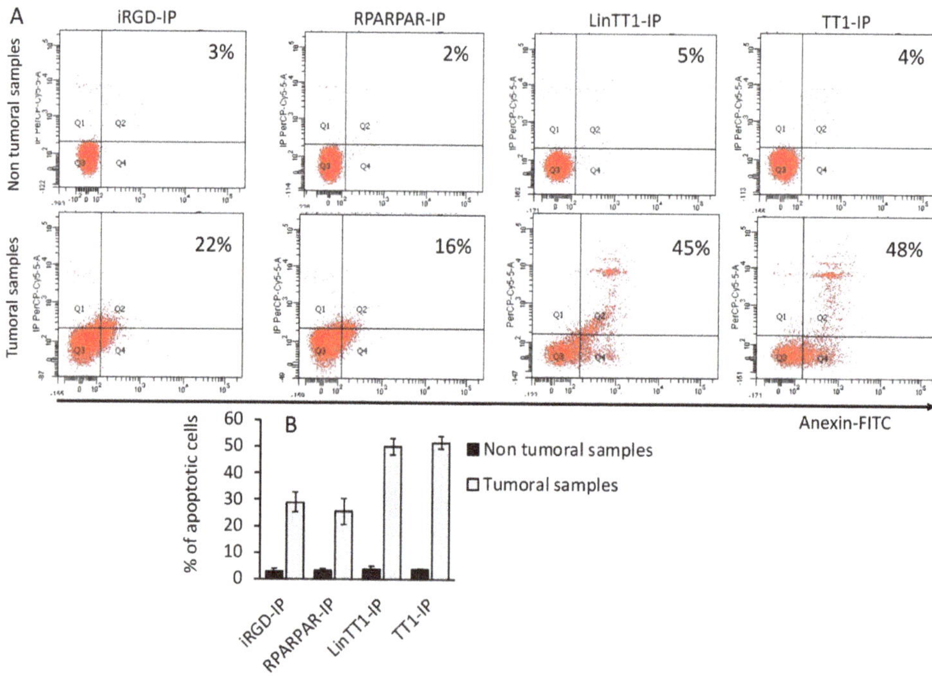

Figure 6. TPP-IPs induce apoptosis in HCC. Hepatocytes isolated from HCC (patient 7) or non-malignant tumors (patient 1) were cultured 12 h with 25 μM and the resulting apoptosis was estimated by annexin-V-FITC staining using flow cytometry. (**A**) Flow cytometry plots from one non-tumoral (patient 1) and one tumoral sample (patient 7). (**B**) Quantification of the percentage of apoptotic cells from three non-tumoral and three tumoral samples. Error bars = ± standard error.

4. Discussion

Liver cancer remains a global health challenge, and its incidence is growing worldwide. It is estimated that by 2025, liver cancer will affect one million people annually [37,38]. The most common form of liver cancer is hepatocellular carcinoma (HCC), which accounts for ~90% of cases. Approximately 25% of HCC tumors present mutations, but they remain undruggable [39,40]. The histology-based definition of the morphological heterogeneity of liver cancer has been modified in an effort to employ personalized therapies for patient treatment [27].

The type of HCC treatment depends on tumor stage, patient performance, and the hepatic functional reserve. The pathogenesis of HCC is a complex multistage process, where angiogenesis plays an important role. For patients with advanced disease, only a handful of kinase inhibitors are approved for therapy, such as cabozantinib, regorafenib, lenvatinib, or sorafenib [41–45]. Anti-angiogenic agents, as well as some monoclonal antibodies, are also approved for use in HCC treatment.

Several therapeutic approaches to specifically target tumoral cells have been investigated. Interfering peptides are emerging as promising therapeutic agents that block intracellular protein–protein interactions [46,47]. The serine/threonine phosphatase PP2A is frequently altered in cancer, either in terms of expression levels or activation [48–50]. The physiological inhibitor of PP2A, i.e., the oncoprotein SET, engages with the catalytic subunit of PP2A to block its activation. Competitive interfering peptides able to block the PP2A/SET interaction can therefore restore PP2A activity [51,52]. We have generated four TPP-IPs able to specifically penetrate tumoral hepatocytes and B-cells and induce apoptosis of malignant cells [26]. Here, the four peptides (iRGD-IP, RPARPAR-IP, LinTT1-IP, and

TT1-IP) that block the PP2A/SET interaction were able to penetrate tumoral hepatocytes isolated from HCC, but crucially, they were not internalized by non-malignant tumors.

Different parameters have been used to define an aggressiveness score. Recent publications define the aggressiveness score based on four clinical parameters, i.e., tumor size, multifocality, presence of portal vein thrombus, and blood alpha-fetoprotein levels [53,54]. Here we defined an aggressiveness score based on the six parameters indicated in Table 1. Given these criteria, we classified the patients into three groups: non-aggressive (non-malignant tumors), moderately aggressive (score up to five), and highly aggressive (score of six and higher). The results show that there was a correlation between the number of TPP receptors expressed by tumoral cells, level of TPP-IP internalization, and HCC aggressiveness.

Treatment of HCC cells with the bi-functional peptides tested here led to a higher level of apoptosis in HCC cells than in non-malignant samples. The penetration-induced apoptosis was mediated by the associated IP and the specific tumoral penetration. Internalization of the TPP-IPs is the result of a multistep mechanism. First, the bi-functional peptides are associated to their primary receptors (αvβ3/5 integrins for iRGD, p32 for LinTT1, and TT1and NRP-1 for RPARPAR). After proteolytic cleavage by tumoral proteases, they bind to the NRP-1 receptor, triggering cellular internalization. One possible explanation for the lower internalization of iRGD-IP, TT1-IP, and LinTT1-IP compared to RPARPAR-IP could be that after the incubation times used here, only a fraction of these peptides gets cleaved to expose the CendR motif. The involvement of several tumor-dependent steps renders this mechanism highly selective toward tumor cells expressing integrin, p32, and NRP-1 receptors [7,10,14–16]. There are several lines of evidence showing that NRP-1 mediates angiogenesis and that increased NRP-1 expression correlates with a decrease in tumor progression, angiogenesis, and immune evasion [55,56]. Overexpression of NRP-1 in vitro and in vivo correlates with decreased tumor vascularization and apoptosis, suggesting a direct correlation between level of NRP-1 expression and aggressiveness of the tumor [55,56]. Similarly, the level of p32 expression by tumoral cells and tissues has been associated with cancer progression and metastasis in several cancers, such as thyroid, pancreatic, gastric, and lung cancer [57].

Uncontrolled tumor cell proliferation and escape from apoptosis play an important role in HCC growth, which makes inhibition of cancer proliferation and induction of apoptosis a crucial target for HCC treatment. Patients with late-stage HCC currently have to rely on systemic chemotherapy [58]. However, the prognosis of patients undergoing chemotherapy for HCC is severely compromised by the toxic side effects of the drugs and by the emergence of drug-resistant HCC tumors [59]. Consequently, there is a real need to search for new targets to treat liver tumors.

Protein phosphatase PP2A and its physiological inhibitor are implicated in HCC as well as other types of cancers. PP2A is a tumor suppressor that negatively regulates many signaling pathways associated with cancer progression by dephosphorylating crucial proteins in these pathways, such as Wnt/β-catenin, PI3K, MAPK, and so on [60,61].

PP2A phosphatase inhibitors have been shown to have therapeutic effects against HCC in clinical trials [62,63], suggesting that PP2A is a promising target for HCC treatment. Unfortunately, the extensive constitutive expression of PP2A in normal tissues, as well as the many PP2A partners and signaling pathways in which PP2A is involved, have bottlenecked the efforts to exploit PP2A as a target for therapeutic intervention. In addition, in clinical trials, phosphatase inhibitors exert a toxic effect against normal hepatic tissue [48–50,64], indicating that the therapeutic efficacy of PP2A inhibitors depends on precise cancer-targeted delivery systems. We have addressed this challenge by developing bi-specific peptides composed of a TPP and an IP module. Peptide drug conjugates are gaining importance in cancer therapy. For example, PEPAXTO® (melphalan flufenamide), a peptide drug conjugate that enters cells by passive diffusion and releases the drug via the action of intracellular aminopeptidases, has been recently approved for clinical use. The advantage of our TPP-IPs over other peptide drug conjugates like PEPAXTO is their high specificity for malignant cells. The TPPs target receptors that are highly expressed in tumoral cells and, once inside the cell, they only dissociate the pool of PP2A associated to SET, without any effect on the free PP2A and SET partners.

Aberrant expression of SET has been reported in other cancers such as leukemia, breast cancer, and colon, liver, and lung carcinoma [65–69]. The oncogenic role of SET in HCC was first suggested by Fukukawa et al. [70] who demonstrated that SET expression is highly upregulated in progressive HCC, indicating that SET may be involved in HCC development. Furthermore, SET activity is associated with development of resistance to chemotherapies [71–73]. Results obtained with patient primary cells support the oncogenic role of SET in HCC, suggesting that SET may serve as a novel biomarker to guide treatment in patients with HCC [69].

In conclusion, we report a selective tumoral internalization and apoptotic effect of peptides with potential clinical applications in liver cancer. The correlation between TPP receptor expression levels, TPP-IP internalization levels, and tumor aggressiveness score suggests that TPP receptor expression could serve as a marker of HCC aggressiveness.

Author Contributions: E.S. and F.C. performed the experiment and provided samples; O.S. provided samples; L.S.-G., T.T. and A.R. wrote, reviewed, and edited the manuscript; P.T. analyzed the data. All authors have read and agreed to the published version of the manuscript.

Funding: This work was supported by Inserm. T.T. and L.S.G. were supported by the European Regional Development Fund (Project No. 2014-2020.4.01.15-0012) and the Estonian Research Council (grants PRG230 and EAG79).

Institutional Review Board Statement: The study was conducted according to the Declaration of Helsinki and approved by the Ethics Committee (CODECOH DC-2019-3261, Commission Nationale Informatique et Liberté, (CNIL) 1929196vO). Date of the approval: 10 January 2019. All the patients gave informed consent as per French law.

Informed Consent Statement: Informed consent was obtained from all subjects involved in this study.

Data Availability Statement: Data are available upon request.

Acknowledgments: We thank G. Langsley and G. Snounou for editing the manuscript.

Conflicts of Interest: The authors declare no conflict of interest.

References

1. Xie, H.G.; Frueh, F.W. Pharmacogenomics steps toward personalized medicine. *Future Med.* **2005**, *2*, 325–337. [CrossRef] [PubMed]
2. Watters, J.W.; McLeod, H.L. Cancer pharmacogenomics: Current and future applications. *Biochim. Biophys. Acta* **2003**, *1603*, 99–111. [CrossRef]
3. Hanahan, D.; Weinberg, R.A. Hallmarks of cancer: The next generation. *Cell* **2011**, *144*, 646–674. [CrossRef] [PubMed]
4. Seebacher, N.A.; Stacy, A.E.; Porter, G.M.; Merlot, A.M. Clinical development of targeted and immune based anti-cancer therapies. *J. Exp. Clin. Cancer Res.* **2019**, *38*, 156. [CrossRef]
5. Man, S.; Luo, C.; Yan, M.; Zhao, G.; Ma, L.; Gao, W. Treatment for liver cancer: From sorafenib to natural products. *Eur. J. Med. Chem.* **2021**, *224*, 113690. [CrossRef]
6. Wildner, G. Tumors, tumor therapies, autoimmunity and the eye. *Autoimmun. Rev.* **2021**, *20*, 102892. [CrossRef]
7. Teesalu, T.; Sugahara, K.N.; Ruoslahti, E. Tumor-penetrating peptides. *Front. Oncol.* **2013**, *3*, 216. [CrossRef]
8. Ruoslahti, E. Tumor penetrating peptides for improved drug delivery. *Adv. Drug Deliv. Rev.* **2017**, *110*, 3–12. [CrossRef]
9. Zanuy, D.; Kotla, R.; Nussinov, R.; Teesalu, T.; Sugahara, K.N.; Aleman, C.; Haspel, N. Sequence dependence of C-end rule peptides in binding and activation of neuropilin-1 receptor. *J. Struct. Biol.* **2013**, *182*, 78–86. [CrossRef]
10. Teesalu, T.; Sugahara, K.N.; Kotamraju, V.R.; Ruoslahti, E. C-end rule peptides mediate neuropilin-1-dependent cell, vascular, and tissue penetration. *Proc. Natl. Acad. Sci. USA* **2009**, *106*, 16157–16162. [CrossRef]
11. Willmore, A.M.; Simon-Gracia, L.; Toome, K.; Paiste, P.; Kotamraju, V.R.; Molder, T.; Sugahara, K.N.; Ruoslahti, E.; Braun, G.B.; Teesalu, T. Targeted silver nanoparticles for ratiometric cell phenotyping. *Nanoscale* **2016**, *8*, 9096–9101. [CrossRef]
12. Wonder, E.; Simon-Gracia, L.; Scodeller, P.; Majzoub, R.N.; Kotamraju, V.R.; Ewert, K.K.; Teesalu, T.; Safinya, C.R. Competition of charge-mediated and specific binding by peptide-tagged cationic liposome-DNA nanoparticles in vitro and in vivo. *Biomaterials* **2018**, *166*, 52–63. [CrossRef]
13. Sugahara, K.N.; Teesalu, T.; Karmali, P.P.; Kotamraju, V.R.; Agemy, L.; Girard, O.M.; Hanahan, D.; Mattrey, R.F.; Ruoslahti, E. Tissue-penetrating delivery of compounds and nanoparticles into tumors. *Cancer Cell* **2009**, *16*, 510–520. [CrossRef]

14. Fogal, V.; Zhang, L.; Krajewski, S.; Ruoslahti, E. Mitochondrial/cell-surface protein p32/gC1qR as a molecular target in tumor cells and tumor stroma. *Cancer Res.* **2008**, *68*, 7210–7218. [CrossRef]
15. Simon-Gracia, L.; Scodeller, P.; Fuentes, S.S.; Vallejo, V.G.; Rios, X.; San Sebastian, E.; Sidorenko, V.; Di Silvio, D.; Suck, M.; De Lorenzi, F.; et al. Application of polymersomes engineered to target p32 protein for detection of small breast tumors in mice. *Oncotarget* **2018**, *9*, 18682–18697. [CrossRef]
16. Paasonen, L.; Sharma, S.; Braun, G.B.; Kotamraju, V.R.; Chung, T.D.; She, Z.G.; Sugahara, K.N.; Yliperttula, M.; Wu, B.; Pellecchia, M.; et al. New p32/gC1qR Ligands for Targeted Tumor Drug Delivery. *Chembiochem* **2016**, *17*, 570–575. [CrossRef]
17. Sugahara, K.N.; Scodeller, P.; Braun, G.B.; de Mendoza, T.H.; Yamazaki, C.M.; Kluger, M.D.; Kitayama, J.; Alvarez, E.; Howell, S.B.; Teesalu, T.; et al. A tumor-penetrating peptide enhances circulation-independent targeting of peritoneal carcinomatosis. *J. Control. Release* **2015**, *212*, 59–69. [CrossRef]
18. Saalik, P.; Lingasamy, P.; Toome, K.; Mastandrea, I.; Rousso-Noori, L.; Tobi, A.; Simon-Gracia, L.; Hunt, H.; Paiste, P.; Kotamraju, V.R.; et al. Peptide-guided nanoparticles for glioblastoma targeting. *J. Control. Release* **2019**, *308*, 109–118. [CrossRef]
19. Hunt, H.; Simon-Gracia, L.; Tobi, A.; Kotamraju, V.R.; Sharma, S.; Nigul, M.; Sugahara, K.N.; Ruoslahti, E.; Teesalu, T. Targeting of p32 in peritoneal carcinomatosis with intraperitoneal linTT1 peptide-guided pro-apoptotic nanoparticles. *J. Control. Release* **2017**, *260*, 142–153. [CrossRef]
20. Simon-Gracia, L.; Hunt, H.; Teesalu, T. Peritoneal Carcinomatosis Targeting with Tumor Homing Peptides. *Molecules* **2018**, *23*, 1190. [CrossRef]
21. Simon-Gracia, L.; Hunt, H.; Scodeller, P.; Gaitzsch, J.; Kotamraju, V.R.; Sugahara, K.N.; Tammik, O.; Ruoslahti, E.; Battaglia, G.; Teesalu, T. iRGD peptide conjugation potentiates intraperitoneal tumor delivery of paclitaxel with polymersomes. *Biomaterials* **2016**, *104*, 247–257. [CrossRef]
22. Diaz Bessone, M.I.; Simon-Gracia, L.; Scodeller, P.; Ramirez, M.L.A.; Lago Huvelle, M.A.; Soler-Illia, G.; Simian, M. iRGD-guided tamoxifen polymersomes inhibit estrogen receptor transcriptional activity and decrease the number of breast cancer cells with self-renewing capacity. *J. Nanobiotechnol.* **2019**, *17*, 120. [CrossRef]
23. Scodeller, P.; Asciutto, E.K. Targeting Tumors Using Peptides. *Molecules* **2020**, *25*, 808. [CrossRef]
24. Simon-Gracia, L.; Sidorenko, V.; Uustare, A.; Ogibalov, I.; Tasa, A.; Tshubrik, O.; Teesalu, T. Novel Anthracycline Utorubicin for Cancer Therapy. *Angew. Chem. Int. Ed. Engl.* **2021**, *60*, 17018–17027. [CrossRef]
25. Ikemoto, H.; Lingasamy, P.; Anton Willmore, A.M.; Hunt, H.; Kurm, K.; Tammik, O.; Scodeller, P.; Simon-Gracia, L.; Kotamraju, V.R.; Lowy, A.M.; et al. Hyaluronan-binding peptide for targeting peritoneal carcinomatosis. *Tumor Biol.* **2017**, *39*, 1010428317701628. [CrossRef]
26. Simon-Gracia, L.; Savier, E.; Parizot, C.; Brossas, J.Y.; Loisel, S.; Teesalu, T.; Conti, F.; Charlotte, F.; Scatton, O.; Aoudjehane, L.; et al. Bifunctional Therapeutic Peptides for Targeting Malignant B Cells and Hepatocytes: Proof of Concept in Chronic Lymphocytic Leukemia. *Adv. Ther.* **2020**, *3*, 2000131. [CrossRef]
27. Lu, X.Y.; Xi, T.; Lau, W.Y.; Dong, H.; Zhu, Z.; Shen, F.; Wu, M.C.; Cong, W.M. Hepatocellular carcinoma expressing cholangiocyte phenotype is a novel subtype with highly aggressive behavior. *Ann. Surg. Oncol.* **2011**, *18*, 2210–2217. [CrossRef]
28. Llovet, J.M.; Kelley, R.K.; Villanueva, A.; Singal, A.G.; Pikarsky, E.; Roayaie, S.; Lencioni, R.; Koike, K.; Zucman-Rossi, J.; Finn, R.S. Hepatocellular carcinoma. *Nat. Rev. Dis. Primers* **2021**, *7*, 6. [CrossRef]
29. Iguchi, T.; Aishima, S.; Sanefuji, K.; Fujita, N.; Sugimachi, K.; Gion, T.; Taketomi, A.; Shirabe, K.; Maehara, Y.; Tsuneyoshi, M. Both fibrous capsule formation and extracapsular penetration are powerful predictors of poor survival in human hepatocellular carcinoma: A histological assessment of 365 patients in Japan. *Ann. Surg. Oncol.* **2009**, *16*, 2539–2546. [CrossRef]
30. Decaens, T.; Roudot-Thoraval, F.; Badran, H.; Wolf, P.; Durand, F.; Adam, R.; Boillot, O.; Vanlemmens, C.; Gugenheim, J.; Dharancy, S.; et al. Impact of tumour differentiation to select patients before liver transplantation for hepatocellular carcinoma. *Liver. Int.* **2011**, *31*, 792–801. [CrossRef]
31. Ziol, M.; Pote, N.; Amaddeo, G.; Laurent, A.; Nault, J.C.; Oberti, F.; Costentin, C.; Michalak, S.; Bouattour, M.; Francoz, C.; et al. Macrotrabecular-massive hepatocellular carcinoma: A distinctive histological subtype with clinical relevance. *Hepatology* **2018**, *68*, 103–112. [CrossRef] [PubMed]
32. Duvoux, C.; Roudot-Thoraval, F.; Decaens, T.; Pessione, F.; Badran, H.; Piardi, T.; Francoz, C.; Compagnon, P.; Vanlemmens, C.; Dumortier, J.; et al. Liver transplantation for hepatocellular carcinoma: A model including alpha-fetoprotein improves the performance of Milan criteria. *Gastroenterology* **2012**, *143*, 986–994. [CrossRef] [PubMed]
33. Akiba, J.; Nakashima, O.; Hattori, S.; Naito, Y.; Kusano, H.; Kondo, R.; Nakayama, M.; Tanikawa, K.; Todoroki, K.; Umeno, Y.; et al. The expression of arginase-1, keratin (K) 8 and K18 in combined hepatocellular-cholangiocarcinoma, subtypes with stem-cell features, intermediate-cell type. *J. Clin. Pathol.* **2016**, *69*, 846–851. [CrossRef] [PubMed]
34. Nishida, T.; Kataoka, H. Glypican 3-Targeted Therapy in Hepatocellular Carcinoma. *Cancers* **2019**, *11*, 1339. [CrossRef]
35. Sempoux, C.; Chang, C.; Gouw, A.; Chiche, L.; Zucman-Rossi, J.; Balabaud, C.; Bioulac-Sage, P. Benign hepatocellular nodules: What have we learned using the patho-molecular classification. *Clin. Res. Hepatol. Gastroenterol.* **2013**, *37*, 322–327. [CrossRef]
36. Moudi, B.; Heidari, Z.; Mahmoudzadeh-Sagheb, H. Study of liver in HBV-related hepatocellular carcinoma: Stereology shows quantitative differences in liver structure. *Eur. J. Histochem.* **2018**, *62*, 2950. [CrossRef]
37. Llovet, J.M.; Zucman-Rossi, J.; Pikarsky, E.; Sangro, B.; Schwartz, M.; Sherman, M.; Gores, G. Hepatocellular carcinoma. *Nat. Rev. Dis. Primers* **2016**, *2*, 16018. [CrossRef]
38. Villanueva, A. Hepatocellular Carcinoma. *N. Engl. J. Med.* **2019**, *380*, 1450–1462. [CrossRef]

39. Schulze, K.; Imbeaud, S.; Letouze, E.; Alexandrov, L.B.; Calderaro, J.; Rebouissou, S.; Couchy, G.; Meiller, C.; Shinde, J.; Soysouvanh, F.; et al. Exome sequencing of hepatocellular carcinomas identifies new mutational signatures and potential therapeutic targets. *Nat. Genet.* **2015**, *47*, 505–511. [CrossRef]
40. Llovet, J.M.; Montal, R.; Sia, D.; Finn, R.S. Molecular therapies and precision medicine for hepatocellular carcinoma. *Nat. Rev. Clin. Oncol.* **2018**, *15*, 599–616. [CrossRef]
41. Abou-Alfa, G.K.; Meyer, T.; Cheng, A.L.; El-Khoueiry, A.B.; Rimassa, L.; Ryoo, B.Y.; Cicin, I.; Merle, P.; Chen, Y.; Park, J.W.; et al. Cabozantinib in Patients with Advanced and Progressing Hepatocellular Carcinoma. *N. Engl. J. Med.* **2018**, *379*, 54–63. [CrossRef]
42. Haber, P.K.; Puigvehi, M.; Castet, F.; Lourdusamy, V.; Montal, R.; Tabrizian, P.; Buckstein, M.; Kim, E.; Villanueva, A.; Schwartz, M.; et al. Evidence-based management of HCC: Systematic review and meta-analysis of randomized controlled trials (2002–2020). *Gastroenterology* **2021**, *161*, 879–898. [CrossRef]
43. Bruix, J.; Qin, S.; Merle, P.; Granito, A.; Huang, Y.H.; Bodoky, G.; Pracht, M.; Yokosuka, O.; Rosmorduc, O.; Breder, V.; et al. Regorafenib for patients with hepatocellular carcinoma who progressed on sorafenib treatment (RESORCE): A randomised, double-blind, placebo-controlled, phase 3 trial. *Lancet* **2017**, *389*, 56–66. [CrossRef]
44. Kim, J.J.; McFarlane, T.; Tully, S.; Wong, W.W.L. Lenvatinib Versus Sorafenib as First-Line Treatment of Unresectable Hepatocellular Carcinoma: A Cost-Utility Analysis. *Oncologist* **2019**, *25*, 512–519. [CrossRef]
45. Llovet, J.M.; Ricci, S.; Mazzaferro, V.; Hilgard, P.; Gane, E.; Blanc, J.F.; de Oliveira, A.C.; Santoro, A.; Raoul, J.L.; Forner, A.; et al. Sorafenib in advanced hepatocellular carcinoma. *N. Engl. J. Med.* **2008**, *359*, 378–390. [CrossRef]
46. Ahmad, A.; Ahmad, E.; Rabbani, G.; Haque, S.; Arshad, M.; Khan, R.H. Identification and design of antimicrobial peptides for therapeutic applications. *Curr. Protein Pept. Sci.* **2012**, *13*, 211–223. [CrossRef]
47. Rabbani, G.; Baig, M.H.; Ahmad, K.; Choi, I. Protein-protein Interactions and their Role in Various Diseases and their Prediction Techniques. *Curr. Protein Pept. Sci.* **2018**, *19*, 948–957. [CrossRef]
48. Haesen, D.; Sents, W.; Lemaire, K.; Hoorne, Y.; Janssens, V. The Basic Biology of PP2A in Hematologic Cells and Malignancies. *Front. Oncol.* **2014**, *4*, 347. [CrossRef]
49. Ciccone, M.; Calin, G.A.; Perrotti, D. From the Biology of PP2A to the PADs for Therapy of Hematologic Malignancies. *Front. Oncol.* **2015**, *5*, 21. [CrossRef]
50. Kalev, P.; Sablina, A.A. Protein phosphatase 2A as a potential target for anticancer therapy. *Anticancer Agents Med. Chem.* **2011**, *11*, 38–46. [CrossRef]
51. Switzer, C.H.; Cheng, R.Y.; Vitek, T.M.; Christensen, D.J.; Wink, D.A.; Vitek, M.P. Targeting SET/I(2)PP2A oncoprotein functions as a multi-pathway strategy for cancer therapy. *Oncogene* **2011**, *30*, 2504–2513. [CrossRef]
52. Neviani, P.; Harb, J.G.; Oaks, J.J.; Santhanam, R.; Walker, C.J.; Ellis, J.J.; Ferenchak, G.; Dorrance, A.M.; Paisie, C.A.; Eiring, A.M.; et al. PP2A-activating drugs selectively eradicate TKI-resistant chronic myeloid leukemic stem cells. *J. Clin. Investig.* **2013**, *123*, 4144–4157. [CrossRef]
53. Carr, B.I.; Guerra, V. A Hepatocellular Carcinoma Aggressiveness Index and Its Relationship to Liver Enzyme Levels. *Oncology* **2016**, *90*, 215–220. [CrossRef]
54. Carr, B.I.; Guerra, V.; Giannini, E.G.; Farinati, F.; Ciccarese, F.; Rapaccini, G.L.; Di Marco, M.; Benvegnu, L.; Zoli, M.; Borzio, F.; et al. A Liver Index and its Relationship to Indices of HCC Aggressiveness. *J. Integr. Oncol.* **2016**, *5*, 178. [CrossRef]
55. Miao, H.Q.; Lee, P.; Lin, H.; Soker, S.; Klagsbrun, M. Neuropilin-1 expression by tumor cells promotes tumor angiogenesis and progression. *FASEB J.* **2000**, *14*, 2532–2539. [CrossRef]
56. Jubb, A.M.; Strickland, L.A.; Liu, S.D.; Mak, J.; Schmidt, M.; Koeppen, H. Neuropilin-1 expression in cancer and development. *J. Pathol.* **2012**, *226*, 50–60. [CrossRef]
57. Saha, P.; Datta, K. Multi-functional, multicompartmental hyaluronan-binding protein 1 (HABP1/p32/gC1qR): Implication in cancer progression and metastasis. *Oncotarget* **2018**, *9*, 10784–10807. [CrossRef]
58. El-Serag, H.B. Hepatocellular carcinoma. *N. Engl. J. Med.* **2011**, *365*, 1118–1127. [CrossRef]
59. Asghar, U.; Meyer, T. Are there opportunities for chemotherapy in the treatment of hepatocellular cancer? *J. Hepatol.* **2012**, *56*, 686–695. [CrossRef]
60. Bos, C.L.; Kodach, L.L.; van den Brink, G.R.; Diks, S.H.; van Santen, M.M.; Richel, D.J.; Peppelenbosch, M.P.; Hardwick, J.C. Effect of aspirin on the Wnt/beta-catenin pathway is mediated via protein phosphatase 2A. *Oncogene* **2006**, *25*, 6447–6456. [CrossRef]
61. Meng, G.; Wang, W.; Chai, K.; Yang, S.; Li, F.; Jiang, K. Combination treatment with triptolide and hydroxycamptothecin synergistically enhances apoptosis in A549 lung adenocarcinoma cells through PP2A-regulated ERK, p38 MAPKs and Akt signaling pathways. *Int. J. Oncol.* **2015**, *46*, 1007–1017. [CrossRef] [PubMed]
62. Wang, G.S. Medical uses of mylabris in ancient China and recent studies. *J. Ethnopharmacol.* **1989**, *26*, 147–162. [CrossRef]
63. Li, W.; Xie, L.; Chen, Z.; Zhu, Y.; Sun, Y.; Miao, Y.; Xu, Z.; Han, X. Cantharidin, a potent and selective PP2A inhibitor, induces an oxidative stress-independent growth inhibition of pancreatic cancer cells through G2/M cell-cycle arrest and apoptosis. *Cancer Sci.* **2010**, *101*, 1226–1233. [CrossRef] [PubMed]
64. Kinch, M.S. An overview of FDA-approved biologics medicines. *Drug Discov. Today* **2015**, *20*, 393–398. [CrossRef]
65. Carlson, S.G.; Eng, E.; Kim, E.G.; Perlman, E.J.; Copeland, T.D.; Ballermann, B.J. Expression of SET, an inhibitor of protein phosphatase 2A, in renal development and Wilms' tumor. *J. Am. Soc. Nephrol.* **1998**, *9*, 1873–1880. [CrossRef]
66. Chae, H.; Lim, J.; Kim, M.; Park, J.; Kim, Y.; Han, K.; Lee, S.; Min, W.S. Phenotypic and genetic characterization of adult T-cell acute lymphoblastic leukemia with del(9)(q34);SET-NUP214 rearrangement. *Ann. Hematol.* **2012**, *91*, 193–201. [CrossRef]

67. Cristobal, I.; Garcia-Orti, L.; Cirauqui, C.; Cortes-Lavaud, X.; Garcia-Sanchez, M.A.; Calasanz, M.J.; Odero, M.D. Overexpression of SET is a recurrent event associated with poor outcome and contributes to protein phosphatase 2A inhibition in acute myeloid leukemia. *Haematologica* **2012**, *97*, 543–550. [CrossRef]
68. Cristobal, I.; Rincon, R.; Manso, R.; Carames, C.; Zazo, S.; Madoz-Gurpide, J.; Rojo, F.; Garcia-Foncillas, J. Deregulation of the PP2A inhibitor SET shows promising therapeutic implications and determines poor clinical outcome in patients with metastatic colorectal cancer. *Clin. Cancer Res.* **2015**, *21*, 347–356. [CrossRef]
69. Hung, M.H.; Chen, Y.L.; Chu, P.Y.; Shih, C.T.; Yu, H.C.; Tai, W.T.; Shiau, C.W.; Chen, K.F. Upregulation of the oncoprotein SET determines poor clinical outcomes in hepatocellular carcinoma and shows therapeutic potential. *Oncogene* **2016**, *35*, 4891–4902. [CrossRef]
70. Fukukawa, C.; Shima, H.; Tanuma, N.; Ogawa, K.; Kikuchi, K. Up-regulation of I-2(PP2A)/SET gene expression in rat primary hepatomas and regenerating livers. *Cancer Lett.* **2000**, *161*, 89–95. [CrossRef]
71. Agarwal, A.; MacKenzie, R.J.; Pippa, R.; Eide, C.A.; Oddo, J.; Tyner, J.W.; Sears, R.; Vitek, M.P.; Odero, M.D.; Christensen, D.J.; et al. Antagonism of SET using OP449 enhances the efficacy of tyrosine kinase inhibitors and overcomes drug resistance in myeloid leukemia. *Clin. Cancer Res.* **2014**, *20*, 2092–2103. [CrossRef]
72. Hung, M.H.; Wang, C.Y.; Chen, Y.L.; Chu, P.Y.; Hsiao, Y.J.; Tai, W.T.; Chao, T.T.; Yu, H.C.; Shiau, C.W.; Chen, K.F. SET antagonist enhances the chemosensitivity of non-small cell lung cancer cells by reactivating protein phosphatase 2A. *Oncotarget* **2016**, *7*, 638–655. [CrossRef]
73. Huang, C.Y.; Hung, M.H.; Shih, C.T.; Hsieh, F.S.; Kuo, C.W.; Tsai, M.H.; Chang, S.S.; Hsiao, Y.J.; Chen, L.J.; Chao, T.I.; et al. Antagonizing SET Augments the Effects of Radiation Therapy in Hepatocellular Carcinoma through Reactivation of PP2A-Mediated Akt Downregulation. *J. Pharmacol. Exp. Ther.* **2018**, *366*, 410–421. [CrossRef]

pharmaceutics

MDPI

Article

B3Pred: A Random-Forest-Based Method for Predicting and Designing Blood–Brain Barrier Penetrating Peptides

Vinod Kumar [1,2], Sumeet Patiyal [1], Anjali Dhall [1], Neelam Sharma [1] and Gajendra Pal Singh Raghava [1,*]

1 Department of Computational Biology, Indraprastha Institute of Information Technology, Okhla 110020, India; vinodporiya032@gmail.com (V.K.); sumeetp@iiitd.ac.in (S.P.); anjalid@iiitd.ac.in (A.D.); neelams@iiitd.ac.in (N.S.)
2 Bioinformatics Centre, CSIR-Institute of Microbial Technology, Sector-39A, Chandigarh 160036, India
* Correspondence: raghava@iiitd.ac.in; Tel.: +91-011-26907444

Abstract: The blood–brain barrier is a major obstacle in treating brain-related disorders, as it does not allow the delivery of drugs into the brain. We developed a method for predicting blood–brain barrier penetrating peptides to facilitate drug delivery into the brain. These blood–brain barrier penetrating peptides (B3PPs) can act as therapeutics, as well as drug delivery agents. We trained, tested, and evaluated our models on blood–brain barrier peptides obtained from the B3Pdb database. First, we computed a wide range of peptide features. Then, we selected relevant peptide features. Finally, we developed numerous machine-learning-based models for predicting blood–brain barrier peptides using the selected features. The random-forest-based model performed the best with respect to the top 80 selected features and achieved a maximal 85.08% accuracy with an AUROC of 0.93. We also developed a webserver, B3pred, that implements our best models. It has three major modules that allow users to predict/design B3PPs and scan B3PPs in a protein sequence.

Keywords: blood–brain barrier; penetrating peptides; machine learning techniques; drug delivery; prediction server

Citation: Kumar, V.; Patiyal, S.; Dhall, A.; Sharma, N.; Raghava, G.P.S. B3Pred: A Random-Forest-Based Method for Predicting and Designing Blood–Brain Barrier Penetrating Peptides. *Pharmaceutics* **2021**, *13*, 1237. https://doi.org/10.3390/pharmaceutics13081237

Academic Editors: Prisca Boisguérin and Sébastien Deshayes

Received: 7 June 2021
Accepted: 14 July 2021
Published: 11 August 2021

Publisher's Note: MDPI stays neutral with regard to jurisdictional claims in published maps and institutional affiliations.

Copyright: © 2021 by the authors. Licensee MDPI, Basel, Switzerland. This article is an open access article distributed under the terms and conditions of the Creative Commons Attribution (CC BY) license (https://creativecommons.org/licenses/by/4.0/).

1. Introduction

The blood–brain barrier (BBB) is the primary barrier between the brain's interstitial fluid and the blood. It is the connection between the central nervous system (CNS) and the peripheral nervous system (PNS) [1–4]. The neurovascular unit (NVU) is the structural and functional unit of the BBB, formed by neurons, macrophages, endothelial cells, astrocytes, and pericytes [5] (Figure 1). The NVU regulates the biochemical environment between the blood and the brain, which is essential for neural function. The endothelial cells of the NVU allow the entry or exit of molecules, such as glucose, amino acids, and proteins/peptides, into or from the CNS [6–8]. In the last few decades, researchers have made many attempts to develop drug delivery systems that can deliver drugs into the brain. Despite advances made by the scientific community in developing drug delivery systems, it is still challenging to penetrate the BBB [9].

In the past, researchers have attempted to develop peptide/protein-based drug delivery vehicles. In this approach, a major challenge is to identify peptides that can penetrate the BBB [10]. In addition, researchers are exploring peptide-based therapeutics to treat CNS-associated diseases, including neurodegenerative disorders such as Parkinson's disease, Alzheimer's disease [11,12], and glioblastoma [13]. This means that peptides can be used as therapeutic agents as well as drug delivery vehicles. In recent studies, numerous peptides, such as shuttle peptides [14], self-assembled peptides [15], and peptide-decorated nanoparticles [16], have been used for efficient drug delivery into the brain. Some neuropeptides are utilized as potential therapeutic targets against many neurological diseases, such as epilepsy [17,18], depression [19,20], and neuroimmune disorders [21]. Due to the low toxicity of these peptides, they may act as potential peptide-based drug candidates against neurological diseases. The major limitation of these peptide-based drugs is low bioavailability, short half-life [22], and weak penetration of the BBB [23]. For example, tumor homing peptides (THPs) [24] and cell-penetrating peptides (CPPs) [25] can be used as drug delivery vehicles [26,27]. The tumor homing peptides need a carrier to cross the BBB, while selected CPPs can directly pass through the BBB [28].

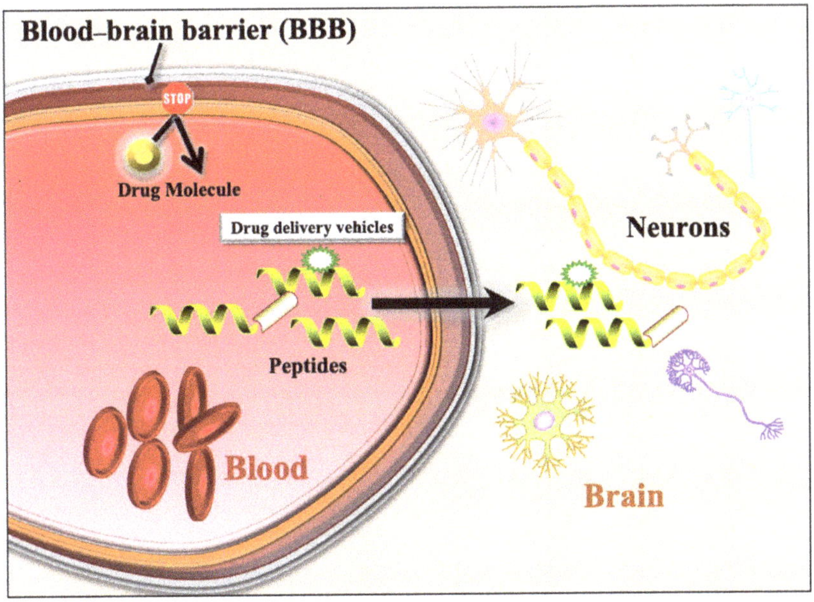

Figure 1. A schematic diagram shows inhibition of drug molecules entry from blood to brain due to Blood–brain barrier. It also shows entry of drug molecules from blood to brain with the support of Blood–brain barrier penetrating peptides.

The cell-penetrating peptides are short peptides which act as molecular delivery vehicles, and are able to deliver various therapeutic molecules inside a cell [29,30]. There are CPPs that can even cross the blood–brain barrier, which are called blood–brain barrier penetrating peptides (B3PPs). These B3PPs can be used to deliver several cargo molecules (e.g., peptides/proteins, siRNA, plasmid DNA) into the brain [31–34]. Mainly, these peptides are obtained from naturally occurring proteins/peptides such as signal peptides, RNA/DNA-binding proteins, viral proteins, and antimicrobial peptides [35]. Several studies have shown that B3PPs may be synthesized chemically or designed with rDNA technology [36–38] to enhance the stability and half-life of the B3PPs [39]. In the past, several methods have been developed for predicting cell-penetrating peptides, such as cellPPD, SkipCPP-Pred, CPPred-RF, KELM-CPPpred, CellPPDMod, and CPPred-FL [40–45]. In addition, various methods have been developed for predicting chemical-based drug delivery vehicles to cross the blood–brain barrier [46–48]. In contrast, a limited attempt has been made to develop methods to predict B3PPs. Recently, Dai et al. developed an in silico method, BBPpred, to identify B3PPs [49].

In this study, we have developed a computational tool named "B3Pred" for predicting B3PPs with high reliability and precision. This method has the ability to classify BBPs vs. non-BBPs and CPPs vs. BBPs; it uses a large dataset for training and validation. We used three datasets, i.e., Dataset_1 (269 B3PPs and 269 CPPs), Dataset_2 (269 B3PPs, and 269 non-B3PPs), and Dataset_3 (269 B3PPs and 2690 non-B3PPs), for training and validation. We have used more than 9000 descriptors/features for the generation of the prediction models using several machine learning techniques, such as RF, DT, LR, XGB, SVM, and GBM.

2. Materials and Methods

2.1. Dataset Collection

In this study, we collected 465 blood–brain barrier penetrating peptides (B3PPs) from the B3Pdb database (https://webs.iiitd.edu.in/raghava/b3pdb/, accessed on 22 July 2020) [50]. We considered B3PPs having a length between 6 and 30 amino acid (AA) residues, inclusive. For the positive dataset, we collected 269 unique B3PPs. The major challenge of this type of study is to generate an authenticated negative dataset. We used three negative datasets in this study. Firstly, we collected unique 269 cell-penetrating peptides (CPPs) [51], other than B3PPs, and called them non-B3PPs or negative Dataset_1. In negative Dataset_2, we randomly generated 269 non-B3PPs from the Swiss-Prot database [52]. Our third negative dataset is ten times larger than the positive dataset, i.e., 2690 unique non-B3PPs randomly generated using the Swiss-Prot database. Finally, we combined the three datasets, i.e., Dataset_1 (269 B3PPs and 269 CPPs), Dataset_2 (269 B3PPs, and 269 non-B3PPs), and Dataset_3 (269 B3PPs and 2690 non-B3PPs).

2.2. Amino Acid Composition

Amino acid composition (AAC) analysis of peptides helped us to find out whether there were any amino acid compositional similarities/differences in different types of peptides. We compared the amino acid composition of B3PPs, CPPs, and randomly generated peptides. The following equation is used to calculate AAC:

$$AAC_i = \frac{AAR_i}{TNR} \times 100 \quad (1)$$

where AAC_i and AAR_i are the percentage composition and number of residues of type i in a peptide, respectively. TNR is the total number of residues in a peptide [53].

2.3. Two Sample Logo

The Two Sample Logo (TSL) tool was used to identify the amino acid preference at a specific position in the peptide sequences [54]. This tool needed an input amino acid sequence vector of fixed length, since the minimum size of peptides in all datasets was five residues; hence, we selected five residues from the N-terminal, and five amino acids from the C-terminal, of the peptide sequences. To create a fixed input vector, the N-terminus side residues and C-terminus residues were grouped together to generate a sequence of 10 amino acid residues. We used the 10-residue sequences generated from our dataset peptides to develop TSLs. To build these Two Sample Logos, we used all B3PPs and all non-B3PPs from the three different negative datasets.

2.4. Generation of Peptide Features

In order to calculate a wide range of features from the protein or peptide sequences, we used the Pfeature package [55]. Pfeature is used to generate thousands of features/descriptors. We computed the composition-based module of Pfeature to calculate >9000 descriptors of peptide sequences for positive and negative datasets. This module calculated fifteen types of features (AAC, DPC, RRI, DDOR, SE, SER, SEP, CTD, CeTD, PAAC, APAAC, QSO, TPC, ABC, and SOCN). The input vector of 9189 descriptors was used further for feature selection and machine learning purposes (Supplementary Table S1).

2.5. Feature Selection

This study used the SVC-L1 feature selection technique to extract an essential set of features from all the datasets. We chose the SVC-L1 method because it is much faster than other feature selection methods [56]. This method applies the L1 penalty to select a relevant set of features, after selecting the non-zero coefficients. SVC-L1 mainly considers regularization and the loss function. During the optimization process, the L1 regularization generates a sparse matrix by choosing some model features. The other important parameter used in this technique is the "C" parameter; its value is directly proportional to the selected features. The smaller the value of "C", the fewer the number of features determined by the method. We chose the default value (i.e., 0.01) of the "C" parameter [57]. Using SVC-L1, 73 important features were identified from the 9189 features for Dataset_1 (B3PPs and CPPs peptides) and Dataset_2 (B3PPs and balanced non-B3PPs). Similarly, 145 features were selected for Dataset_3 (i.e., B3PPs and random non-B3PPs).

2.6. Feature Ranking

After selecting an important set of features, we ranked the features based on their importance in classification. The Feature-selector method is based on a decision-tree-like algorithm and uses the Light Gradient Boosting Machine (LightGBM) method [58]. It computes the rank of each feature based on the feature that is used to split the dataset across all the trees. Further, the top-most ranked features for each dataset were used in different machine learning techniques for the classification of B3PPs and non-B3PPs.

2.7. Machine Learning Techniques

We used several machine learning algorithms to classify B3PPs and non-B3PPs. In this study, we implemented decision tree (DT), random forest (RF), Logistic Regression (LR), k-nearest neighbors (KNN), Gaussian Naive Bayes (GNB), XGBoost (XGB), and Support Vector Classifier (SVC) machine learning classifiers. The different classification methods were implemented with the help of a python-based library known as Scikit-learn [59]. DT algorithms work based on non-parametric supervised learning models. The major aim of the classifier is to identify the output instance by learning various decision rules, provided in the form of input data [60]. The GNB method is a probabilistic classifier and builds on Bayes' theorem. It is based on the assumption that the consecutive variable of every group follows the Gaussian (or normal) distribution [61]. Random forest is an ensemble-based classifier, which predicts a single tree as a response variable by training the number of decision trees. It also controls the overfitting of the models [62]. The LR technique is used to train the logistic/logit model, which gives the likelihood of an event happening. It applies a logistic function to predict the response variable or occurrence of a class [63]. The KNN method is an instance-based classifier. It usually collects the instances of the training dataset. Its prediction is based on the maximum number of votes given to a particular class which is closest to the nearest neighbor data point [64]. The XGB classifier uses the scalable tree boosting algorithm, in which an iterative approach is used for the prediction of the final output [65]. The SVC is developed on the library of support vector machines. It usually fits the data points provided as input features and provides the most suitable fit of a hyperplane that categorizes the data into two classes [66].

2.8. Cross-Validation Techniques

We used internal and external validation techniques to assess the performance of our classification models. In the past, several methods used 80:20 splitting of the complete dataset for training and validation [67,68]. In the current study, we implemented a similar strategy to evaluate our classification models. For each dataset, 80% of the data were used for training, and the remaining 20% were used for external validation. We applied 5-fold cross-validation techniques on the training dataset; this is called internal validation. In internal validation, training data are equally divided into five sets/folds in which four folds were used for training, and the fifth fold is used for testing the model (Supplementary Table S2–S4). This process is repeated five times so that each set is used once for testing. The final performance is computed by taking the average of the performance on the five sets. In the case of external validation, the performance of the best model on the training dataset was evaluated on a validation, or independent, dataset.

2.9. Performance Evaluation Parameters

We used standard evaluation parameters to compute the performance of the classification models. Threshold-dependent and -independent parameters were used in this study. The performance of the models was calculated using threshold-dependent parameters, such as sensitivity (Sens), accuracy (Acc), and specificity (Spec). Area Under the Receiver Operating Characteristic (AUROC) curve, a threshold-independent parameter, was used to measure the models' performance. AUROC generates a curve by plotting sensitivity against (1-specificity) on various thresholds. Threshold-dependent parameters were computed using the given equations:

$$Sensitivity = \frac{TP}{TP + FN} \times 100 \tag{2}$$

$$Specificity = \frac{TN}{TN + FP} \times 100 \tag{3}$$

$$Accuracy = \frac{TP + TN}{TP + FP + TN + FN} \times 100 \tag{4}$$

where *TP*, *FP*, *TN*, and *FN* are true positive, false positive, true negative, and false negative predictions, respectively.

2.10. Webserver Implementation

We developed a webserver named "B3Pred" (https://webs.iiitd.edu.in/raghava/b3pred/, accessed on 22 February 2021) to identify blood–brain barrier penetrating peptides and non-B3PPs. We used HTML5, JAVA, CSS3, and PHP scripts to develop the front-end and back-end of the webserver. The B3Pred server is compatible with all the latest devices, such as mobiles, tablets, iMacs, and desktop computers. It mainly incorporates the predict, design, and protein scan modules.

3. Results

3.1. Amino Acid Composition Analysis

The acid composition of B3PPs, CPPs, and random peptides is shown by a graph (Figure 2); the compositional difference is clearly visible. Arginine is highest in CPPs and B3PPs, which shows that it plays a crucial role in the penetration of peptides into cells. Tyrosine, an aromatic amino acid, is high in B3PPs as compared to other types of peptides. The unique amino acids proline and glycine are prevalent in B3PPs, which contrasts with other types of peptides.

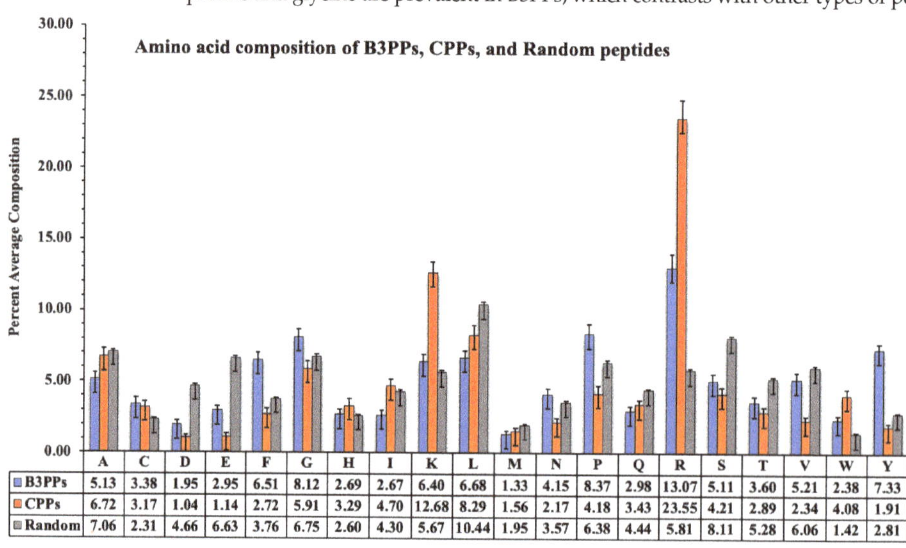

	A	C	D	E	F	G	H	I	K	L	M	N	P	Q	R	S	T	V	W	Y
B3PPs	5.13	3.38	1.95	2.95	6.51	8.12	2.69	2.67	6.40	6.68	1.33	4.15	8.37	2.98	13.07	5.11	3.60	5.21	2.38	7.33
CPPs	6.72	3.17	1.04	1.14	2.72	5.91	3.29	4.70	12.68	8.29	1.56	2.17	4.18	3.43	23.55	4.21	2.89	2.34	4.08	1.91
Random	7.06	2.31	4.66	6.63	3.76	6.75	2.60	4.30	5.67	10.44	1.95	3.57	6.38	4.44	5.81	8.11	5.28	6.06	1.42	2.81

Figure 2. A bar graph to represent percentage amino acid composition of blood–brain barrier penetrating peptides (B3PPs), cell-penetrating peptides (CPPs), and random peptides.

3.2. Amino Acid Position Analysis

The preferential amino acid position is denoted in Figure 3, which was generated with the help of Two Sample Logo software. The preferred position of amino acids can be seen in the figure; tyrosine, glycine, and arginine are more prominent in the first three positions in B3PPs. The Two Sample Logos suggest that tyrosine, glycine, arginine, and lysine are more preferred throughout the B3PPs.

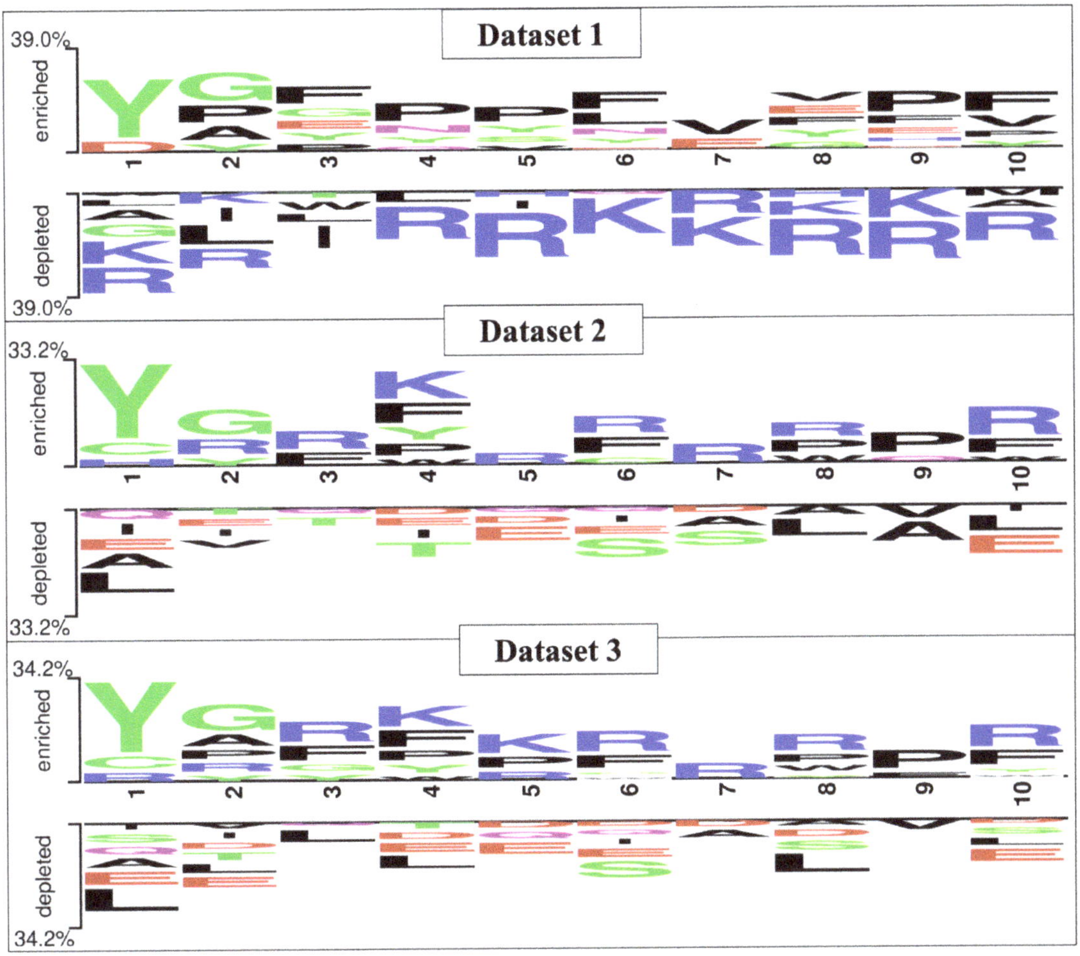

Figure 3. Two Sample Logo (TSL) depiction of all the three datasets (i.e., Dataset_1, Dataset_2, and Dataset_3), preferred positions for amino acids can be seen in the TSLs.

3.3. B3PPs Prediction Methods on Different Datasets

B3PPs prediction models were built using various machine learning techniques, such as random forest (RF), XG Boosting (XGB), Logistic Regression (LR), Support Vector Classifier (SVC), k-nearest neighbor (KNN), Gaussian Naive Bayes (GNB), and decision tree (DT) on various datasets. The best model was implemented in the webserver and standalone software. As we created three different datasets for the prediction of B3PPs, we generated 9189 peptide features by using Pfeature. These peptide features on each dataset were scrutinized and reduced by an SVC-L1 feature selection technique. The feature selection technique highlighted 73 features of Dataset_1, 73 features of Dataset_2, and 145 features of Dataset_3. After selecting features for the datasets, we developed prediction methods using different machine learning techniques. In order to classify B3PPs and CPPs, we developed models on Dataset_1, which contains 269 B3PPs and 269CPPs. Our random forest model achieved maximum performance using 73 selected features. Our RF-based method obtained an 85.12% accuracy with an AUROC of 0.92 on the training dataset, and an 84.25% accuracy with an AUROC of 0.89 on the validation dataset. KNN performed the worst and obtained a 65.58% accuracy with an AUROC of 0.74 on the training dataset, and a 50.92% accuracy with AUROC of 0.64 on the validation dataset (Table 1).

Table 1. The performance of classification models developed on Dataset_1 for discriminating B3PPs and CPPs; models were developed using different machine learning techniques.

Methods	Training Dataset_1					Validation Dataset_1				
	Sens	Spec	Acc	AUROC	MCC	Sens	Spec	Acc	AUROC	MCC
RF	86.04	84.18	82.09	0.90	0.64	75.92	87.03	81.48	0.88	0.63
XGB	81.39	82.32	80.93	0.88	0.62	79.63	88.89	84.25	0.88	0.68
LR	82.79	83.25	83.48	0.90	0.67	81.48	87.03	84.26	0.91	0.69
SVC	83.25	82.79	81.86	0.88	0.64	74.07	92.59	83.33	0.91	0.67
KNN	66.51	64.65	65.58	0.74	0.32	48.18	77.77	62.93	0.72	0.27
GNB	84.18	82.32	80	0.86	0.61	53.70	94.44	74.07	0.86	0.52
DT	78.14	75.34	73.49	0.79	0.47	74.07	70.37	72.22	0.76	0.44

We developed classification models on Dataset_2 to classify B3PPs and non-B3PPs using different machine learning algorithms. Our RF-based model performed better than other models and achieved an 82.09% accuracy with an AUROC of 0.90 on the training dataset, and an 81.48% accuracy with an AUROC of 0.88 on the validation dataset (Table 2).

Table 2. The performance of classification models developed on Dataset_2 for discriminating B3PPs and non-B3PPs; models were developed using different machine learning techniques.

Methods	Training Dataset_2					Validation Dataset_2				
	Sens	Spec	Acc	AUROC	MCC	Sens	Spec	Acc	AUROC	MCC
RF	80.57	84.18	82.09	0.90	0.64	75.92	87.03	81.48	0.88	0.63
XGB	80.46	81.39	80.93	0.88	0.62	79.63	88.89	84.25	0.88	0.68
LR	80.46	86.52	83.48	0.90	0.67	81.48	87.03	84.26	0.91	0.69
SVC	79.07	84.65	81.86	0.88	0.64	74.07	92.59	83.33	0.91	0.67
KNN	50.23	80.93	65.58	0.74	0.32	48.18	77.77	62.93	0.72	0.27
GNB	72.55	87.44	80	0.86	0.61	53.70	94.44	74.07	0.86	0.52
DT	73.02	73.95	73.49	0.79	0.47	74.07	70.37	72.22	0.76	0.44

Finally, classification models were developed on Dataset_3 for discriminating B3PPs and randomly generated non-B3PPs. Our RF-based model achieved the best performance with respect to the top 80 features (Supplementary Table S5). The performance of the RF model was an 85.25% accuracy with an AUROC of 0.93 on the training dataset, and an 82.93% accuracy with an AUROC of 0.90 on the validation dataset. It was the highest-performing among all the methods on all the datasets, so we incorporated this RF model into our webserver for the prediction of the B3PPs (Table 3).

Table 3. The performance of classification models developed on Dataset_3 for discriminating B3PPs and CPPs; models were developed using different machine learning techniques.

Methods	Training Dataset_3					Validation Dataset_3				
	Sens	Spec	Acc	AUROC	MCC	Sens	Spec	Acc	AUROC	MCC
RF	86.97	85.08	85.25	0.93	0.51	81.48	83.08	82.93	0.90	0.44
XGB	72.55	93.82	91.88	0.92	0.58	72.22	92.00	90.20	0.892	0.52
LR	80.93	89.73	88.93	0.92	0.54	83.33	89.40	88.85	0.93	0.55
SVC	80.00	84.75	84.32	0.90	0.45	85.18	82.15	82.43	0.90	0.45
KNN	83.72	80.76	81.03	0.88	0.43	79.63	78.44	78.54	0.84	0.37
GNB	80.46	75.74	76.20	0.84	0.35	83.33	72.67	73.65	0.86	0.34
DT	85.11	65.00	66.83	0.82	0.30	64.82	63.20	63.40	0.72	0.20

We also computed the performance of the models in terms of AUROC on Dataset_3; the models were developed using different machine learning techniques. As shown in Figure 4A, the RF-based model achieved the highest AUROC of 0.93 on the training dataset. As shown in Figure 4B, the SVC-based model achieved the maximal AUROC of 0.92 on the validation dataset (Figure 4).

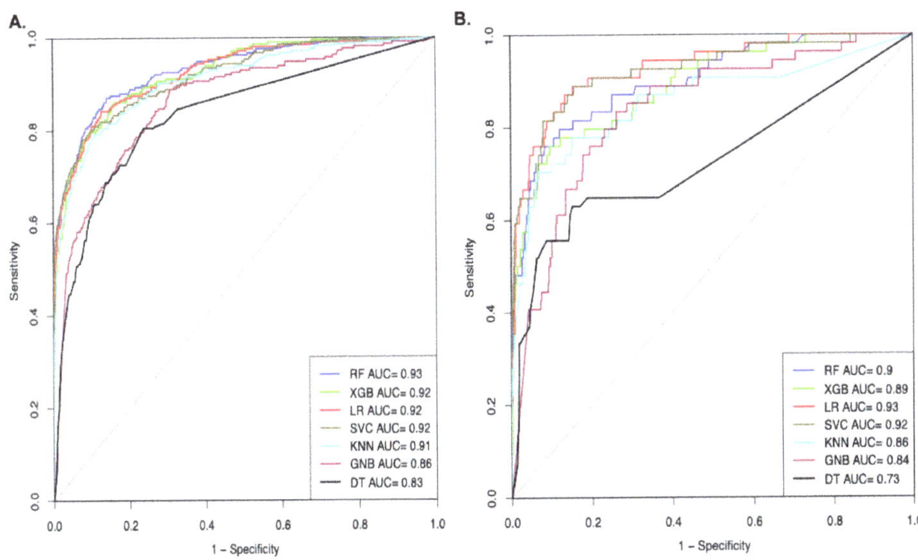

Figure 4. AUROC plot shows the performance of models on Dataset_3 developed using top selected features (**A**). AUROC curve for the training dataset (**B**). AUROC curve for validation dataset.

3.4. Webserver and Standalone Software

One of the major objectives of this study is to facilitate the scientific community in discovering B3PP-based drug delivery vehicles that can deliver cargo into brain tissues. Thus, we developed a standalone software as well as a web-based service to assist the researcher in finding new B3PPs or designing efficient B3PPs. Our webserver, B3Pred, has three major modules: predict, design, and scan. The predict module of B3pred allows users to predict B3PPs in a set of protein sequences submitted by the user. It allows users to select models developed on any dataset used in this study (Figure 5). The design module of B3pred was developed to discover the most promiscuous B3PPs for a given peptide. This module first generates all possible analogs of a peptide, then predicts the score for each analog. It also allows users to sort analogs, based on their score, and to select the best analog of a peptide. The scan module provides the facility to identify the B3PPs region in the user's query protein. It allows the user to select the length of the peptide segment to be scanned in the protein sequence they submit. In addition to this web-based service, we also developed standalone software for searching B3PPs at a large scale, including searching B3PPs at the genome level.

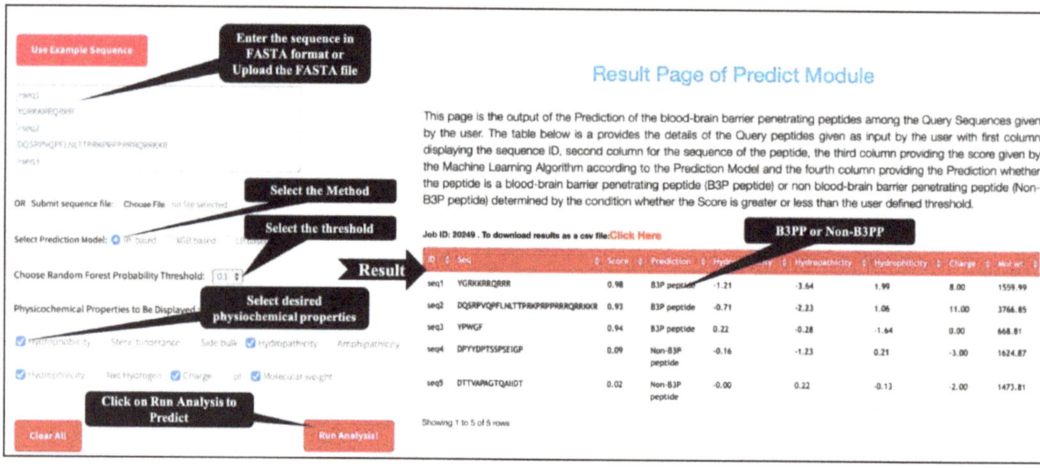

Figure 5. Depiction of predict module present in B3Pred webserver.

3.5. Comparison with the Existing Method

It is crucial to compare this newly developed method with existing methods to understand its benefits and drawbacks. BBPpred has been developed to predict B3PPs, which is trained on 100 B3PPs and 100 non-B3PPs, and the model is tested on only 19 B3PPs and 19 non-B3PPs. On the other hand, B3Pred is trained and tested on three different datasets: Dataset_1 contains 269 B3P peptides and 269 CPPs; Dataset_2 comprises 269 B3P peptides and 269 non-B3P peptides randomly generated using the Swiss-Prot database; and Dataset_3 accommodates 269 B3P peptides and 2690 non-B3P peptides randomly generated using the Swiss-Prot database. In terms of performance, BBPpred achieved a maximal AUROC of 0.87, whereas B3Pred achieved AUROCs of 0.92, 0.90, and 0.93 on Dataset_1, Dataset_2, and Dataset_3, respectively. BBPpred only provides the prediction facility; on the other hand, B3Pred provides a prediction, design, and scan facility. In addition, B3Pred is also available as standalone software, so that users can run it on their local machine at a large scale.

4. Discussion and Conclusions

The blood–brain barrier (BBB) is the natural guard of the brain, which inhibits unwanted molecules from crossing into brain tissue [69]. Unfortunately, neurological disease prevalence has increased tremendously in the last few decades. Thus, there is a need to discover new drugs that can be used to treat brain-associated diseases such as Alzheimer's disease and Parkinson's disease. Due to advancements in technology, researchers can discover drugs to treat these disorders in vitro. One of the major hurdles in treating brain-associated disease is delivering drugs into brain tissue, as the blood–brain barrier inhibits these drug molecules from reaching this tissue [70]. The transportation or delivery of the therapeutic molecules across the barriers of the brain is the major bottleneck in treating brain tumors and CNS diseases [71].

Several in silico methods have been developed to predict and improve the delivery of therapeutic molecules that circumvent the BBB. A study has shown that D-Ala-Peptide

T-amide (DAPTA), or peptide T is an antiviral peptide that can cross the blood–brain barrier. Intranasal Peptide T can be obtained from the envelope protein of the human immunodeficiency virus (HIV). This peptide shows antiviral properties, usually inhibits chemokine (CCR5) receptors, and also acts as a B3PP [72,73]. Researchers have also found that AH-D, an amphipathic α-helical BBB-penetrating peptide, can act as a therapeutic agent for deadly viruses. It is used as a direct antiviral agent (DAA) to inhibit specific viral proteins. A recent study has suggested that potential antiviral AH-D is a target against deadly viruses, such as chikungunya virus, Zika, dengue, and yellow fever, with different inhibitory and cytotoxic concentrations [74–77]. These studies show that such peptides can be helpful in viral infections, along with any neurological complications that arise due to these viruses. These peptides can be used as therapeutic substitutes for antiviral drugs which are unable to cross the brain. This may help in controlling the neurological complications that arise due to COVID-19 [78].

In the present scenario, there is the utmost need to develop an efficient prediction tool that can accurately predict the peptides that have the property of penetrating through the blood–brain barrier. To facilitate the researchers working in this area, we proposed a method named B3pred for predicting B3PPs. We have also developed a free webserver, named B3pred, and have incorporated various modules to predict, design, scan for and analyze B3PPs. We believe that our method will help in the accurate prediction of B3PPs and aid the scientific community working in this area.

Supplementary Materials: The following are available online at https://www.mdpi.com/1999-4923/13/8/1237/s1, Table S1: Description of all the 9189 features calculated using composition-based module of Pfeature; Table S2: Fold-wise performance of various machine learning algorithms on Dataset_1; Table S3: Fold-wise performance of various machine learning algorithms on Dataset_2; Table S4: Fold-wise performance of various machine learning algorithms on Dataset_3; Table S5: Top 80 features selected after implementation of SVC-L1 with their importance score calculated using feature-selector python library

Author Contributions: Collected and processed the datasets, V.K. and S.P.; created the prediction models, V.K. and S.P.; analyzed and interpreted the results, V.K., S.P., A.D., N.S. and G.P.S.R.; developed the webserver, V.K. and S.P.; prepared the manuscript, V.K., S.P., A.D. and G.P.S.R.; coordinated the project, G.P.S.R. All authors have read and agreed to the published version of the manuscript.

Funding: Authors received the funding in terms of fellowships and financial support from the following funding agencies. Kumar V. received from University Grant Commission (UGC), Dhall A. and Sharma N. from Department of Science and Technology (DST), and Patiyal S. from Department of Biotechnology, Govt. of India.

Institutional Review Board Statement: Not applicable.

Informed Consent Statement: Not applicable.

Data Availability Statement: The datasets are available at https://webs.iiitd.edu.in/raghava/b3pred/download.php (accessed on 7 July 2021).

Acknowledgments: Authors are thankful to funding agencies, University Grant Commission (UGC), Department of Science and Technology (DST) and Department of Biotechnology, Govt. of India for financial support and fellowships.

Conflicts of Interest: The authors declare no conflict of interest.

References

1. Abbott, N.J.; Rönnbäck, L.; Hansson, E. Astrocyte-endothelial interactions at the blood-brain barrier. *Nat. Rev. Neurosci.* **2006**, *7*, 41–53. [CrossRef]
2. Kniesel, U.; Wolburg, H. Tight junctions of the blood-brain barrier. *Cell. Mol. Neurobiol.* **2000**, *20*, 57–76. [CrossRef]
3. Fenstermacher, J.; Gross, P.; Sposito, N.; Acuff, V.; Pettersen, S.; Gruber, K. Structural and Functional Variations in Capillary Systems within the brain. *Ann. N. Y. Acad. Sci.* **1988**, *529*, 21–30. [CrossRef] [PubMed]

4. Rhea, E.M.; Banks, W.A. Role of the Blood-Brain Barrier in Central Nervous System Insulin Resistance. *Front. Neurosci.* **2019**, *13*. [CrossRef] [PubMed]
5. Muoio, V.; Persson, P.B.; Sendeski, M.M. The neurovascular unit—Concept review. *Acta Physiol.* **2014**, *210*, 790–798. [CrossRef] [PubMed]
6. Oldendorf, W.H. Brain uptake of radiolabeled amino acids, amines, and hexoses after arterial injection. *Am. J. Physiol.* **1971**, *221*, 1629–1639. [CrossRef]
7. Tietz, S.; Engelhardt, B. Brain barriers: Crosstalk between complex tight junctions and adherens junctions. *J. Cell Biol.* **2015**, *209*, 493–506. [CrossRef] [PubMed]
8. Pardridge, W.M. Blood-brain barrier delivery. *Drug Discov. Today* **2007**, *12*, 54–61. [CrossRef]
9. Islam, Y.; Leach, A.G.; Smith, J.; Pluchino, S.; Coxon, C.; Sivakumaran, M.; Downing, J.E.; Fatokun, A.A.; Teixidò, M.; Ehtezazi, T. Peptide based drug delivery systems to the brain. *Nano Express* **2020**, *1*, 012002. [CrossRef]
10. Banks, W.A. Peptides and the blood-brain barrier. *Peptides* **2015**, *72*, 16–19. [CrossRef]
11. Funke, S.A.; Willbold, D. Peptides for Therapy and Diagnosis of Alzheimer's Disease. *Curr. Pharm. Des.* **2012**, *18*, 755–767. [CrossRef]
12. Baig, M.H.; Ahmad, K.; Saeed, M.; Alharbi, A.M.; Barreto, G.E.; Ashraf, G.M.; Choi, I. Peptide based therapeutics and their use for the treatment of neurodegenerative and other diseases. *Biomed. Pharmacother.* **2018**, *103*, 574–581. [CrossRef] [PubMed]
13. Raucher, D. Tumor targeting peptides: Novel therapeutic strategies in glioblastoma. *Curr. Opin. Pharmacol.* **2019**, *47*, 14–19. [CrossRef] [PubMed]
14. Oller-Salvia, B.; Sánchez-Navarro, M.; Giralt, E.; Teixidó, M. Blood-brain barrier shuttle peptides: An emerging paradigm for brain delivery. *Chem. Soc. Rev.* **2016**, *45*, 4690. [CrossRef] [PubMed]
15. Wu, L.-P.; Ahmadvand, D.; Su, J.; Hall, A.; Tan, X.; Farhangrazi, Z.S.; Moghimi, S.M. Crossing the blood-brain-barrier with nanoligand drug carriers self-assembled from a phage display peptide. *Nat. Commun.* **2019**, *10*, 1–16. [CrossRef] [PubMed]
16. Nosrati, H.; Tarantash, M.; Bochani, S.; Charmi, J.; Bagheri, Z.; Fridoni, M.; Abdollahifar, M.-A.; Davaran, S.; Danafar, H.; Manjili, H.K. Glutathione (GSH) Peptide Conjugated Magnetic Nanoparticles As Blood-Brain Barrier Shuttle for MRI-Monitored Brain Delivery of Paclitaxel. *ACS Biomater. Sci. Eng.* **2019**, *5*, 1677–1685. [CrossRef]
17. Solbrig, M.V.; Koob, G.F. Epilepsy, CNS viral injury and dynorphin. *Trends Pharmacol. Sci.* **2004**, *25*, 98–104. [CrossRef] [PubMed]
18. Balasubramaniam, A. Clinical potentials of neuropeptide Y family of hormones. *Am. J. Surg.* **2002**, *183*, 430–434. [CrossRef]
19. Claes, S.J. Corticotropin-releasing hormone (CRH) in psychiatry: From stress to psychopathology. *Ann. Med.* **2004**, *36*, 50–61. [CrossRef]
20. Ströhle, A.; Holsboer, F. Stress Responsive Neurohormones in Depression and Anxiety. *Pharmacopsychiatry* **2003**, *36*, 207–214.
21. Li, C.; Wu, X.; Liu, S.; Zhao, Y.; Zhu, J.; Liu, K. Roles of Neuropeptide Y in Neurodegenerative and Neuroimmune Diseases. *Front. Neurosci.* **2019**, *13*, 869. [CrossRef]
22. Dwibhashyam, V.S.N.M.; Nagappa, A. Strategies for enhanced drug delivery to the central nervous system. *Indian J. Pharm. Sci.* **2008**, *70*, 145–153.
23. Reese, T.S.; Karnovsky, M.J. Fine structural localization of a blood-brain barrier to exogenous peroxidase. *J. Cell Biol.* **1967**, *34*, 207–217. [CrossRef] [PubMed]
24. Kapoor, P.; Singh, H.; Gautam, A.; Chaudhary, K.; Kumar, R.; Raghava, G.P.S. TumorHoPe: A database of tumor homing peptides. *PLoS ONE* **2012**, *7*, e35187. [CrossRef]
25. Gautam, A.; Singh, H.; Tyagi, A.; Chaudhary, K.; Kumar, R.; Kapoor, P.; Raghava, G.P.S. CPPsite: A curated database of cell penetrating peptides. *Database* **2012**, *2012*, bas015. [CrossRef]
26. Sharma, A.; Kapoor, P.; Gautam, A.; Chaudhary, K.; Kumar, R.; Chauhan, J.S.; Tyagi, A.; Raghava, G.P.S. Computational approach for designing tumor homing peptides. *Sci. Rep.* **2013**, *3*, 1607. [CrossRef] [PubMed]
27. Gautam, A.; Sharma, M.; Vir, P.; Chaudhary, K.; Kapoor, P.; Kumar, R.; Nath, S.K.; Raghava, G.P.S. Identification and characterization of novel protein-derived arginine-rich cell-penetrating peptides. *Eur. J. Pharm. Biopharm.* **2015**, *89*, 93–106. [CrossRef] [PubMed]
28. Shergalis, A.; Bankhead, A.; Luesakul, U.; Muangsin, N.; Neamati, N. Current challenges and opportunities in treating glioblastomas. *Pharmacol. Rev.* **2018**, *70*, 412–445. [CrossRef]
29. Stalmans, S.; Wynendaele, E.; Bracke, N.; Gevaert, B.; D'Hondt, M.; Peremans, K.; Burvenich, C.; De Spiegeleer, B. Chemical-Functional Diversity in Cell-Penetrating Peptides. *PLoS ONE* **2013**, *8*, e71752. [CrossRef] [PubMed]
30. Stalmans, S.; Bracke, N.; Wynendaele, E.; Gevaert, B.; Peremans, K.; Burvenich, C.; Polis, I.; De Spiegeleer, B. Cell-penetrating peptides selectively cross the blood-brain barrier in vivo. *PLoS ONE* **2015**, *10*, e0139652. [CrossRef] [PubMed]
31. Yamano, S.; Dai, J.; Hanatani, S.; Haku, K.; Yamanaka, T.; Ishioka, M.; Takayama, T.; Yuvienco, C.; Khapli, S.; Moursi, A.M.; et al. Long-term efficient gene delivery using polyethylenimine with modified Tat peptide. *Biomaterials* **2014**, *35*, 1705–1715. [CrossRef]
32. Huwyler, J.; Wu, D.; Pardridge, W.M. Brain drug delivery of small molecules using immunoliposomes. *Proc. Natl. Acad. Sci. USA* **1996**, *93*, 14164–14169. [CrossRef]
33. Knight, A.; Carvajal, J.; Schneider, H.; Coutelle, C.; Chamberlain, S.; Fairweather, N. Non-viral neuronal gene delivery mediated by the H(C) fragment of tetanus toxin. *Eur. J. Biochem.* **1999**, *259*, 762–769. [CrossRef] [PubMed]
34. El-Andaloussi, S.; Holm, T.; Langel, U. Cell-Penetrating Peptides: Mechanisms and Applications. *Curr. Pharm. Des.* **2005**, *11*, 3597–3611. [CrossRef] [PubMed]

35. Milletti, F. Cell-penetrating peptides: Classes, origin, and current landscape. *Drug Discov. Today* **2012**, *17*, 850–860. [CrossRef] [PubMed]
36. Stewart, K.M.; Horton, K.L.; Kelley, S.O. Cell-penetrating peptides as delivery vehicles for biology and medicine. *Org. Biomol. Chem.* **2008**, *6*, 2242–2255. [CrossRef]
37. Mueller, J.; Kretzschmar, I.; Volkmer, R.; Boisguerin, P. Comparison of cellular uptake using 22 CPPs in 4 different cell lines. *Bioconjug. Chem.* **2008**, *19*, 2363–2374. [CrossRef] [PubMed]
38. Meade, A.J.; Meloni, B.P.; Mastaglia, F.L.; Knuckey, N. The application of cell penetrating peptides for the delivery of neuroprotective peptides/proteins in experimental cerebral ischaemia studies. *J. Exp. Stroke Transl. Med.* **2009**, *2*, 22–40. [CrossRef]
39. Mathur, D.; Prakash, S.; Anand, P.; Kaur, H.; Agrawal, P.; Mehta, A.; Kumar, R.; Singh, S.; Raghava, G.P.S. PEPlife: A Repository of the Half-life of Peptides. *Sci. Rep.* **2016**, *6*, 36617. [CrossRef]
40. Gautam, A.; Chaudhary, K.; Kumar, R.; Sharma, A.; Kapoor, P.; Tyagi, A.; Raghava, G.P.S. In silico approaches for designing highly effective cell penetrating peptides. *J. Transl. Med.* **2013**, *11*, 74. [CrossRef]
41. Wei, L.; Tang, J.; Zou, Q. SkipCPP-Pred: An improved and promising sequence-based predictor for predicting cell-penetrating peptides. *BMC Genom.* **2017**, *18*, 742. [CrossRef]
42. Wei, L.; Xing, P.; Su, R.; Shi, G.; Ma, Z.S.; Zou, Q. CPPred-RF: A Sequence-based Predictor for Identifying Cell-Penetrating Peptides and Their Uptake Efficiency. *J. Proteome Res.* **2017**, *16*, 2044–2053. [CrossRef]
43. Pandey, P.; Patel, V.; George, N.V.; Mallajosyula, S.S. KELM-CPPpred: Kernel Extreme Learning Machine Based Prediction Model for Cell-Penetrating Peptides. *J. Proteome Res.* **2018**, *17*, 3214–3222. [CrossRef]
44. Kumar, V.; Agrawal, P.; Kumar, R.; Bhalla, S.; Usmani, S.S.; Varshney, G.C.; Raghava, G.P.S. Prediction of cell-penetrating potential of modified peptides containing natural and chemically modified residues. *Front. Microbiol.* **2018**, *9*, 725. [CrossRef]
45. Qiang, X.; Zhou, C.; Ye, X.; Du, P.-F.; Su, R.; Wei, L. CPPred-FL: A sequence-based predictor for large-scale identification of cell-penetrating peptides by feature representation learning. *Brief. Bioinform.* **2018**, *21*, 11–23. [CrossRef] [PubMed]
46. Shaker, B.; Yu, M.-S.; Song, J.S.; Ahn, S.; Ryu, J.Y.; Oh, K.-S.; Na, D. LightBBB: Computational prediction model of blood–brain-barrier penetration based on LightGBM. *Bioinformatics* **2021**, *37*, 1135–1139. [CrossRef] [PubMed]
47. Carpenter, T.; Kirshner, D.A.; Lau, E.Y.; Wong, S.E.; Nilmeier, J.P.; Lightstone, F.C. A Method to Predict Blood-Brain Barrier Permeability of Drug-Like Compounds Using Molecular Dynamics Simulations. *Biophys. J.* **2014**, *107*, 630–641. [CrossRef] [PubMed]
48. Mensch, J.; Oyarzabal, J.; Mackie, C.; Augustijns, P. In vivo, in vitro and in silico methods for small molecule transfer across the BBB. *J. Pharm. Sci.* **2009**, *98*, 4429–4468. [CrossRef] [PubMed]
49. Dai, R.; Zhang, W.; Tang, W.; Wynendaele, E.; Zhu, Q.; Bin, Y.; De Spiegeleer, B.; Xia, J. BBPpred: Sequence-Based Prediction of Blood-Brain Barrier Peptides with Feature Representation Learning and Logistic Regression. *J. Chem. Inf. Model.* **2021**, *61*, 525–534. [CrossRef]
50. Kumar, V.; Patiyal, S.; Kumar, R.; Sahai, S.; Kaur, D.; Lathwal, A.; Raghava, G.P.S. B3Pdb: An archive of blood–brain barrier-penetrating peptides. *Brain Struct. Funct.* **2021**. [CrossRef]
51. Agrawal, P.; Bhalla, S.; Usmani, S.S.; Singh, S.; Chaudhary, K.; Raghava, G.P.S.; Gautam, A. CPPsite 2.0: A repository of experimentally validated cell-penetrating peptides. *Nucleic Acids Res.* **2016**, *44*, D1098–D1103. [CrossRef] [PubMed]
52. Boutet, E.; Lieberherr, D.; Tognolli, M.; Schneider, M.; Bansal, P.; Bridge, A.; Poux, S.; Bougueleret, L.; Xenarios, I.; Boutet, E. Uniprotkb/swiss-prot, the manually annotated section of the uniprot knowledgebase: How to use the entry view. *Methods Mol. Biol.* **2016**, *1374*, 23–54. [PubMed]
53. Agrawal, P.; Raghava, G.P.S. Prediction of Antimicrobial Potential of a Chemically Modified Peptide From Its Tertiary Structure. *Front. Microbiol.* **2018**, *9*, 2551. [CrossRef] [PubMed]
54. Vacic, V.; Iakoucheva, L.M.; Radivojac, P. Two Sample Logo: A graphical representation of the differences between two sets of sequence alignments. *Bioinformatics* **2006**, *22*, 1536–1537. [CrossRef]
55. Pande, A.; Patiyal, S.; Lathwal, A.; Arora, C.; Kaur, D.; Dhall, A.; Mishra, G.; Kaur, H.; Sharma, N.; Jain, S.; et al. Computing wide range of protein/peptide features from their sequence and structure. *bioRxiv* **2019**. [CrossRef]
56. Tang, J.; Alelyani, S.; Liu, H. Feature selection for classification: A review. *Data Classif Algorithms Appl.* **2014**, 37–64. Available online: https://www.cvs.edu.in/upload/feature_selection_for_classification.pdf (accessed on 22 February 2021).
57. Chang, K.-W.; Hsieh, C.-J.; Lin, C.-J. LIBLINEAR: A Library for Large Linear Classification Rong-En Fan Xiang-Rui Wang. *J. Mach. Learn. Res.* **2008**, *9*, 1871–1874.
58. Ke, G.; Meng, Q.; Finley, T.; Wang, T.; Chen, W.; Ma, W.; Ye, Q.; Liu, T.-Y. LightGBM: A Highly Efficient Gradient Boosting Decision Tree. *Adv. Neural Inf. Process. Syst.* **2017**, *30*, 3146–3154.
59. Pedregosa, F.; Varoquaux, G.; Gramfort, A.; Michel, V.; Thirion, B.; Grisel, O.; Blondel, M.; Prettenhofer, P.; Weiss, R.; Dubourg, V.; et al. Scikit-learn: Machine Learning in Python. *J. Mach. Learn. Res.* **2011**, *12*, 2825–2830.
60. Webb, G.I. Decision Tree Grafting From the All-Tests-But-One Partition. In Proceedings of the Sixteenth International Joint Conference on Artificial Intelligence, IJCAI 99, Stockholm, Sweden, 31 July–6 August 1999.
61. Zhang, H. Exploring conditions for the optimality of naïve bayes. *Int. J. Pattern Recognit. Artif. Intell.* **2005**, *19*, 183–198. [CrossRef]
62. Geurts, P.; Ernst, D.; Wehenkel, L. Extremely randomized trees. *Mach. Learn.* **2006**, *63*, 3–42. [CrossRef]
63. Tolles, J.; Meurer, W.J. Logistic regression: Relating patient characteristics to outcomes. *J. Am. Med. Assoc.* **2016**, *316*, 533–534. [CrossRef] [PubMed]

64. Mucherino, A.; Papajorgji, P.J.; Pardalos, P.M. Data Mining in Agriculture. *Dyn. Disasters* **2009**, *34*. [CrossRef]
65. Chen, T.; Guestrin, C. XGBoost: A Scalable Tree Boosting System. In Proceedings of the 22nd ACM SIGKDD International Conference on Knowledge Discovery and Data Mining, San Francisco, CA, USA, 13–17 August 2016.
66. Chang, C.-C.; Lin, C.-J. LIBSVM: A Library for Support Vector Machines. *ACM Trans. Intell. Syst. Technol.* **2011**, *2*, 1–27. [CrossRef]
67. Kumar, V.; Kumar, R.; Agrawal, P.; Patiyal, S.; Raghava, G.P.S. A Method for Predicting Hemolytic Potency of Chemically Modified Peptides From Its Structure. *Front. Pharmacol.* **2020**, *11*, 54. [CrossRef] [PubMed]
68. Dhall, A.; Patiyal, S.; Sharma, N.; Usmani, S.S.; Raghava, G.P.S. Computer-aided prediction and design of IL-6 inducing peptides: IL-6 plays a crucial role in COVID-19. *Brief. Bioinform.* **2021**, *22*, 936–945. [CrossRef] [PubMed]
69. Thangudu, S.; Cheng, F.-Y.; Su, C.-H. Advancements in the Blood-Brain Barrier Penetrating Nanoplatforms for Brain Related Disease Diagnostics and Therapeutic Applications. *Polymers* **2020**, *12*, 55. [CrossRef] [PubMed]
70. He, Q.; Liu, J.; Liang, J.; Liu, X.; Li, W.; Liu, Z.; Ding, Z.; Tuo, D. Towards Improvements for Penetrating the Blood-Brain Barrier-Recent Progress from a Material and Pharmaceutical Perspective. *Cells* **2018**, *7*, 24. [CrossRef]
71. Banks, W.A. From blood-brain barrier to blood-brain interface: New opportunities for CNS drug delivery. *Nat. Rev. Drug Discov.* **2016**, *15*, 275–292. [CrossRef]
72. Polianova, M.T.; Ruscetti, F.W.; Pert, C.B.; Tractenberg, R.E.; Leoung, G.; Strang, S.; Ruff, M.R. Antiviral and immunological benefits in HIV patients receiving intranasal peptide T (DAPTA). *Peptides* **2003**, *24*, 1093–1098. [CrossRef]
73. Barrera, C.M.; Kastin, A.J.; Banks, W.A. D-[Ala1]-peptide T-Amide is transported from blood to brain by a saturable system. *Brain Res. Bull.* **1987**, *19*, 629–633. [CrossRef]
74. Jackman, J.A.; Costa, V.V.; Park, S.; Real, A.L.C.V.; Park, J.H.; Cardozo, P.L.; Ferhan, A.R.; Olmo, I.G.; Moreira, T.P.; Bambirra, J.L.; et al. Therapeutic treatment of Zika virus infection using a brain-penetrating antiviral peptide. *Nat. Mater.* **2018**, *17*, 971–977. [CrossRef] [PubMed]
75. Cho, N.-J.; Dvory-Sobol, H.; Xiong, A.; Cho, S.-J.; Frank, C.W.; Glenn, J.S. Mechanism of an amphipathic α-helical peptide's antiviral activity involves size-dependent virus particle lysis. *ACS Chem. Biol.* **2009**, *4*, 1061–1067. [CrossRef] [PubMed]
76. Zou, J.; Shi, P.Y. Targeting vesicle size. *Nat. Mater.* **2018**, *17*, 955–956. [CrossRef]
77. Boldescu, V.; Behnam, M.; Vasilakis, N.; Klein, C.D. Broad-spectrum agents for flaviviral infections: Dengue, Zika and beyond. *Nat. Rev. Drug Discov.* **2017**, *16*, 565–586. [CrossRef]
78. Mao, X.Y.; Jin, W.L. The COVID-19 Pandemic: Consideration for Brain Infection. *Neuroscience* **2020**, *437*, 130–131. [CrossRef]

MDPI
St. Alban-Anlage 66
4052 Basel
Switzerland
www.mdpi.com

Pharmaceutics Editorial Office
E-mail: pharmaceutics@mdpi.com
www.mdpi.com/journal/pharmaceutics

Disclaimer/Publisher's Note: The statements, opinions and data contained in all publications are solely those of the individual author(s) and contributor(s) and not of MDPI and/or the editor(s). MDPI and/or the editor(s) disclaim responsibility for any injury to people or property resulting from any ideas, methods, instructions or products referred to in the content.

www.ingramcontent.com/pod-product-compliance
Lightning Source LLC
Chambersburg PA
CBHW040840250926
51850CB00007BA/1361
* 9 7 8 3 7 2 5 8 1 2 5 8 5 *